Biosensors
Materials and Applications

Edited by

Inamuddin[1,2,3], Tauseef Ahmad Rangreez[4], Mohd Imran Ahamed[5] and Abdullah M. Asiri[1,2]

[1]Centre of Excellence for Advanced Materials Research, King Abdulaziz University, Jeddah 21589, Saudi Arabia

[2]Chemistry Department, Faculty of Science, King Abdulaziz University, Jeddah 21589, Saudi Arabia

[3]Department of Applied Chemistry, Faculty of Engineering and Technology, Aligarh Muslim University, Aligarh-202 002, India

[4]Department of Chemistry, National Institute of Technology, Srinagar, Jammu and Kashmir-190006, India

[5]Department of Chemistry, Faculty of Science, Aligarh Muslim University, Aligarh-202 002, India

Published by **Materials Research Forum LLC**
Millersville, PA 17551, USA

Published as part of the book series
Materials Research Foundations
Volume 47 (2019)
ISSN 2471-8890 (Print)
ISSN 2471-8904 (Online)

Print ISBN 978-1-64490-012-3
eBook ISBN 978-1-64490-013-0

Distributed worldwide by

Materials Research Forum LLC
105 Springdale Lane
Millersville, PA 17551
USA
http://www.mrforum.com

Manufactured in the United States of America
10 9 8 7 6 5 4 3 2 1

Table of Contents

Preface

Modern societies have become extremely cautious towards the maintenance of health standards. The countries worldwide are making every effort to make health care safe, easily available and affordable. The same is evident from the large increase of grants/funds allocated to health care. The financial assistance supports various health programmes like extending healthy life, reducing illness, disability, disorders and discomfort and developing procedures, means and advanced devices for proper health care. These efforts have started to bear fruit as is clear from the overall increase in the average healthy life span of humans in countries like Japan, Switzerland, Australia, UK and USA. In order to maintain the highest standards of health care, it is imperative to have standard procedures and devices that help in the detection of all those analytes, molecules which are detrimental to the well being and human health. Also maintaining the quality and safety standards require easy and fast detection of all those substances which may cause spoilage and or contamination of food. Over the last decade, biosensors have evolved as a major field of research. These have been used in several areas viz. health care, food and environment. These biosensors have revolutionized the point of care devices as these are highly specific and sensitive, low cost, compact in size, provide fast response and easy to operate. The affordability is an important aspect related to biosensors as these are used in several critical fields. A large number of biosensors have been developed till date but the cost component has limited their use. Biosensors have simplified the screening of population groups. The fast and rapid results available from biosensors have proven to be a boon to emergency/critical care sector of health care as these not only save time but also provide important data rapidly and accurately. This book is indented to explore various aspect of biosensors such as detection of environmental contaminants, disease-causing pathogens, genetic materials, tumor cells, cancer, infectious diseases, *in vivo* monitoring of key molecules.

We are very thankful to the authors and their co-authors for their contribution. We also want to thank the copyright holders and publishers who have granted permission to use their figures and tables in this book. Although special care has been taking to get permission for the figures, tables and schemes used in this book, we would like to apologise if any copyright has been infringed unknowingly.

Chapter 1 intends to describe the role of aptasensors for the detection of environmental contaminants and disease-causing pathogens that pose a threat to the health care system.

Chapter 2 deals with some insight into molecularly imprinted polymers and their application as biosensors for the detection of genetic materials.

Chapter 3 discusses recent advances of functional metal nanoparticles such as the gold and silver based biosensor technology for biomarkers detection. The basic structure of a biosensor device and its working principle are also discussed. These functional nanoparticles are targeted to specific tumour cells for detection as well as imaging.

Chapter 4 focuses on the properties of layered double hydroxide, fabrication of layered double hydroxide based biosensors and their electroanalytical applications for sensing of various analytes.

Chapter 5 focuses on various electrochemical nano-biosensors established for the detection of commonly occurring cancers (lung, breast, prostate and colorectal). It also emphasizes the detailed biosensor fabrication strategies utilized for the detection of biomarkers related to particular cancer types. Recently developed electrochemical techniques based on different bio-recognition elements (antibody, enzyme, nucleic acid, aptamer, phage and lectin) and nanomaterials have been explored with their advantages and limitations.

Chapter 6 summarizes the role of polymer-based nanoparticles for combating the infectious diseases.

Chapter 7 discusses an overview of the types of nanocarriers employed in diversified nanoparticulated systems based on their theranostic application, beneficial as well as deleterious impacts, present status and future prospects are also discussed.

Chapter 8 reviews the main issues (sensitivity, selectivity, tolerance to fluctuations in oxygen levels, biocompatibility, long-term functional stability, etc.) involved in the design phase and the application of biosensors for *in vivo* monitoring of key molecules in health applications.

Inamuddin[1,2,3], Tauseef Ahmad Rangreez[4], Mohd Imran Ahamed[5] and Abdullah M. Asiri[1,2]

[1]Centre of Excellence for Advanced Materials Research, King Abdulaziz University, Jeddah 21589, Saudi Arabia

[2]Chemistry Department, Faculty of Science, King Abdulaziz University, Jeddah 21589, Saudi Arabia.

[3]Department of Applied Chemistry, Faculty of Engineering and Technology, Aligarh Muslim University, Aligarh-202 002, India

[4]Department of Chemistry, National Institute of Technology, Srinagar, Jammu and Kashmir-190006, India

[5]Department of Chemistry, Faculty of Science, Aligarh Muslim University, Aligarh-202 002, India

Biosensors: Materials and Applications
Materials Research Foundations 47 (2019) 1-50

Materials Research Forum LLC
doi: http://dx.doi.org/10.21741/9781644900130-1

Chapter 1

Applications of Aptasensors in Health Care

Abhijeet Dhiman[1,2*], Harleen Kaur[3*] Chanchal Kumar[1], Yusra Ahmad[2], Tarun Kumar Sharma[4**]

[1]Department of Biotechnology, All India Institute of Medical Sciences (AIIMS), New Delhi, India

[2]Faculty of Pharmacy, Uttarakhand Technical University (UTU), Dehradun, Uttarakhand, India.

[3]Astrazeneca, USA

[4]Centre for Biodesign and Diagnostics, Translational Health Science and Technology Institute (THSTI), Faridabad, Haryana, India

*Contributed equally to this work, **tarun@thsti.res.in

Abstract

Access to quality diagnostics is vital to maintain a healthy life as rapid and accurate diagnostic tests can guide the clinician to appropriately and timely manage diseases. However, a major limitation in the diagnostic sector is that it depends heavily on the supply of antibodies. These antibodies are used as molecular recognition elements in a variety of diagnostic assays but they usually displayed a high level of batch-to-batch variation thus pose a great threat in maintaining the quality of a diagnostic test. To overcome this limitation in recent years chemical rivals of antibody called aptamers have emerged. Aptamers are ssDNA or RNA molecule that can acquire a typical 2D or 3D structure to recognize its target with high affinity and specificity. In the last few years, aptamers have attained an impressive growth by showing their utility in detection of a variety of analytes ranging from small molecules to protein to the whole cell. It would not be surprising if aptamers will replace antibodies in diagnostic assays in coming years. This chapter is intended to highlight the importance and utility of aptasensors in health care.

Keywords

Aptamer, Aptasensors, Health, Diagnostics, Pathogen

Contents

1.   Introduction

With the advancement in technology the development of highly sensitive, reliable, rapid, inexpensive and easy-to-use assays to detect small molecules, various contaminants, biomarkers and pathogens is utterly important for food safety, environmental analysis and clinical diagnosis [1]. Generally, chromatographic techniques are used for the detection of molecules, pathogen or disease biomarkers due to its highly sensitive and selective detection but despite these advantages it requires highly skilled technicians for the operation and they are not suitable for screening analysis [2,3]. However, mass spectrometry is used as the "gold standard" for the detection of small molecules [4] but unfortunately its association and dependence on sophisticated instruments limits its application. In the case of pathogenic bacterial cultures, it is considered as a "gold standard" but due to its high turnaround time results in delayed diagnosis and hence delayed treatment. Therefore, there is an urgent and unmet need for the development of rapid, cost-effective and point-of-care (POC) devices for the diagnosis of analytes (small molecules and pathogens) [5–8].

Aptamers are single-stranded (ss) DNA or RNA molecules that can be generated by a process of *in vitro* evolution called Systematic Evolution of Ligands by exponential enrichment (SELEX) [9]. The SELEX was first reported in 1990 by three independent groups i.e. Ellington and Szostak, Robertson and Joyce, and Tuerk and Gold, separately [9–11]. Aptamer form 2D and 3D shapes to bind their targets with high affinity and specificity. Owing to the speed in selection, synthesis and scale up, these molecules have spurred great interest in the health care sector as a diagnostic tool [12–17]. As compared to antibodies, aptamers have numerous advantages including but not limited to *in vitro* evolution ease in functionalization with nanoparticles and fluorophores, room temperature storage, no batch-to-batch variation, high specificity, slow off-rates and high stability with an ability to replace antibodies in all possible diagnostic formats. These attributes make them as an attractive diagnostic tool for the detection of a plethora of diverse targets [12–18]. Moreover, in some studies aptamers were compared with single-chain variable fragments of the antibody, in which aptamers surpass the antibodies due to high affinity and excellent regenerability [19]. Affinity of aptamers can be further enhanced by creating dimeric or multivalent aptamers reported first time by Hianik et al. and Hasegawa et al. [20,21]. On the contrary, Kaur et al. [22] adopted a stem-loop truncation strategy to truncate the aptamer structure in order to achieve improve binding affinity and specificity of the aptamer. Aptamers are able to achieve an impressive growth rate as a diagnostic tool as it is easily adaptable to various formats and platforms thus contributing to the global diagnostic market [23]. Various aptamer technological formats and platforms have been developed for the growing demands in the global market in the area such as diagnostics, therapeutics and biosensors [23]. The global demand for the *in vitro* diagnostics market was valued at US $64.02 billion in 2017 and is expected to reach US $87.93 billion by 2023 [13,24,25]. The exact contribution of the aptamers based diagnostics industry has not been estimated till date, however, the aptamer industry is swiftly growing and estimated to reach US $244.93 million by 2020 from US $107.56 million in 2015. With this market value of the aptamer, more than 50 industries are engaging themselves globally into this business with a primary focus on diagnostics or therapeutics. Herein, in this chapter, we mainly focus on recent advancements in aptamer-based biosensors for the detection of small molecules, bacterial, viral, protozoan and the whole cell, with their analytical performances, unique features and their potential application in the health care system.

In recent time, aptamer-based biosensor technology has captured great attention for the detection and diagnosis of harmful molecules and biomarkers of infectious diseases due to its ability to provide better sensitivity, specificity, rapidness and cost-effective measurement [26]. Aptamer-based biosensor are called aptasensor in which aptamer act as a recognition element [27]. Biosensors consist of two main components **(Figure 1)**, a biological receptor element which is able to specifically interact with a target molecule and a transducer component which include electrical or electrochemical, optical and mass sensors through these interaction, the recognition event can be converted into a detectable signal [5]. For any successful biosensor device, the recognition molecule plays an important role as the sensitivity and specificity entirely depends on it [28].

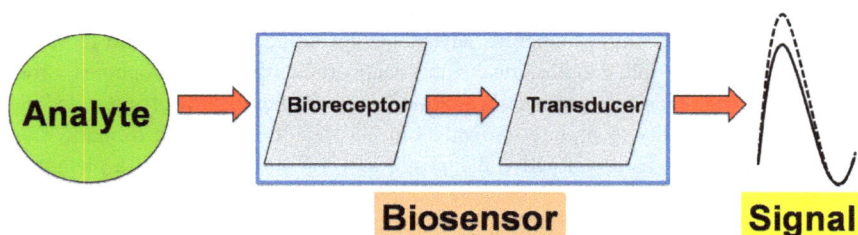

Figure 1: Schematic illustration of a biosensor.

To achieve this, a variety of recognition molecules like antibodies, enzymes and nucleic acid (aptamers) are employed to a given target with enhanced affinity and specificity for the development of fast, reliable and sensitive biosensors. However, the use of antibodies as immunosensors and enzymes as enzyme sensor has not gain much recognition when compared to aptasensor due to various advantages associated with aptamers like high stability with minimal cross-reactivity, ease in modifications, large dynamic detection range, and a higher resistance to denaturation and prolonged shelf life [29–31]. Moreover, aptamer evinces structural switching in presence of its cognate target that makes them an ideal molecular recognition element for electrochemical sensing platforms. Therefore, all these properties of aptamer-based aptasensor make it a first choice to be impressively used in a variety of applications for clinical, environmental monitoring, medical, and as well as in food analysis [32–36].

2.   Aptamer-based sensing platform

Various readout mechanisms have been carried out by the aptasensor from label-free methods, such as quartz crystal microbalance (QCM) measurements and surface plasmon resonance (SPR) [37], to label based aptasensor methods such as electrochemistry [38,39], colorimetry [40,41], fluorescence [42–45], chemiluminescence [46,47], mass sensitivity [48], and amplification of nucleic acids [49] and field effect transistors which all together enhanced the advancement of this field. The three most widely used aptasensors for the analysis of small molecules i.e., fluorescence, colorimetry and electrochemistry is depicted in **Figure 2A, B and C** respectively. Among all these biosensors, whether it is optic (fluorescent, colorimetric, chemiluminescent, and SPR),

Biosensors: Materials and Applications   Materials Research Forum LLC
Materials Research Foundations **47** (2019) 1-50   doi: http://dx.doi.org/10.21741/9781644900130-1

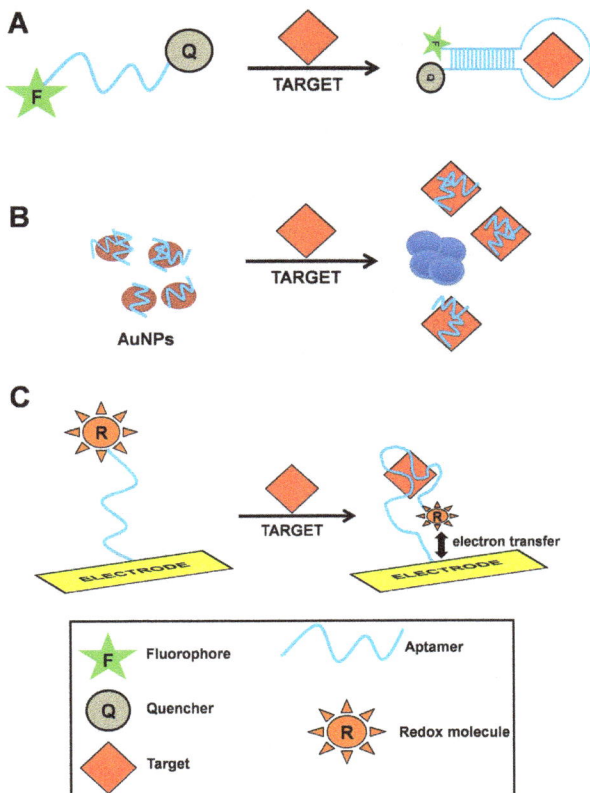

Figure 2: Aptamer-based three different formats of biosensors. (A) Fluorescence, (B) Colorimetry and (C) Electrochemistry aptasensors.

electric (amperometric, impedimetric and ECL) or mass-based (piezoelectric), the optical and electrical aptasensor are the most powerful and widely used analytical tools due to their various advantages such as high sensitivity, easy-to-use chemistry, low cost, easy handling, minimal requirement of sophisticated instrument , real-time and high-frequency monitoring [5,50,51]. On the other hand, aptasensor based on mass have not been extensively used as an analytical tool as compared to others due to lack of proper

Biosensors: Materials and Applications   Materials Research Forum LLC
Materials Research Foundations **47** (2019) 1-50   doi: http://dx.doi.org/10.21741/9781644900130-1

immobilization of aptamer on the crystal surface and the viscous drag experience in the liquid phase. The transducer associated with electrochemical aptasensor based on the principle of change in an electric signal i.e., the immobilized biomolecule (aptamer) and target (analyte) produce or consume ions or electrons, which change the measurable electrical properties of the solution, such an electric current or potential. Change in the electric signal can be readout as different parameters such as current (amperometric), the potential (potentiometric), impedance (impedimetric) or conductance (conductometric). Recently, various researchers around the world have developed electrochemical aptasensors using different strategies like amperometric, potentiometric, cyclic voltammetry (CV), differential pulse voltammetry (DPV) and impedimetric sensor for the detection of diverse analytes ranging from small molecules (saxitoxin, bisphenol A, tetracycline in milk, Ochratoxin A) to whole cell (pathogenic bacteria) [52–58].

Amperometric is a widely used electrochemical detection method which involves constant potential and monitoring the change in current is associated with the oxidation/reduction of an electroactive molecule (eg: Methylene blue, MB) [1], which are involved in the recognition process. The recorded current response is proportional to the concentration of the target to be analyzed in the sample [59–61].

Potentiometric sensors are based on either solid-state or solvent polymeric membrane ion-selective electrodes [62] and become a well-established tool for analysis of small molecules such as ions [63,64]. Moreover, the nature of the ion selective membrane determines the selectivity of the electrode [59]. The electrodes play a major role in this type of solid-state sensor as surface potential change take place on electrodes through the target-induced adsorption, desorption, uptake of ions and charged molecules on the surfaces [63,65–67]. With continuous advancement in the field to improve the sensitivity and selectivity of the sensors, researchers are designing and introducing new materials, synthetic receptors, and various specific recognition molecules. In 2012, dithizone modified gold nanoparticles based potentiometric sensors have been developed by Michalska et al. [68] for the determination of copper ions in which the cations diffuse into the gold nanoparticle layer and the interaction take place with dithizone leads to change in the surface charge. The potentiometric aptasensor has been extensively adopted as a promising transducer due to its simple device design, rapid response, field-portable, ease of use, low cost and its application has a wide variety of range from small molecules to whole-cell detection. Likewise, Rius et al. [69] has developed an aptamer-based potentiometric sensor using functionalized single-walled carbon nanotubes (SWCN) as a transducing material for the detection of ultra-low bacterial cfus (colony forming units). Recently, Wei Qin's [58] group also shown the detection of an endocrine disrupter small molecule Bisphenol A (BPA), using the for mentioned strategy but instead of SWCN

Biosensors: Materials and Applications   Materials Research Forum LLC
Materials Research Foundations **47** (2019) 1-50   doi: http://dx.doi.org/10.21741/9781644900130-1

they used carboxylated multiwall carbon nanotubes (CNTs), on which the aptamers (polyanion) were immobilized layer-by-layer onto the electrode surface.

Voltammetric biosensors use different voltammetric techniques such as Differential Pulse Voltammetry (DPV), Cyclic Voltammetry (CV), Square Wave Voltammetry (SWV) and Alternating Current Voltammetry (ACV) [70]. The unique properties of this sensor are, its scanning potential over the set potential range where both the current and potential are measured and recorded [70]. Similarly, the amperometric sensor also recorded current response is proportional to the concentration of the target to be analyzed in the sample. Moreover, multiple analytes (target) with different potential peaks are detected in a single scan which makes it a simple, rapid and time-saving sensor [71].

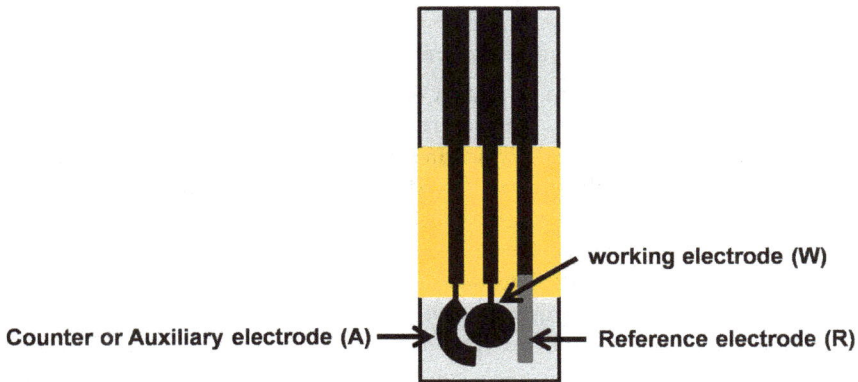

Figure 3: Schematic representation of Screen Printed Electrode (SPE) where R is Reference electrode, W is Working electrode and A is Auxillary electrode.

The choice of the electrode is one of the most important contributing factors in the development of biosensors whether it is amperometric, potentiometric or voltammetric aptasensors as the electrochemical reaction occurring on the surface of the electrode. Generally, a three electrodes system is adopted in aptasensor namely the reference electrode (R), a counter or auxiliary electrode (A) and a working electrode (W), also known as the sensing or redox electrode **(Figure 3)** [72]. The 'R' electrode, commonly composed of Ag/AgCl, is kept at a distance from the reaction site in order to maintain a known and stable potential. The 'W' electrode serves as the transduction element in the biochemical reaction, while the 'A' electrode established the connection to the electrolytic solution so that a current can be applied to the W electrode [73]. Usually, carbon and noble metals such as gold, platinum, silver and stainless steel are used as working electrode surface due to their exceptional electrical and mechanical properties [74,75]. With the advent of screen-printed electrodes (SPE), glassy carbon electrodes (GCE), and carbon paste electrodes (CPE) all of them have gained much appreciation due to their prominent characteristics, such as simple and low-cost fabrication, mass production, and conveniently practical application [74]. Incorporation of these electrodes in aptasensors enables them to use few microlitre volumes of the sample and reagents to be used for the analysis which have a great advantage as in case of clinical diagnosis where the sample volume is a major concern [61,76]. The SPE based biosensors have been widely used for detections of biomolecules [77], pesticides [78], antigen [79] and anion [80].

3.   Immobilization of recognition molecules

The technique of aptamers immobilization onto the surface of transducer or electrode plays a most crucial step in shaping the overall performance of electrochemical aptasensor as the affinity, specificity and the stability of the aptamer towards its target depends on it. The immobilization method which is used in construction of aptasensors are mainly depended on the nature of aptamer and the surface onto which it is adsorbed i.e., electrode/ transducer, the physicochemical properties of the target molecule, and primarily the operating at optimal conditions of the sensor which all contribute in fetching the maximum activity of the sensor [81]. The most successful and widely used methods are immobilization of affinity interactions (streptavidin/avidin/neutravidin-biotin) and covalent interaction via functional groups other commonly used methods are physical adsorption (physisorption) and chemical adsorption (chemisorptions) [81–83].

In affinity interactions **(Figure 4A)**, streptavidin/avidin-biotin interaction has been used, as Streptavidin (homo-tetramers) have an amazingly high affinity via non-covalent interactions for biotin with a dissociation constant (K_d) of $10^{-14}/10^{-15}$mol/L [84]. In comparison to avidin, streptavidin is more frequently used as a linker due to its low non-specific binding than avidin which is related to its near-neutral isoelectric point (pI) [72]. Another favourable feature is that the very low amount of biotin-labelled aptamer is needed for the detection of the target [85,86].

In covalent interaction **(Figure 4B)**, the sensor surface is modified with the reactive groups like amine: $-NH_2$; hydroxyl: -OH; carboxylic acid: -COOH; aldehyde: -CHO) to which the modified aptamer like amino $(-NH_2)$- labelled is covalently linked onto the surface of the electrode which is pre-functionalized with thiolic acids followed by activation of the terminal carboxylic group (-COOH) via with EDC-NHS chemistry (1-ethyl-3-[3-dimethylaminopropyl]-carbodiimide   hydrochloride/N-

Biosensors: Materials and Applications   Materials Research Forum LLC
Materials Research Foundations 47 (2019) 1-50   doi: http://dx.doi.org/10.21741/9781644900130-1

hydroxysulfosuccinimide [82]. Covalent interaction is the most adopted method for the immobilization of the aptamer for two reasons first; it improves the uniformity, density and distribution of the aptamer, as well as reproducibility and homogeneity of the electrode surface which also contributing in high specificity of the aptasensors and allow minimum non-specific adsorption [87,88] and secondly, it also provides a variety of different electrode surfaces such as gold or platinum electrodes, glassy carbon, carbon pasta, silicate and polymer surfaces, gold nanoparticles (AuNPs) and gold nanorods (AuNRs) [89].

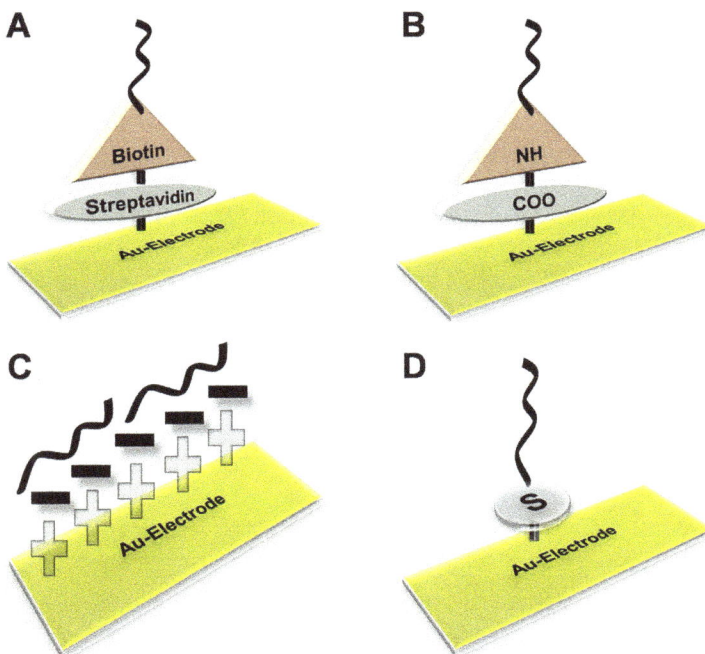

Figure 4: Schematic representation of aptamer immobilization on Au-Electrode surface. (A) Affinity interactions: biotinylated aptamer immobilized via streptavidin-biotin interaction; (B) Covalent interaction: amine-labeled aptamer covalently linked onto the electrode surface which is pre-functionalized with terminal carboxylic group (-COOH); (C) Physical adsorption: aptamer immobilized onto the electrode surface through charge-charge interaction via electrostatic forces; (D) Chemical adsorption: thiolated aptamer immobilized onto the electrode surface via covalent bond between the sulfur and gold atoms.

Biosensors: Materials and Applications   Materials Research Forum LLC
Materials Research Foundations 47 (2019) 1-50   doi: http://dx.doi.org/10.21741/9781644900130-1

In case of physical adsorption **(Figure 4C)**, the aptamers are immobilized to their surface via electrostatic interaction like Vander Waals, hydrophobic interaction, hydrogen bonds, and ionic interactions and no modification of aptamers is required but it has a limitation that the aptamer get easily desorbed from the surface due to the weak interaction on the surface [90,91]. The chemical adsorption **(Figure 4D)** is the most simple immobilization technique in which thiolate aptamer is covalently attached to the thiol groups (-SH) of the gold electrode surfaces in a stable and flexible self-assembled monolayer (SAM) with repeatable aptamer surface layers [88].

4.   Design and strategies of aptasensors

Different design and strategies have been employed on aptamer-based electrochemical sensors depending upon the size of analyte, sensitivity and modification required on the electrodes. Cheng et al. and Han et al. (2010) [92,93] extensively reviewed these strategies based on the differences in the design of the DNA/RNA-modified electrodes. Cheng et al. [92] has classified three different strategies of electrochemical aptasensors namely; configuration change (in which the analyte/target molecule binding induces either an assembly or dissociation of the aptamer); conformation change (in which the analyte/target molecule binding induces a variation in the conformation of the immobilized aptamer strand onto the electrode surface) and conductivity change (in which the analyte/target molecule binding exhibit "switches on" the conductivity of the surface-bound aptamer-DNA strand). Contrarily, Han et al. [93] have classified these strategies into four categories namely, target-induced structure switching mode; a sandwich or sandwich-like mode; target-induced dissociation or displacement mode; and competitive replacement mode. Recently, Rapini et al. [1] has reviewed two main approaches on the basis of the size of the analyte and required sensitivity through an electrochemical aptasensors detection; label (using labelled aptamers) and label-free (using non-labelled aptamers) formats.

In label formats, aptamer are covalently labelled either 3' or 5' ends with a variety of redox or electro-active molecule such as alkaline phosphatase (AP), glucose oxidase (GOD), horseradish peroxidase (HRP) enzymes or ferrocene dyes like neutral red or methylene blue (MB) and more recently nanomaterials such as gold nanoparticles (GNPs) and quantum dots (QDs) were also applied for the biosensing assays [94]. At one end, the aptamer is immobilized on the electrode surface and the other end labelled with redox or electro-active molecule is utilized for the generation of an electrochemical signal which will be amplified upon the formation of aptamer-target-complex [94–97]. To read the signal of this aptamer-target-complex in the form of an electrochemical signal, two strategies have been employed, one is 'switch-on' and the other is 'switch-off'. In case of

Biosensors: Materials and Applications   Materials Research Forum LLC
Materials Research Foundations **47** (2019) 1-50   doi: http://dx.doi.org/10.21741/9781644900130-1

the 'switch-on' strategy, the redox or electro-active molecule labelled aptamer undergo a conformational change upon binding with the target as a result, the distance between the redox molecule and the electrode surface decreases which allow the smooth flow of electron transfer and current signal. On the other hand, in the 'switch-off' strategy, the distance between the redox molecule and the electrode surface increases upon binding with the target in such a way that the labelled aptamer moves away from the electrode surface resulting in the inconsistent flow of electron transfer as a result decrease current flow [98–101].

In label-free formats, the aptamers are devoid of any labelled redox molecules and the aptamer-target interaction is monitored by electrochemical techniques like voltammetry and electrochemical impedance spectroscopy (EIS). In this method, the aptamer is electrostatically associated or diffusively interacted (attraction/repulsion) with electroactive molecules such as ferricyanide and ruthenium complexes $[Fe(CN)_6]^{3-/4-}$ and $Ru(NH_3)_6]^{3+}$ [102]. Moreover, based on the morphology of the analytes, aptasensors can be used in various formats such as sandwich and competitive assays. Generally, a sandwich assay is taking into consideration when the size of the analyte/target molecule is large as multiple recognition sites are required within the target for binding at the same time which is not possible in case of small molecules [103]. In a sandwich assay, two different aptamers are used for the detection of the analyte; the first aptamer is called capture aptamer which is immobilized on the electrode surface (solid phase) onto which the analyte is added. Appropriate washing steps required to make sure that no unbound components are left, the second labelled aptamer called detection aptamer mostly labelled with enzymes is incubated with the solid surface which is further coupled with the desired substrate for amplified detection. Basically, the amount of secondary aptamer bound to the electrode surface (solid phase) produce an electroactive product which is monitored by means of transducer [70]. On the other hand, a competitive assay is used when the analyte/target molecule is small and as the name suggests there is a competition between the two components for the third one. Here, the target is immobilized onto the electrode surface (solid phase), and the aptamer guided against it, is bound to the immobilized target. Sample containing the free target is then added onto the solid phase resulting into the competitive binding of the free target to the aptamer, as a result, the aptamer is released from the immobilized target which could be easily detected by transducers like EIS or SPR. As an alternate strategy, in spite of target, the aptamer can also be immobilized on the electrode surface (solid phase). In this method, prior to analyte detection, the labelled target is bound to the aptamer whereas non-labelled target present in the sample competitively bind to the immobilized aptamer, as a result, the bound target is displaced [104–106].

5.   Application of aptasensor for small molecules

5.1   Aptasensor for pesticide

In the present scenario, the population is increasing day-by-day and the demand to feed them poses a great challenge around the world. To meet growing demand, one has to improve the quality of crops yields and to achieve this task the crop should be free of insects, pests and diseases. According to UN Food and Agriculture Organization (FAO), nearly 40% of crops were destroyed while standing in the field and about 6-7% after harvesting due to pests and insects in developing countries [107–109]. The use of pesticide in preventing and controlling the crops from these losses has made chemical pesticides an integral part of the crop production. Pesticides are very efficient in the eradication of different insects and are easy in synthesis with low production cost [110,111]. It is estimated that nearly 3 million tons of pesticides are used worldwide out of which 45% is consumed in Europe alone, 25% in America and the remaining for the rest of the world [112,113]. Organophosphorus (OPs) is one of the most common classes of pesticides used worldwide [114] and is linked to severe issues related to human health and environmental problems. When Pesticides comes to direct contact with human beings it causes a harmful impact on human bodies and may be responsible for various diseases including but not limited to cancer, kidney, liver and respiratory damage, improper functioning of nervous and reproductive system. Its impact is not only limited to humans only but it also affects the zoonotic and aquatic lives as it starts accumulating in water bodies, soil and agriculture food products which results in the disturbance of the ecosystem. Overall, the situation is like 'on the horns of a dilemma' as on one hand, we have to fulfil the demand of the growing population and in doing so; pesticide play an important role in agriculture as without it around 70% of crop production could have been lost to pests [112] while its need is essential its global environmental effect should be minimized. Widespread application of pesticides (Organophosphorus) in the field results into an accumulation of pesticide residues in food products and water bodies which ultimately lead to severe complication to humans and animals. Considering these facts, various analytical and conventional techniques have been employed for the detection of toxic pesticides in aqueous solution which include gas chromatography-mass spectrometry (GC-MS) [115–118], high performance liquid chromatography (HPLC) [119–122], enzyme-linked immunosorbent assay (ELISA) [123–125] and electrophoresis [126]. However, all these methods are well-used in terms of reliability and relatively low limit of detection but in spite of this, they are restricted to be used in outdoor field settings and accompanied with numerous short-comings. These methods are time consuming, sophisticated, required expensive instrumentation, complex sample preparation and highly trained personnel which all limits their application on-site [127–

129]. Therefore, in order to overcome these drawbacks for the detection of pesticides researchers are now developing and designing modern approaches which are more reliable, rapid, simple, selective, cost-effective, user-friendly, etc. [1].

Recently, the aptamer-based biosensor has gathered much attention of researchers around the world due to its wide range of application in the biological, agricultural and medical field. As we know, aptamers are short single-stranded DNA/RNA which binds to its target with high affinity, specificity and sensitivity by acquiring three-dimensional structures [17]. Moreover, due to its unique features like small size, ease of synthesis, negligible immunogenicity, ease in modification, low cost production, chemical and thermal stability made this molecule as the first choice to be used in biosensor for the detection of antibiotics, food allergens, bacterial toxins, pesticides, toxic heavy metals, environmental pollutants, different class of infectious pathogens [130]. Aptamers can be generated to a wide variety of targets such as ions, drugs, proteins, peptides, viruses, bacteria, whole cell and tissues. Therefore, aptamer emerge as a promising recognition molecule in biosensor over antibody and enzyme which has the limitation of short shelf life, poor chemical and physical stability [131]. Due to these properties, aptamer-based biosensor provide an easy, sensitive, rapid, low cost, reproducible and reusable platform for the detection of different pestiferous substance [132]. Various aptasensors have been developed based on different principles such as fluorescent [133], chemiluminescent [134] and chromatographic [135] but aptamer-based electrochemical sensors are the most widely used aptasensors for the detection of pesticides [136].

Recently, a novel Electrochemical Impedance Spectroscopy (EIS) based aptasensor which is a highly sensitive, label-free electrochemical method has been developed by Madianos et al. [137] for the detection of herbicide atrazine, which is known to play a major role in carcinogenesis, birth defects and disturb the endocrine system due to which it has been banned by the European Union since 2004. The group has made platinum nanoparticles (Pt NPs) based microwires and integrated them between the interdigitated electrodes (IDEs) in a bridge-like manner. The Pt NPs microwires were chemically functionalized which allow efficient covalent immobilization of aptamer against atrazine onto the sensor surfaces. By employing this strategy, they achieved remarkably highly sensitive and selective detection of atrazine with a linear range of 100 pM to 1 μM and a limit of detection (LoD) of 10 pM [137]. The same group also detected another common and efficient pesticide acetamiprid, which inhibit the catalytic activity of acetylcholinesterase (AChE), an essential enzyme for the proper functioning of central and peripheral nervous systems [138] and is also reported to have a role in causing erectile dysfunction in human males [139] by using the same strategy and they were able to achieve a good response with a linear range of 10 pM to 100 nM with a lowest LoD of

1 pM [137]. In similar fashion, i.e., on EIS based aptasensor, three independent research groups, Fan et al., Fei et al., and Jiang et al. [140–142] have been able to detect acetamiprid with LoD of 1×10^{-9}M, 1.7×10^{-14}M and 3.3 x 10^{-14}M respectively. Rapini et al., [143] used a slightly different approach for the detection of acetamiprid which was based on a competitive format. In this format, a thiolated aptamer against acetamiprid was chemisorbed on the pre-immobilized with gold nanoparticles surface of graphite screen-printed electrodes (SPE). Followed by a solution containing acetamiprid and a fixed amount of biotinylated complementary aptamer was then applied and the competitive reaction was allowed to take place between acetamiprid and complementary biotinylated aptamer onto the electrode surface followed by the addition of streptavidin-alkaline phosphatase conjugate. The enzymatic electroactive product was then detected by Differential Pulse Voltammetry (DPV). Using DPV a LoD of 0.086 µM was determined. Weerathunge et al. [144] exploited a novel approach where the peroxidase-like NanoZyme activity of gold nanoparticles (GNPs) was used in combination with acetamiprid specific aptamer for the highly sensitive and rapid detection of acetamiprid based on a colorimetric biosensing assay. In the absence of acetamiprid, the peroxidase activity of the pristine GNP was inhibited by shielding its surface through the adsorption of acetamiprid-specific aptamer and hence no color change is observed on addition of 3,3,5,5-tetramethylbenzidine (TMB) substrate. Conversely, in the presence of acetamiprid, the peroxidase activity of the pristine GNP was resumed as the aptamer exhibited target responsive structural changes which further allowed its desorption from the GNP surface resulting in the color change of colorless TMB to the purplish-blue product in the presence of H_2O_2 **(Figure 5)**. The output of this NanoZyme activity of gold nanoparticles (GNPs) can either be directly visualized in the form of color change of the peroxidase reaction product or can be quantified using UV-visible absorbance spectroscopy. By employing this strategy, the author was able to achieve the detection of as low as 0.1 ppm within an assay time of 10 min.

Carbendazim (CBZ) is another class of pesticide (fungicide) used extensively in the field containing highly stable benzimidazole ring due to which it remains in the soil for a long time and get absorbed by the plants, leaves and fruits and ultimately entered into the food chain [145]. CBZ has a more or less similar effect as acetamiprid and causes infertility in males, hormonal imbalance and disruption of endocrine system [146,147]. Recently, Eissa et al. [148] developed label-free aptasensor for CBZ detection using EIS. The aptamer generated against CBZ was coupled with the redox molecule $[Fe(CN)_6]^{3-/4-}$ which undergoes conformation change upon binding with CBZ due to this the electron transfer change was measured by EIS. Through this, the aptasensor exhibited rapid response (30 min) with high selectivity and sensitivity with a detection limit of 8.2 pg/ml.

Biosensors: Materials and Applications   Materials Research Forum LLC
Materials Research Foundations **47** (2019) 1-50   doi: http://dx.doi.org/10.21741/9781644900130-1

The aptasensors developed by these authors have also been applied to real-life sample such as wastewater, fruit juices, cucumbers, tomatoes, etc. for the detection of pesticides analysis.

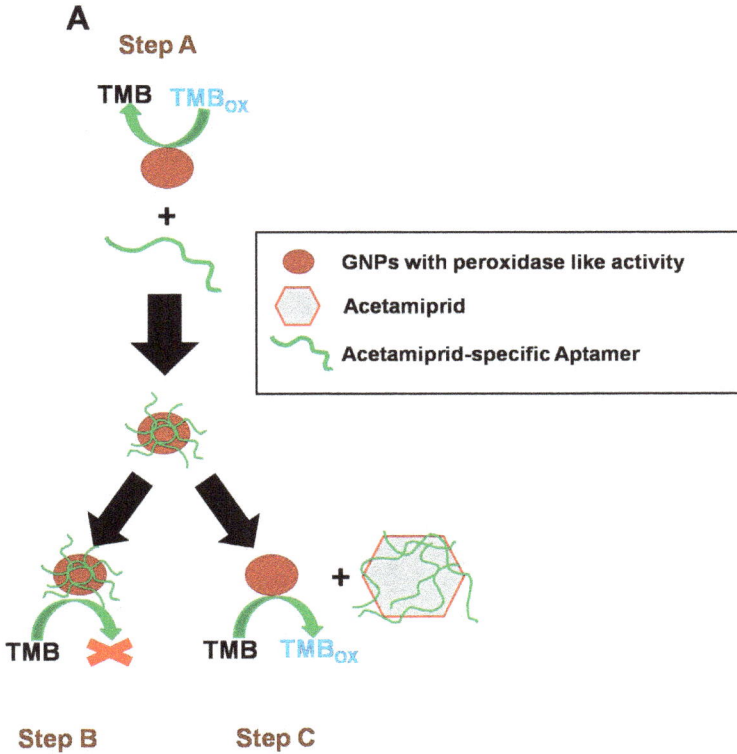

Figure 5: Schematic representation showing inhibition of NanoZyme activity of GNPs using an acetamiprid-specific aptamer (Step A) shows intrinsic peroxidase-like activity of pristine GNP that gets inhibited after conjugating with acetamiprid-specific aptamer (Step B), in the presence of acetamiprid the aptamer undergoes target responsive structural changes and forms a supramolecular complex with acetamiprid resulting in free GNP to resume its peroxidase like activity (Step C).

5.2   Aptasensor for small molecules (Cocaine and Adenosine)

Cocaine another small molecule generally known as coke is mostly used as an illegal abusive drug globally due to its physiological and psychological related health issue among people [149]. 2014 report suggests that globally around 1.5 million cocaine users are of age 12 or above (0.6% of the population) which were relatively stable in 2009 according to National Survey on Drug Use and Health (NSDUH) in the United States [150]. Cocaine is regarded as a powerful stimulant of the central nervous system (CNS) which increases the dopamine level and inhibits the reuptake of neurotransmitter at the synapse [151]. Abusive and an overdose of cocaine leads to serious health issues like cardiac arrest, anxiety, an expanse of human immunodeficiency and organ failure [152,153]. Moreover, addiction of cocaine is a serious issue among the adolescent population [154]. Therefore, detecting and quantify cocaine has great importance in forensic, law enforcement clinical and medical diagnosis. Several analytical methods are available for its detection which are time-consuming, high-cost, sophisticated instrumentations and require extensive sample preparation such as high performance liquid chromatography (HPLC), Ion mobility spectrometry (IMS), gas chromatography-mass spectrometry (GCMS), capillary electrophoresis (CE), radioimmunoassay and enzyme-linked immunosorbent assay (ELISA) [155–161]. In this respect, aptasensor showed a significant contribution due to their obvious reasons mentioned above. Roushani et al. [162] developed highly sensitive and specific aptasensor based on the EIS method for the detection of cocaine using for the first time novel nanocomposite of dendrimer and silver nanoparticles (AgNPs). In this study, they modified the SPE with a nanocomposite of dendrimer and AgNPs, onto which cocaine-binding aptamer was immobilized, due to which a complex of cocaine-aptamer was formed on the electrode surface. Consequently, folding and conformational change happened in the aptamer structure that result into the change in the signal which was recorded as an impedimetric signal. Using this method, they determined the LoD of 333 mol L^{-1}. Hashemi et al. [163] developed label-free aptasensor for cocaine detection. The assay was based on customized SPE with three-dimensional magnetic reduced graphene oxide(3D-MRGO)/polyaniline(PA)/gold nanoparticle(AuNP) nanocomposite for EIS detection of cocaine. The developed assay determined the cocaine detection limit of 0.029 nM and in order to achieve the real application of the sensor, the author also analyze the cocaine in urine and serum samples which found to be satisfactory. In a novel approach, Shen et al. [164] developed electrochemical- enzyme-based aptasensor for the detection of cocaine. They have designed a cocaine aptamer and cleaved them into two fragments, one fragment was thiol labeled for the immobilization on gold electrode surface while the other was biotinylated at its 3'-end before being conjugated with streptavidin. Afterward,

Biosensors: Materials and Applications   Materials Research Forum LLC
Materials Research Foundations **47** (2019) 1-50   doi: http://dx.doi.org/10.21741/9781644900130-1

the two fragments form a supramolecular structure in presence of cocaine to which streptavidin molecule was conjugated that provide an anchor to biotinylated circular DNA. This was followed by the addition of nucleotides and phi29 DNA polymerase this initiated the rolling circle amplification (RCA) produce micrometer-long single-strand DNA containing hundreds of tandem-repeats which could bind to the biotinylated detection probes. Streptavidin-alkaline phosphatase was then added to capture the biotinylated detection probes which catalyzed the α-naphthyl phosphate (α-NP) substrate to produce an enzymatic product and was detected by DPV. Quantitative detection limit of cocaine via this approach was found to be 1.3 nM. Several other aptasensors based on electrochemical [165–167], colorimetric [168] and fluorescent [169,170] assays have been developed which also exhibited their potential to quantitatively determine cocaine in real biological samples.

Adenosine is the most common chemical entity found in almost all living systems. Being an endogenous nucleoside, it has multiple impactful pharmacological and physiological functions [171] which include stimulation of angiogenesis, regulate the blood flow in CNS and arteries, regulate immune responses, inhibit inflammatory reactions, the supply of efficient myocardial oxygen and also modulates neurotransmission [172–174]. Adenosine is also an essential intermediate molecule and takes part in the synthesis of ATP, adenine, vidarabine and adenylic acid [175,176]. Therefore, determination of adenosine level under optimized conditions is crucial to further investigate physiological functions, behaviour and also related to heart and brain physiology [177]. For routine analysis of adenosine various analytical techniques have been employed like high-performance liquid chromatography (HPLC) [178,179], GC/LC-MS [180], capillary electrophoresis [181] and radioimmunoassay [182] but due to their cumbersome sample preparation requirement, heavy instrumentation and trained personal requirement makeing them difficult for on-site and real-time detection. Aptasensor, on the other hand, delivered rapid, sensitive, and cost-effective detection for small molecules like adenosine.

5.3   Aptasensor for Lysozyme

Recently, Deng et al. [183] developed a bifunctional electrochemical-based aptasensor for highly sensitive detection of lysozyme (protein). Lysozyme is the natural antibiotic of the human body as it possesses tremendous lytic activity against polysaccharide (bacterial cell wall) and also exhibits antiviral activity [183]. Lysozyme is also a potent biomarker of rheumatoid arthritis and certain other diseases, therefore, its quantitative detection is important [184–186]. In the present assay, two aptamers against adenosine and lysozyme were immobilized on the modified gold electrode surface to form DNA-DNA duplex. This was followed by the addition of an electroactive redox molecule $[Ru(NH_3)_6]^{3+}$ which could interact with a negatively charged backbone of DNA. In the presence of adenosine and lysozyme, the aptamers undergo structure switching to form DNA-adenosine complex and DNA- lysozyme complex respectively from DNA-DNA duplex as a result lysozyme aptamer dehybridized and released into the solution. In either case, the change in the peak current of $[Ru(NH_3)_6]^{3+}$ would decrease and monitored by cyclic voltammertic (CV). Moreover, the sensitivity of the aptasensor were used to enhance by using DNA functionalized Au nanoparticles (DNA-AuNPs) as a result more electroactive molecule $[Ru(NH_3)_6]^{3+}$ could be added. The developed aptasensor exhibited LoD of 0.02 nM for adenosine and 0.01 μg/mL for lysozyme [187]. Shahdost-fard et al. [188] designed a label-free approach for the detection of adenosine. Herein, they used nanocomposites containing multi walled carbon nanotubes, ionic liquid and chitosan (MWCNTs-IL-CHIT) for the modification of glassy carbon electrode (GCE) onto which amine labeled capture-aptamer (12-mer) was covalently immobilized via linking agent glutaraldehyde (GA). Afterward, adenosine-specific aptamer (32-mer) was immobilized on the electrode surface through hybridization methylene blue (MB) as the redox probe. As a concentration of adenosine increases, the peak current decreases linearly due to the formation of aptamer-adenosine complex and consequently the aptamer is released from the electrode surface. LoD of 150 pM was achieved by means of DPV using this strategy subsequently the author has also applied the aptasensor for measuring adenosine concentration in real samples like drug formulation and blood serum. More or less in a similar approach, Wu et al. [189] also proposed an aptasensor in which GCE modified with Au-GS was decorated with 3'-amine labeled capture aptamer (A1) and thionine (TH). 3'-SH labeled adenosine aptamer (A2) was adsorbed onto palladium/copper MWCNTs (PdCu@MWCNTs)-conjugated with multiple bienzymes, glucose oxidase (GOx) and horseradish peroxidase (HRP). This complex of A2/PdCu@MWCNTs/HRP/GO$_X$ was then immobilized on the electrode surface via the hybridization between A1 and A2. In this assay, change in TH peak current was monitored by square wave voltammetry (SWV) through multiple HRP-catalyzed TH-mediated electro reduction of H_2O_2 which was generated by GOx. In presence of adenosine, the TH peak current decreases due to the release of adenosine and A2/PdCu@MWCNTs/HRP/GO$_X$ from the modified electrode. The developed aptasensor could able to detect the adenosine with LoD of 2.5 nM. Moreover, the application of this aptasensor was also tested in human serum samples showing promising results. Wang et al. [190] developed label-free aptasensor based on EIS for ultrasensitive detection of adenosine with dual back fillers with superior LoD when compared with single back fillers aptasensor. In this assay, they modified the gold electrode surface by co-assembling the thiolated aptamer, dithiothreitol (DTT) and 6-mercaptohexanol (MCH) as a result a complex was formed of Au/aptamer-DTT/MCH. The electro-active redox molecule $[Fe(CN)^6]^{3-/4-}$ was used as a probe consequently an interfacial electron transfer resistance (Ret) of the aptasensor was increased with adenosine concentration. The change in Ret (ΔRet) was monitored by EIS was found to be linear against the logarithm of adenosine concentration over the range from 0.05 pM to 17 pM. The detection limit of adenosine with this strategy was determined as 0.02 pM and that of aptasensor decorated with Au/aptamer/MCH and Au/aptamer-DTT alone was found to be 0.03 nM and 0.2 pM respectively.

5.4   Application of aptasensor for bacterial, viral and protozoan

Infectious diseases are the invasion of disease-causing organisms such as bacterial, viral or protozoal agents in human bodies that may cause disease (or infection) or even mortality. Infectious diseases are highly contagious and can easily be transmitted through food, air and water. This has a major economic impact. Food and water borne pathogens have been a cause of a large number of infectious diseases worldwide and most prevalent in low-income countries. According to the World Health Organization (WHO) 2015 report, lower respiratory infections are among the top 10 leading causes of deaths across all income groups [191]. It is important to contain these pathogens and detect their presence in food and water before they enter the human body and cause a serious outbreak. Early detection, identification and quantification of microbial pathogens is very critical for public health and ensuring food safety. Microbial-culture based tests and immunological assays are commonly used for diagnosis of infectious diseases [192]. These methods are either time consuming or require special conditions for storage and handling of monoclonal antibodies to prevent their denaturation and trained personnel necessity results in increased cost of analysis. Thus, the development of fast, reliable, easy to use and cost-effective pathogen detection method is a necessity. Due to non-immunogenic nature, cheaper to synthesize, easy to modify with high affinity and high specificity for their target make aptamers are a promising alternative to immunological methods. Several aptasensors have been developed for the detection of bacterial, viral and protozoan proteins. Different strategies have been adopted for the whole cell detection of bacteria using quantum dots and gold nanoparticles. Aptamer-quantum dots based diagnostic assay was developed to detect *Bacillus thuringiensis* (BT) spores [193]. The aptamer-QD conjugate was incubated with BT spores for 30 min followed by filtration and collection of the spores conjugated to aptamer-QDs and fluorescence measurement. A series of control samples such as samples without spores, QDs not conjugated to aptamers and *Bacillus globigii* (BG) spores were also used to test the specificity of the assay. The assay can detect BT concentration of about 10^3 CFU/ml and can differentiate BT and BG spores at concentrations above 10^5 CFU/ml. In another strategy, aptasensor using gold nanoparticles (AuNPs) has been developed for the detection of enteropathogenic bacteria such as *Escherichia coli* (*E. coli*) O157:H7 and *Salmonella typhimurium* [194]. The aptamer was used as a target recognition element and unmodified AuNPs served as a signal transducer element. Compared to the existing methods, no labelling of aptamers or modification of AuNPs is required which significantly reduces the cost. The aptamer-AuNPs solution was mixed with the target bacteria and several control bacteria's which were characterized using UV vis spectroscopy or visual observation. The detection of the target bacteria bound to the aptamer was accomplished by red-to-purple colour change upon high-salt conditions by bacteria-induced aggregation of the aptasensor. The reported sensitivity of the assay is as low as 10^5 CFU/ml of target bacteria within 20 min or less, with 100% specificity.

Some of the life-threatening common viral diseases includes human immunodeficiency virus (HIV-1), hepatitis C virus (HCV) and avian flu variants e.g. H5N1. According to WHO, there will be approximately 36.7 million people living with HIV at the end of 2016 and 71 million people globally have chronic hepatitis C infection (HCV) [195,196]. Viral diseases represent an important portion of global public health concerns with thousands of deaths annually. Effective and faster identification of viral pathogens is needed to help prevent the transmission, efficient disease management control and transmission. The conventional viral detection diagnostic method e.g. Enzyme-Linked Immunosorbent Assay (EIA) and supplementary nucleic acid testing (NAT) methods have limitations due to generating false-positive and false-negative results, high cost and labour-intensive thus more effective and improved viral detection methods are required [197,198]. Lee et al. [199] generated a biosensor that specifically detects the core antigen of the hepatitis C virus (HCV) proteins. High affinity and high specificity RNA aptamers were selected against the HCV core antigen using the SELEX procedure. Two aptamers with highest binding affinity (K_d =142nM and K_d =224nM) showed high specificity with the core antigen only and not with NS5 antigen, which is another HCV-expressed antigen and other proteins such as HCV NS3, E1/E2 antigens, HIV p24 antigens, and BSA proteins. Using the two high-affinity aptamers, an aptamer-based biosensor was developed by immobilizing the selected high-affinity RNA aptamers within the 96 well plate using sol-gel based immobilization method. The 96-well format biosensor was not only used for detection of the purified protein but to human sera to determine its potential in real patient's samples. The results demonstrate that this aptamer-based biosensor prototype was able to specifically detect pure recombinant core antigen and serum of the HCV infected patients and has potential application for early diagnosis of HCV infection. A simple quartz crystal microbalance (QCM) construct developed by Wang and colleagues was used for rapid detection of avian influenza virus (AIV) H5N1 [200]. The selected high affinity and high specificity aptamer against AIV H5N1 surface protein were hybridized with ssDNA forming a cross-linker in the polymer hydrogel. The aptamer hydrogel was immobilized on the gold surface of the QCM sensor remaining shrunk in the absence of the H5N1 virus. When exposed to the target virus, the binding reaction between the aptamer and the virus led to the dissolution of the aptamer-ssDNA linkage, resulting in the swelling of the hydrogel. The swelling of the hydrogel was monitored in terms of decreased resonance frequency by the QCM sensor. Among the three polymeric hydrogels used as a sensor coating material in the study, hydrogel III coated aptasensor demonstrated the highest sensitivity of 0.0128 HAU (HA unit) and high specificity for the target H5N1 with total detection time of 30 minutes. Tat (Trans-Activator of Transcription) is a 86 amino acid long trans-activator protein that binds to the trans-activation response region (TAR-1) and regulates viral transcription in HIV-1 [201]. The Tat protein of HIV-1 appears to be a suitable target for the detection of HIV-1. Tombelli et al. [202] generated two aptamer-based biosensors using quartz crystal microbalance (QCM) and surface plasmon resonance (SPR) techniques for the detection of HIV-1 Tat protein. For both QCM and SPR techniques, the immobilized aptamer was allowed to interact with the Tat protein at different concentrations followed by washing with the buffer to remove the unbound protein. This detection approach demonstrated high reproducibility for the immobilization step and the binding step and high sensitivity for the Tat protein with the two biosensors with the negative control Rev protein responding only for the 17% (QCM) and 25% (SPR) with respect to Tat at the same concentration.

Another important field of aptamer biosensor applications is the detection of protozoan infections. Protozoan infections contribute significantly to the burden of infectious diseases worldwide particularly the developing world. In 2010, an estimated 350 million cases of illness were reported that were caused by one of the enteric protozoans infection (*Cryptosporidium* spp., *Entamoeba histolytica*, and *Giardia* spp.) resulting in nearly 33,000 deaths [203]. In 2016, there were roughly 216 million malaria cases and an estimated 445,000 malaria deaths worldwide [204]. Considering the significant health burden caused by protozoan parasites dictates a necessity to search for novel approaches of fast and reliable diagnostics and treatment. The existing diagnostic methods for malarial pathogen detection have several limitations such as the microscopic examination of blood films is a time-consuming method, lengthy polymerase chain reaction (PCR) assays limits the usefulness of PCR in routine clinical practice and rapid diagnostic test (RDT) has inadequate sensitivity and specificity due to low concentrations of parasites [204–207]. Due to the limitations of the existing methods, rapid and accurate diagnostic methods are required for the detection of malaria pathogen. Seonghwan et al. [207] selected high affinity and high specificity DNA aptamers against *Plasmodium* lactate dehydrogenase (pLDH), which is a biomarker for malaria and shows a high expression level in both the sexual and asexual stages of parasites. The selected pLDH aptamers could bind to both recombinant *Plasmodium vivax* LDH (PvLDH) and *Plasmodium falciparum* LDH (PfLDH) with high sensitivity (K_d = 16.8–49.6 nM). The selected aptamer with lower K_d value referred as pL1 aptamer in the study was used to design an aptasensor using electrochemical impedance spectroscopy (EIS). Different concentrations of PvLDH (1 pM, 10 pM, 100 pM, and 1 nM) were applied to the pL1 aptamer monolayer on the gold electrode, and change in charge transfer resistance (Rct) was recorded. As the concentration of PvLDH increased, the Rct values successively increased. A similar pattern was observed with PfLDH as well. The calculated detection limits were 108.5 fM for PvLDH and 120.1 fM for PfLDH. The aptasensor selectively detected *Plasmodium* vivax lactate dehydrogenase (PvLDH) and *Plasmodium* falciparum lactate dehydrogenase (PfLDH) proteins but also native pLDH proteins in the infected blood samples in this work thereby demonstrating its potential as a useful tool for malaria diagnosis. *Cryptosporidium* is a common intestinal protozoan parasite found in humans and animals and is associated with diarrhea cases and outbreaks in developed countries. It is a leading cause of waterborne disease among humans in the United States and approximately 8% of domestically acquired incidents of cryptosporidiosis are food-borne in the United States [208,209]. The spread of *Cryptosporidium* spp. is caused when oocytes of *Cryptosporidium* spp. shed with feces of the host and are immediately infective to the subsequent hosts. Besides water, the routes of transmission of cryptosporidiosis include foodborne, person-to-person (i.e., the fecal-oral route) and zoonotic [210]. Unfortunately, there are not many standardized techniques available for the testing of *Cryptosporidium* spp. in clinical samples. The existing diagnostic tests such as immunofluorescence microscopy and the polymerase chain reaction (PCR) are very useful in testing food samples. However, these tests have few limitations, most notably, complex sample pre-treatment, inadequate specificity when the parasite concentration is low in the foods, expensive instrumentation, and presence of PCR inhibitors [211]. In recent years, electrochemical aptasensors emerged as a promising alternative method for pathogen detection [212]. Electrochemical aptasensors are simple, economical and convenient for on-field applications. In 2015, Asma et al. [213] identified 14 high affinity and high specificity aptamer clones against *Cryptosporidium parvum* (C.*parvum)* oocytes after 10 rounds of selection using cell-SELEX technique. An inexpensive electrochemical aptasensor was developed in the study to screen each of the 14 aptamer clones against C.*parvum* oocyte by detecting the aptamer-oocytes binding using square wave voltammetry (SWV) measurements. The limit of detection of the aptasensor was as low as approximately 100 oocytes with the ability to discriminate between oocytes of C. *parvum* and *Giardia duodenalis*, henceforth suggesting the developed aptasensor a better

Biosensors: Materials and Applications   Materials Research Forum LLC
Materials Research Foundations **47** (2019) 1-50   doi: http://dx.doi.org/10.21741/9781644900130-1

alternative to current methods of detection such as immunofluorescence microscopy and PCR.

5.5   Application of aptasensor for non-infectious disease

Cancer is a major health problem and one of the leading causes of death in the United States and across the world. In 2012, there were 14.1 million new cases and 8.2 million cancer-related deaths worldwide [214]. Despite the diagnostic improvements, the early stage detection of cancer cells with high sensitivity and selectivity is a key challenge in cancer diagnosis, prevention and treatment and considered to be the most effective way to improve the survival rate [215,216]. The existing techniques for the detection of cancer cells include microarray, immunophenotyping by flow cytometry and amplification of malignant cell mutations by PCR [217–219]. However, these methods suffer from time-consuming, false-positivity and false-negative results, and inadequate efficiency. Therefore, the development of new techniques that can rapidly identify and characterize the cancer cells has been of great importance. In order to improve the sensitivity and specificity, different aptamer-based electrochemical methods have been developed for the detection of cancer cells and for sensitive protein detection [220,221]. T-cell acute lymphoblastic leukaemia (T-ALL) is a type of acute leukaemia that progresses quickly affecting the lymphoid-cell-producing stem cells, in particular, the T lymphocytes and accounts for approximately 15% and 25% of ALL pediatric and adult patients respectively [222]. Xiao et al. [223] developed a label-free graphene-oxide (GO)-based aptasensor that exhibits high sensitivity and high selectivity to the presence of T-cell acute lymphoblastic leukemia cells (rare CCRF-CEM cells). The principle of the aptasensor used in this study was based on the cell-triggered cyclic enzymatic signal amplification (CTCESA) based fluorescence. In the absence of the target cells, the GO-based FÖrster resonance energy transfer (FRET) process quench the fluorescence and hairpin aptamer probes (HAPs) and dye-labelled linker DNAs stably coexisted in solution. However, in the presence of the target cells, the specific binding of hairpin aptamer probes (HAPs) to the target cells resulted in conformational change and cleavage of linker DNAs with terminal labelled dyes which resulted in the fluorescence due to the inability of cleaved linker DNA's binding with the GOs and prevent the FRET process. This fluorescent based assay demonstrated high selectivity and sensitivity of the GO-based aptasensor for the target cells with cell concentration ranging from 50 to 10^5 cells. This aptasensor demonstrated a detection limit of 25 cells, which is 20 times lower than the detection limit of existing fluorescence aptasensors. Due to its high sensitivity and selectivity and it being easy to operate, this GO-based aptasensor can be used as a cost-effective approach for early cancer diagnosis. In another study, the aptamer-based electrochemical sensor was developed for detection of different potential protein targets in plasma of lung cancer patients (LCP) and healthy people (HP) using square wave voltammetry (SWV. A high affinity and high specificity DNA aptamer ($K_d = 38nM$) was generated against lung cancer tissue and circulating tumor cells in blood referred as LC-18 in the study was thiolated on the 5'-end and immobilized onto a gold electrode to construct the aptasensor and used for electrochemical measurements. Blood samples taken from patient's vein were incubated with the masking DNA (1ng/µL) for 60 min prior to incubation with aptasensor for another 60 min. SWV for both LCP and HP blood plasma samples were analyzed by the aptasensor. Change in the current (Δ I) for the LCP group was 41.6 µA and for the HP group was 17.0 µA was monitored. Furthermore, the sensitivity of the aptasensor was enhanced by 100 times by sandwich formation with silica-coated iron oxide magnetic beads grafted with hydrophobic C8 and C4 alkyl that binds to the target proteins thereby promoting the reduction of the active surface of the electrode and henceforth decrease of the current. The aptasensor was stable against nucleases in the plasma for over 5 hours. The results from this work demonstrated the electrochemical aptasensor ability to detect cancer-related targets in crude blood plasma of lung cancer patients.

Diabetes is a chronic disease where the pancreas doesn't make enough insulin hormone to metabolize the glucose resulting in elevated levels of glucose in the blood. The prevalence of diabetes has increased rapidly over the past 20 years and is one of the leading causes of premature morbidity and mortality worldwide. According to CDC's Division of Diabetes Translation, 23.4 million people had diagnosed diabetes in 2015, compared to only 1.6 million in 1958 [224]. The most commonly adopted clinical method to monitor the progression of diabetes is the combined measurement of blood glucose and glycosylated haemoglobin (HAb1c). However, these methods are often unreliable in patients with hemolytic anaemia, thalassemia and abnormal haemoglobin where haemoglobin production or red blood survival is affected in the patient's body [225]. For better detection and monitoring of diabetes, Apiwat and his colleagues [226] selected DNA aptamers that could measure the glycosylated human serum albumin (GHSA), which is an intermediate marker for diabetes mellitus. The identified high binding affinity and specificity DNA aptamer against GHSA protein was modified and fluorescently labelled to conjugate with a monolayer of graphene oxide (GO) to form GO-aptamer sensor. Studies have demonstrated that graphene oxide (GO) can interact with ssDNA via π–π stacking and being an excellent fluorescence quencher can quench the fluorescence of the fluorescently labelled DNA [227,228]. On binding to the GO, the aptamer fluorescence signal would quench and in the presence of the GHSA protein, the aptamer detached from the GO surface and bind to the protein resulting in fluorescent signal recovery. The limit of detection of the aptasensor was 50 µg/mL and demonstrated very high specificity for GHSA protein in PBS buffer. The aptamer bound to GO was stable for up to 16 hrs in human serum and DNase at both 37 °C and 25 °C. In addition, the aptasensor platform was able to detect the GHSA protein in clinical serum samples and has the potential for screening and monitoring diabetes mellitus. In a similar study, the 23-mer long glycated albumin (GA) aptamer was selected based on the sensitivity of the aptamer for GA as shown in the previous study by Apiwat et al. [226] and aptamer length suitable for FRET phenomenon. GA is a ketoamine formed by non-enzymatic glycation of human serum albumin through a series of reactions and modifications. GA is determined to serve as a better indicator for diabetes mellitus than other markers in diabetic patients on dialysis or hematologic diseases [229,230]. Ghosh and colleagues [229] designed an optical sensor comprised of GA aptamer, quantum dots and gold nanoparticles for the detection of GA. The aptasensor worked on the principle of fluorescence resonance energy transfer (FRET) phenomenon based on quantum dot-gold nanoparticle interactions. The photoluminescence intensity of the aptasensor increased with the increase in GA concentration (0 nM to 14500 nM) because of the unfolding of aptamer molecule to bind the GA protein and hence pushing the quantum dot and Au nanoparticle quench further apart. The sensor demonstrated limit of detection of 1.008 nM, which is significantly lower than what is reported in the previous studies [230–232]. The GA showed higher quenching efficiency compared to other control proteins such as HSA, transferrin, and IgG at the same concentration demonstrating the specificity of the aptamer and aptasensor to GA. The aptasensor has the potential to be used for efficient diagnosis and monitoring of diabetes mellitus in conjunction with the traditional method of glucose level monitoring.

6.   Future prospects

Aptamers represent a unique class of molecules that are larger than the small molecules and smaller than the antibodies. As mentioned before, aptamers display multiple advantages as biorecognition elements in sensor development when compared to antibodies. Some of the aptamers exhibit similar binding affinities of antibodies and outperform the antibodies in some instances. Being non-immunogenic, aptamers can be selected in conditions similar to the real conditions which are particularly useful for environmental and food aptasensors, easy to label or chemically modify as required and can be subjected to repeated cycles of denaturation and regeneration. However, aptamers susceptibility to nucleases attack may limit the clinical use of aptasensors. Unmodified aptamers especially RNA based susceptible to nuclease attack and have a very short lifetime in biological fluids (typically <10 min). Aptamers contain several sites such as sugar moiety, phosphodiester region where chemical modification can be introduced without interrupting its activity, for example, capping the 3' end rather than on 5' end or both the ends of the aptamer to enhance aptamer stability against serum nucleases [233]. Modifications to the sugars such as 2'-fluoro (2'-F) or 2'-amino (2'-NH$_2$) ribose groups on the pyrimidine residues reduce the affinity of nuclease for DNA degradation [234]. Further, the L-enantiomers form of nucleotides also known as spiegelmers makes aptamer nuclease resistant [235].

Aptasensors are small, chemically unchanging, and affordable detection tools. The electrochemical aptasensors have emerged as a promising tool among the other reported biosensors, due to their tremendous sensitivity and specificity for the recognition of target analytes. Recent advances have also witnessed the electrochemical label-free detection, thus overcoming the requirement of labels. A combination of modern electrochemical tools with cutting-edge aptamer technology will provide powerful analytical methods for the efficient detection of a wide range of analytes. Some of the challenges in the development of an analytical system such as high sensitivity, high selectivity, cost-effectivity, rapid, portable and real-time, on-site monitoring may be addressed by the construction of effective electrochemical aptasensors using frontline and advanced technologies. With the recent development in aptamer technology and following the progress of aptasensors, the next few years will witness an avalanche of new exciting electrochemical aptasensors devoted to the healthcare field. Some of the successfully commercialized aptamers available in the market include Ochratoxin A (OTA) sense system by NeoVentures, test strips-based APOLLO Dx technology and SOMAmer reagents by SomaLogic [236]. Overall, the potential of electrochemical aptasensors is immense, and this exciting and challenging area is on the brink of exponential growth. The future development in aptasensors may focus on the design of aptamers against unexplored target analyte, and then their subsequent integration in the electrochemical platform to monitor those target analytes.

References

[1]   R. Rapini, G. Marrazza, Electrochemical aptasensors for contaminants detection in food and environment: Recent advances, Bioelectrochemistry. 118 (2017) 47–61. https://doi.org/10.1016/j.bioelechem.2017.07.004

[2]   J. Aceña, S. Stampachiacchiere, S. Pérez, D. Barceló, Advances in liquid chromatography–high-resolution mass spectrometry for quantitative and qualitative environmental analysis, Anal. Bioanal. Chem. 407 (2015) 6289–6299. https://doi.org/10.1007/s00216-015-8852-6

Biosensors: Materials and Applications   Materials Research Forum LLC
Materials Research Foundations **47** (2019) 1-50   doi: http://dx.doi.org/10.21741/9781644900130-1

[3]   H.C. Liang, N. Bilon, M.T. Hay, Analytical methods for pesticide residues in the
        water environment, Water Environ. Res. 87 (2015) 1923–1937.
        https://doi.org/10.2175/106143015X14338845156542

[4]   S. Sudsakorn, A. Phatarphekar, T. O'Shea, H. Liu, Determination of 1,25-
        dihydroxyvitamin D2 in rat serum using liquid chromatography with tandem mass
        spectrometry, J. Chromatogr. B. 879 (2011) 139–145.
        https://doi.org/10.1016/j.jchromb.2010.11.025

[5]   N. Duan, S. Wu, S. Dai, H. Gu, L. Hao, H. Ye, Z. Wang, Advances in aptasensors
        for the detection of food contaminants, Analyst. 141 (2016) 3942–3961.
        https://doi.org/10.1039/C6AN00952B

[6]   A. Hayat, J.L. Marty, Aptamer based electrochemical sensors for emerging
        environmental pollutants, Front. Chem. 2 (2014).
        https://doi.org/10.3389/fchem.2014.00041

[7]   A. Dhiman, P. Kalra, V. Bansal, J.G. Bruno, T.K. Sharma, Aptamer-based point-
        of-care diagnostic platforms, Sensors Actuators B Chem. 246 (2017) 535–553.
        https://doi.org/10.1016/j.snb.2017.02.060

[8]   S. Kanchi, M.I. Sabela, P.S. Mdluli, Inamuddin, K. Bisetty, Smartphone based
        bioanalytical and diagnosis applications: A review, Biosens. Bioelectron. 102
        (2018) 136–149. https://doi.org/10.1016/j.bios.2017.11.021

[9]   C. Tuerk, L. Gold, Systematic evolution of ligands by exponential enrichment:
        RNA ligands to bacteriophage T4 DNA polymerase, Science (80-.). 249 (1990)
        505–510. https://doi.org/10.1126/science.2200121

[10]   A.D. Ellington, J.W. Szostak, In vitro selection of RNA molecules that bind
        specific ligands, Nature. 346 (1990) 818–822. https://doi.org/10.1038/346818a0

[11]   D.L. Robertson, G.F. Joyce, Selection in vitro of an RNA enzyme that specifically
        cleaves single-stranded DNA, Nature. 344 (1990) 467–468.
        https://doi.org/10.1038/344467a0

[12]   A. Ozer, J.M. Pagano, J.T. Lis, New technologies provide quantum changes in the
        scale, speed, and success of selex methods and aptamer characterization, Mol.
        Ther. - Nucleic Acids. 3 (2014) e183. https://doi.org/10.1038/mtna.2014.34

[13]   M. Blind, M. Blank, Aptamer selection technology and recent advances, Mol.
        Ther. - Nucleic Acids. 4 (2015) e223. https://doi.org/10.1038/mtna.2014.74

[14]   K.M. Ahmad, Y. Xiao, H.T. Soh, Selection is more intelligent than design: improving the affinity of a bivalent ligand through directed evolution, Nucleic Acids Res. 40 (2012) 11777–11783. https://doi.org/10.1093/nar/gks899

[15]   R.S. and T.K.S. Aradhana Chopra, Aptamers as an emerging player in biology, aptamers Synth. Antibodies. 1 (2014) 11

[16]   G. Hybarger, J. Bynum, R.F. Williams, J.J. Valdes, J.P. Chambers, A microfluidic SELEX prototype, Anal. Bioanal. Chem. 384 (2006) 191–198. https://doi.org/10.1007/s00216-005-0089-3

[17]   T.K. Sharma, J.G. Bruno, A. Dhiman, ABCs of DNA aptamer and related assay development, Biotechnol. Adv. 35 (2017) 275–301. https://doi.org/10.1016/j.biotechadv.2017.01.003

[18]   T.K. and R.S. Sharma, Nucleic acid aptamers as an emerging diagnostic tool for animal pathogens., Adv. Anim. Vet. Sci. 2 (2014) 50–55

[19]   V. Crivianu-Gaita, M. Thompson, Aptamers, antibody scFv, and antibody Fab' fragments: An overview and comparison of three of the most versatile biosensor biorecognition elements, Biosens. Bioelectron. 85 (2016) 32–45. https://doi.org/10.1016/j.bios.2016.04.091

[20]   T. Hianik, A. Porfireva, I. Grman, G. Evtugyn, Aptabodies – New type of artificial receptors for detection proteins, Protein Pept. Lett. 15 (2008) 799–805. https://doi.org/10.2174/092986608785203656

[21]   H. Hasegawa, K. Taira, K. Sode, K. Ikebukuro, Improvement of aptamer affinity by dimerization, Sensors. 8 (2008) 1090–1098. https://doi.org/10.3390/s8021090

[22]   H. Kaur, L.-Y.L. Yung, Probing high affinity sequences of dna aptamer against VEGF165, PLoS One. 7 (2012) e31196. https://doi.org/10.1371/journal.pone.0031196

[23]   J. Bruno, Predicting the uncertain future of aptamer-based diagnostics and therapeutics, Molecules. 20 (2015) 6866–6887. https://doi.org/10.3390/molecules20046866

[24]   G. Abel, Current status and future prospects of point-of-care testing around the globe, Expert Rev. Mol. Diagn. 15 (2015) 853–855. https://doi.org/10.1586/14737159.2015.1060126

[25]   K.S. Lam, M. Lebl, V. Krchňák, The "One-Bead-One-Compound" combinatorial library method, Chem. Rev. 97 (1997) 411–448. https://doi.org/10.1021/cr9600114

Biosensors: Materials and Applications   Materials Research Forum LLC
Materials Research Foundations **47** (2019) 1-50   doi: http://dx.doi.org/10.21741/9781644900130-1

[26] O. Lazcka, F.J. Del Campo, F.X. Muñoz, Pathogen detection: A perspective of traditional methods and biosensors, Biosens. Bioelectron. 22 (2007) 1205–1217. https://doi.org/10.1016/j.bios.2006.06.036

[27] C.K. O'Sullivan, Aptasensors – the future of biosensing?, Anal. Bioanal. Chem. 372 (2002) 44–48. https://doi.org/10.1007/s00216-001-1189-3

[28] I.E. Tothill, Biosensors for cancer markers diagnosis, Semin. Cell Dev. Biol. 20 (2009) 55–62. https://doi.org/10.1016/j.semcdb.2009.01.015

[29] T. Tang, J. Deng, M. Zhang, G. Shi, T. Zhou, Quantum dot-DNA aptamer conjugates coupled with capillary electrophoresis: A universal strategy for ratiometric detection of organophosphorus pesticides, Talanta. 146 (2016) 55–61. https://doi.org/10.1016/j.talanta.2015.08.023

[30] T. Mairal, V. Cengiz Özalp, P. Lozano Sánchez, M. Mir, I. Katakis, C.K. O'Sullivan, Aptamers: molecular tools for analytical applications, Anal. Bioanal. Chem. 390 (2008) 989–1007. https://doi.org/10.1007/s00216-007-1346-4

[31] R. Rapini, G. Marrazza, Biosensor potential in pesticide monitoring, in: 2016: pp. 3–31. https://doi.org/10.1016/bs.coac.2016.03.016

[32] E.W.M. Ng, D.T. Shima, P. Calias, E.T. Cunningham, D.R. Guyer, A.P. Adamis, Pegaptanib, a targeted anti-VEGF aptamer for ocular vascular disease, Nat. Rev. Drug Discov. 5 (2006) 123–132. https://doi.org/10.1038/nrd1955

[33] M.N. Stojanovic, D.W. Landry, Aptamer-based colorimetric probe for cocaine, J. Am. Chem. Soc. 124 (2002) 9678–9679. https://doi.org/10.1021/ja0259483

[34] M. Famulok, J.S. Hartig, G. Mayer, Functional aptamers and aptazymes in biotechnology, diagnostics, and therapy, Chem. Rev. 107 (2007) 3715–3743. https://doi.org/10.1021/cr0306743

[35] F. Wang, S. Liu, M. Lin, X. Chen, S. Lin, X. Du, H. Li, H. Ye, B. Qiu, Z. Lin, L. Guo, G. Chen, Colorimetric detection of microcystin-LR based on disassembly of orient-aggregated gold nanoparticle dimers, Biosens. Bioelectron. 68 (2015) 475–480. https://doi.org/10.1016/j.bios.2015.01.037

[36] L. Li, B. Li, Y. Qi, Y. Jin, Label-free aptamer-based colorimetric detection of mercury ions in aqueous media using unmodified gold nanoparticles as colorimetric probe, Anal. Bioanal. Chem. 393 (2009) 2051–2057. https://doi.org/10.1007/s00216-009-2640-0

[37] V. Ostatná, H. Vaisocherová, J. Homola, T. Hianik, Effect of the immobilisation of DNA aptamers on the detection of thrombin by means of surface plasmon

resonance, Anal. Bioanal. Chem. 391 (2008) 1861–1869.
https://doi.org/10.1007/s00216-008-2133-6

[38]   I. Willner, M. Zayats, Electronic aptamer-based sensors, Angew. Chemie Int. Ed. 46 (2007) 6408–6418. https://doi.org/10.1002/anie.200604524

[39]   E.E. Ferapontova, E.M. Olsen, K. V. Gothelf, An RNA aptamer-based electrochemical biosensor for detection of theophylline in serum, J. Am. Chem. Soc. 130 (2008) 4256–4258. https://doi.org/10.1021/ja711326b

[40]   K.-M. Song, M. Cho, H. Jo, K. Min, S.H. Jeon, T. Kim, M.S. Han, J.K. Ku, C. Ban, Gold nanoparticle-based colorimetric detection of kanamycin using a DNA aptamer, Anal. Biochem. 415 (2011) 175–181. https://doi.org/10.1016/j.ab.2011.04.007

[41]   T.K. Sharma, R. Ramanathan, P. Weerathunge, M. Mohammadtaheri, H.K. Daima, R. Shukla, V. Bansal, Aptamer-mediated 'turn-off/turn-on' nanozyme activity of gold nanoparticles for kanamycin detection, Chem. Commun. 50 (2014) 15856–15859. https://doi.org/10.1039/C4CC07275H

[42]   N. Rupcich, R. Nutiu, Y. Li, J.D. Brennan, Entrapment of fluorescent signaling DNA aptamers in sol–gel-derived silica, Anal. Chem. 77 (2005) 4300–4307. https://doi.org/10.1021/ac0506480

[43]   C.J. Rankin, E.N. Fuller, K.H. Hamor, S.A. Gabarra, T.P. Shields, A simple fluorescent biosensor for theophylline based on its RNA aptamer, Nucleosides, Nucleotides and Nucleic Acids. 25 (2006) 1407–1424. https://doi.org/10.1080/15257770600919084

[44]   N. Rupcich, R. Nutiu, Y. Li, J.D. Brennan, Solid-phase enzyme activity assay utilizing an entrapped fluorescence-signaling DNA aptamer, Angew. Chemie Int. Ed. 45 (2006) 3295–3299. https://doi.org/10.1002/anie.200504576

[45]   M.N. Stojanovic, P. de Prada, D.W. Landry, Aptamer-based folding fluorescent sensor for cocaine, J. Am. Chem. Soc. 123 (2001) 4928–4931. https://doi.org/10.1021/ja0038171

[46]   X. Wang, J. Zhou, W. Yun, S. Xiao, Z. Chang, P. He, Y. Fang, Detection of thrombin using electrogenerated chemiluminescence based on Ru(bpy)32+-doped silica nanoparticle aptasensor via target protein-induced strand displacement, Anal. Chim. Acta. 598 (2007) 242–248. https://doi.org/10.1016/j.aca.2007.07.050

[47]   K.-H. Leung, L. Lu, M. Wang, T.-Y. Mak, D.S.-H. Chan, F.-K. Tang, C.-H. Leung, H.-Y. Kwan, Z. Yu, D.-L. Ma, A label-free luminescent switch-on assay

for atp using a g-quadruplex-selective Iridium(III) complex, PLoS One. 8 (2013) e77021. https://doi.org/10.1371/journal.pone.0077021

[48]   G. Cappi, F.M. Spiga, Y. Moncada, A. Ferretti, M. Beyeler, M. Bianchessi, L. Decosterd, T. Buclin, C. Guiducci, Label-free detection of tobramycin in serum by transmission-localized surface plasmon resonance, Anal. Chem. 87 (2015) 5278–5285. https://doi.org/10.1021/acs.analchem.5b00389

[49]   E.J. Cho, L. Yang, M. Levy, A.D. Ellington, Using a deoxyribozyme ligase and rolling circle amplification to detect a non-nucleic acid analyte, ATP, J. Am. Chem. Soc. 127 (2005) 2022–2023. https://doi.org/10.1021/ja043490u

[50]   M. Zayats, Y. Huang, R. Gill, C. Ma, I. Willner, Label-free and reagentless aptamer-based sensors for small molecules, J. Am. Chem. Soc. 128 (2006) 13666–13667. https://doi.org/10.1021/ja0651456

[51]   P. Miao, Y. Tang, J. Yin, MicroRNA detection based on analyte triggered nanoparticle localization on a tetrahedral DNA modified electrode followed by hybridization chain reaction dual amplification, Chem. Commun. 51 (2015) 15629–15632. https://doi.org/10.1039/C5CC05499K

[52]   C.K. Dixit, K. Kadimisetty, B.A. Otieno, C. Tang, S. Malla, C.E. Krause, J.F. Rusling, Electrochemistry-based approaches to low cost, high sensitivity, automated, multiplexed protein immunoassays for cancer diagnostics, Analyst. 141 (2016) 536–547. https://doi.org/10.1039/C5AN01829C

[53]   T.H. Le, V.P. Pham, T.H. La, T.B. Phan, Q.H. Le, Electrochemical aptasensor for detecting tetracycline in milk, Adv. Nat. Sci. Nanosci. Nanotechnol. 7 (2016) 015008. https://doi.org/10.1088/2043-6262/7/1/015008

[54]   M. Wei, S. Feng, A signal-off aptasensor for the determination of Ochratoxin A by differential pulse voltammetry at a modified Au electrode using methylene blue as an electrochemical probe, Anal. Methods. 9 (2017) 5449–5454. https://doi.org/10.1039/C7AY01735A

[55]   A. Abbaspour, F. Norouz-Sarvestani, A. Noori, N. Soltani, Aptamer-conjugated silver nanoparticles for electrochemical dual-aptamer-based sandwich detection of staphylococcus aureus, Biosens. Bioelectron. 68 (2015) 149–155. https://doi.org/10.1016/j.bios.2014.12.040

[56]   M. Labib, A.S. Zamay, O.S. Kolovskaya, I.T. Reshetneva, G.S. Zamay, R.J. Kibbee, S.A. Sattar, T.N. Zamay, M. V. Berezovski, Aptamer-based viability

Biosensors: Materials and Applications   Materials Research Forum LLC
Materials Research Foundations **47** (2019) 1-50   doi: http://dx.doi.org/10.21741/9781644900130-1

impedimetric sensor for bacteria, Anal. Chem. 84 (2012) 8966–8969. https://doi.org/10.1021/ac302902s

[57]   L. Hou, L. Jiang, Y. Song, Y. Ding, J. Zhang, X. Wu, D. Tang, Amperometric aptasensor for saxitoxin using a gold electrode modified with carbon nanotubes on a self-assembled monolayer, and methylene blue as an electrochemical indicator probe, Microchim. Acta. 183 (2016) 1971–1980. https://doi.org/10.1007/s00604-016-1836-1

[58]   E. Lv, J. Ding, W. Qin, Potentiometric aptasensing of small molecules based on surface charge change, Sensors Actuators B Chem. 259 (2018) 463–466. https://doi.org/10.1016/j.snb.2017.12.067

[59]   D. Grieshaber, R. MacKenzie, J. Vörös, E. Reimhult, Electrochemical biosensors - sensor principles and architectures, Sensors. 8 (2008) 1400–1458. https://doi.org/10.3390/s80314000

[60]   M. Rahman, P. Kumar, D.-S. Park, Y.-B. Shim, Electrochemical sensors based on organic conjugated polymers, Sensors. 8 (2008) 118–141. https://doi.org/10.3390/s8010118

[61]   N.J. Ronkainen, H.B. Halsall, W.R. Heineman, Electrochemical biosensors, Chem. Soc. Rev. 39 (2010) 1747. https://doi.org/10.1039/b714449k

[62]   E. Bakker, P. Bühlmann, E. Pretsch, Carrier-based ion-selective electrodes and bulk optodes. 1. general characteristics, Chem. Rev. 97 (1997) 3083–3132. https://doi.org/10.1021/cr940394a

[63]   E.G. Harsányi, K. Tóth, E. Pungor, The adsorption of copper ions on the surface of copper(II) sulphide precipitate-based ion-selective electrodes, Anal. Chim. Acta. 152 (1983) 163–171. https://doi.org/10.1016/S0003-2670(00)84906-5

[64]   E. Pungor, K. Tóth, Precipitate-based ion-selective electrodes, in: ion-selective electrodes Anal. Chem., Springer US, Boston, MA, 1978: pp. 143–210. https://doi.org/10.1007/978-1-4684-2592-5_2

[65]   V. Agarwala, M.C. Chattopadhyaya, A heterogeneous precipitate based Mn(II) coated wire ion-selective electrode, Anal. Lett. 22 (1989) 1451–1457. https://doi.org/10.1080/00032718908051611

[66]   E.G. Harsányi, K. Tóth, E. Pungor, Y. Umezawa, S. Fujiwara, Study of the potential response of solid-state chloride electrodes at low concentration ranges, Talanta. 31 (1984) 579–584. https://doi.org/10.1016/0039-9140(84)80173-3

Biosensors: Materials and Applications   Materials Research Forum LLC
Materials Research Foundations **47** (2019) 1-50   doi: http://dx.doi.org/10.21741/9781644900130-1

[67] Y. Tani, H. Eun, Y. Umezawa, A cation selective electrode based on copper(II) and nickel(II) hexacyanoferrates: dual response mechanisms, selective uptake or adsorption of analyte cations, Electrochim. Acta. 43 (1998) 3431–3441. https://doi.org/10.1016/S0013-4686(98)00089-9

[68] E. Woźnica, M.M. Wójcik, M. Wojciechowski, J. Mieczkowski, E. Bulska, K. Maksymiuk, A. Michalska, Dithizone modified gold nanoparticles films for potentiometric sensing, Anal. Chem. 84 (2012) 4437–4442. https://doi.org/10.1021/ac300155f

[69] G.A. Zelada-Guillén, J. Riu, A. Düzgün, F.X. Rius, Immediate detection of living bacteria at ultralow concentrations using a carbon nanotube based potentiometric aptasensor, Angew. Chemie Int. Ed. 48 (2009) 7334–7337. https://doi.org/10.1002/anie.200902090

[70] S.G. Meirinho, L.G. Dias, A.M. Peres, L.R. Rodrigues, Voltammetric aptasensors for protein disease biomarkers detection: A review, Biotechnol. Adv. 34 (2016) 941–953. https://doi.org/10.1016/j.biotechadv.2016.05.006

[71] R.S. Marks, Nanomaterials for water management, Pan Stanford, 2015. https://doi.org/10.1201/b18715

[72] Y.C. Lim, A.Z. Kouzani, W. Duan, Aptasensors: A review, J. Biomed. Nanotechnol. 6 (2010) 93–105. https://doi.org/10.1166/jbn.2010.1103

[73] R.D. and V.K.R. Sanjay Upadhyay, Mukesh K. Sharma, Mahabul Shaik, Senors-A nanotechnological approach for the detection of organophosphorous compounds/pesticides., in: Impact Pestic., 2012: pp. 391–415

[74] J. Lin, H. Ju, Electrochemical and chemiluminescent immunosensors for tumor markers, Biosens. Bioelectron. 20 (2005) 1461–1470. https://doi.org/10.1016/j.bios.2004.05.008

[75] and Y.Y. Zhang, S., G. Wright, Materials and techniques for electrochemical biosensor design and construction, Biosens Bioelectron. 15 (2000) 273–282. https://doi.org/10.1016/S0956-5663(00)00076-2

[76] F. Wei, P. Patel, W. Liao, K. Chaudhry, L. Zhang, M. Arellano-Garcia, S. Hu, D. Elashoff, H. Zhou, S. Shukla, F. Shah, C.-M. Ho, D.T. Wong, Electrochemical Sensor for Multiplex Biomarkers Detection, Clin. Cancer Res. 15 (2009) 4446–4452. https://doi.org/10.1158/1078-0432.CCR-09-0050

[77] M.A.T. Gilmartin, J.P. Hart, D.T. Patton, Prototype, solid-phase, glucose biosensor, Analyst. 120 (1995) 1973. https://doi.org/10.1039/an9952001973

[78] H. Schulze, R. Schmid, T. Bachmann, Rapid detection of neurotoxic insecticides in food using disposable acetylcholinesterase-biosensors and simple solvent extraction, Anal. Bioanal. Chem. 372 (2002) 268–272. https://doi.org/10.1007/s00216-001-1137-2

[79] J. Wang, P.V.A. Pamidi, K.R. Rogers, Sol−gel-derived thick-film amperometric immunosensors, Anal. Chem. 70 (1998) 1171–1175. https://doi.org/10.1021/ac971093e

[80] C.G. Neuhold, J. Wang, X. Cai, K. Kalcher, Screen-printed electrodes for nitrite based on anion-exchanger-doped carbon inks, Analyst. 120 (1995) 2377. https://doi.org/10.1039/an9952002377

[81] S. Balamurugan, A. Obubuafo, S.A. Soper, D.A. Spivak, Surface immobilization methods for aptamer diagnostic applications, Anal. Bioanal. Chem. 390 (2008) 1009–1021. https://doi.org/10.1007/s00216-007-1587-2

[82] L. Zhou, M.H. Wang, J.P. Wang, Z.Z. Ye, Application of biosensor surface immobilization methods for aptamer, Chinese J. Anal. Chem. 39 (2011) 432–438. https://doi.org/10.1016/S1872-2040(10)60429-X

[83] K. Yugender Goud, G. Catanante, A. Hayat, S. M., K. Vengatajalabathy Gobi, J.L. Marty, Disposable and portable electrochemical aptasensor for label free detection of aflatoxin B1 in alcoholic beverages, Sensors Actuators B Chem. 235 (2016) 466–473. https://doi.org/10.1016/j.snb.2016.05.112

[84] N. MichaelGreen, Avidin, Adv. Protein Chem. 29 (1975) 85–133

[85] D. Wu, Y. Wang, Y. Zhang, H. Ma, X. Pang, L. Hu, B. Du, Q. Wei, Facile fabrication of an electrochemical aptasensor based on magnetic electrode by using streptavidin modified magnetic beads for sensitive and specific detection of Hg^{2+}, Biosens. Bioelectron. 82 (2016) 9–13. https://doi.org/10.1016/j.bios.2016.03.061

[86] D.J. Chung, K.C. Kim, S.H. Choi, Electrochemical DNA biosensor based on avidin–biotin conjugation for influenza virus (type A) detection, Appl. Surf. Sci. 257 (2011) 9390–9396. https://doi.org/10.1016/j.apsusc.2011.06.015

[87] R. Monošík, M. Streďanský, E. Šturdík, Biosensors - classification, characterization and new trends, Acta Chim. Slovaca. 5 (2012) 109–120. https://doi.org/10.2478/v10188-012-0017-z

[88] N. Paniel, J. Baudart, A. Hayat, L. Barthelmebs, Aptasensor and genosensor methods for detection of microbes in real world samples, Methods. 64 (2013) 229–240. https://doi.org/10.1016/j.ymeth.2013.07.001

Biosensors: Materials and Applications   Materials Research Forum LLC
Materials Research Foundations **47** (2019) 1-50   doi: http://dx.doi.org/10.21741/9781644900130-1

[89]   R. Das, M.K. Sharma, V.K. Rao, B.K. Bhattacharya, I. Garg, V. Venkatesh, S. Upadhyay, An electrochemical genosensor for Salmonella typhi on gold nanoparticles-mercaptosilane modified screen printed electrode, J. Biotechnol. 188 (2014) 9–16. https://doi.org/10.1016/j.jbiotec.2014.08.002

[90]   T. Hianik, J. Wang, Electrochemical aptasensors - recent achievements and perspectives, Electroanalysis. 21 (2009) 1223–1235. https://doi.org/10.1002/elan.200904566

[91]   A.-E. Radi, Electrochemical aptamer-based biosensors: recent advances and perspectives, Int. J. Electrochem. 2011 (2011) 1–17. https://doi.org/10.4061/2011/863196

[92]   A.K.H. Cheng, D. Sen, H.-Z. Yu, Design and testing of aptamer-based electrochemical biosensors for proteins and small molecules, Bioelectrochemistry. 77 (2009) 1–12. https://doi.org/10.1016/j.bioelechem.2009.04.007

[93]   K. Han, Z. Liang, N. Zhou, Design strategies for aptamer-based biosensors, Sensors. 10 (2010) 4541–4557. https://doi.org/10.3390/s100504541

[94]   B. Prieto-Simón, M. Campàs, J.-L. Marty, Electrochemical aptamer-based sensors, Bioanal. Rev. 1 (2010) 141–157. https://doi.org/10.1007/s12566-010-0010-1

[95]   I. Palchetti, M. Mascini, Electrochemical nanomaterial-based nucleic acid aptasensors, Anal. Bioanal. Chem. 402 (2012) 3103–3114. https://doi.org/10.1007/s00216-012-5769-1

[96]   C. Gao, Q. Wang, F. Gao, F. Gao, A high-performance aptasensor for mercury(<scp>ii</scp>) based on the formation of a unique ternary structure of aptamer–Hg^{2+}–neutral red, Chem. Commun. 50 (2014) 9397–9400. https://doi.org/10.1039/C4CC03275F

[97]   B. Strehlitz, N. Nikolaus, R. Stoltenburg, Protein detection with aptamer biosensors, Sensors. 8 (2008) 4296–4307. https://doi.org/10.3390/s8074296

[98]   X. Wang, S. Dong, P. Gai, R. Duan, F. Li, Highly sensitive homogeneous electrochemical aptasensor for antibiotic residues detection based on dual recycling amplification strategy, Biosens. Bioelectron. 82 (2016) 49–54. https://doi.org/10.1016/j.bios.2016.03.055

[99]   W. Zheng, J. Teng, L. Cheng, Y. Ye, D. Pan, J. Wu, F. Xue, G. Liu, W. Chen, Hetero-enzyme-based two-round signal amplification strategy for trace detection of aflatoxin B1 using an electrochemical aptasensor, Biosens. Bioelectron. 80 (2016) 574–581. https://doi.org/10.1016/j.bios.2016.01.091

[100] G. Catanante, R.K. Mishra, A. Hayat, J.-L. Marty, Sensitive analytical performance of folding based biosensor using methylene blue tagged aptamers, Talanta. 153 (2016) 138–144. https://doi.org/10.1016/j.talanta.2016.03.004

[101] R.Y. Lai, K.W. Plaxco, A.J. Heeger, Aptamer-based electrochemical detection of picomolar platelet-derived growth factor directly in blood serum, Anal. Chem. 79 (2007) 229–233. https://doi.org/10.1021/ac061592s

[102] L. Cui, J. Wu, H. Ju, Label-free signal-on aptasensor for sensitive electrochemical detection of arsenite, Biosens. Bioelectron. 79 (2016) 861–865. https://doi.org/10.1016/j.bios.2016.01.010

[103] L. Fang, Z. Lü, H. Wei, E. Wang, A electrochemiluminescence aptasensor for detection of thrombin incorporating the capture aptamer labeled with gold nanoparticles immobilized onto the thio-silanized ITO electrode, Anal. Chim. Acta. 628 (2008) 80–86. https://doi.org/10.1016/j.aca.2008.08.041

[104] Y.Y. Hua M, Tao M, Wang P, Zhang Y, Wu Z, Chang Y, Label-free electrochemical cocaine aptasensor based on a target-inducing aptamer switching conformation, Anal Sci. 26 (2010) 1265–70

[105] N. de-los-Santos-Álvarez, M.J. Lobo-Castañón, A.J. Miranda-Ordieres, P. Tuñón-Blanco, Modified-RNA aptamer-based sensor for competitive impedimetric assay of neomycin B, J. Am. Chem. Soc. 129 (2007) 3808–3809. https://doi.org/10.1021/ja0689482

[106] N. de-los-Santos-Álvarez, M.J. Lobo-Castañón, A.J. Miranda-Ordieres, P. Tuñón-Blanco, SPR sensing of small molecules with modified RNA aptamers: Detection of neomycin B, Biosens. Bioelectron. 24 (2009) 2547–2553. https://doi.org/10.1016/j.bios.2009.01.011

[107] A.H. Strickland, Plant protection and world crop production By H. H. Cramer Leverkusen: 'Bayer' Pflanzenschutz (1967), pp. 524., Exp. Agric. 5 (1969) 82. https://doi.org/10.1017/S0014479700010036

[108] D. Pimentel, H. Lehman, eds., The pesticide question, Springer US, Boston, MA, 1993. https://doi.org/10.1007/b102353

[109] A.W. EC. OERKE, H.W. Dehne, F. Schönbeck, Crop production and crop protection estimated losses in major food and cash crops, 1994

[110] D. Liu, W. Chen, J. Wei, X. Li, Z. Wang, X. Jiang, A highly sensitive, dual-readout assay based on gold nanoparticles for organophosphorus and carbamate pesticides, Anal. Chem. 84 (2012) 4185–4191. https://doi.org/10.1021/ac300545p

[111] Q. Long, H. Li, Y. Zhang, S. Yao, Upconversion nanoparticle-based fluorescence resonance energy transfer assay for organophosphorus pesticides, Biosens. Bioelectron. 68 (2015) 168–174. https://doi.org/10.1016/j.bios.2014.12.046

[112] E.C. Oerke, Crop losses to pests, J. Agric. Sci. 144 (2006) 31. https://doi.org/10.1017/S0021859605005708

[113] J. Popp, Cost-benefit analysis of crop protection measures, J. Consum. Prot. Food Saf. 6 (2011) S105–S112. https://doi.org/10.1007/s00003-011-0677-4

[114] R. Singh, R. Prasad, G. Sumana, K. Arora, S. Sood, R.K. Gupta, B.D. Malhotra, STD sensor based on nucleic acid functionalized nanostructured polyaniline, Biosens. Bioelectron. 24 (2009) 2232–2238. https://doi.org/10.1016/j.bios.2008.11.030

[115] S. Berijani, Y. Assadi, M. Anbia, M.-R. Milani Hosseini, E. Aghaee, Dispersive liquid–liquid microextraction combined with gas chromatography-flame photometric detection, J. Chromatogr. A. 1123 (2006) 1–9. https://doi.org/10.1016/j.chroma.2006.05.010

[116] E. Watanabe, T. Iwafune, K. Baba, Y. Kobara, Organic solvent-saving sample preparation for systematic residue analysis of neonicotinoid insecticides in agricultural products using liquid chromatography–diode array detection, Food Anal. Methods. 9 (2016) 245–254. https://doi.org/10.1007/s12161-015-0189-4

[117] Z. Shi, S. Zhang, Q. Huai, D. Xu, H. Zhang, Methylamine-modified graphene-based solid phase extraction combined with UPLC-MS/MS for the analysis of neonicotinoid insecticides in sunflower seeds, Talanta. 162 (2017) 300–308. https://doi.org/10.1016/j.talanta.2016.10.042

[118] M. Saitta, G. Di Bella, M.R. Fede, V. Lo Turco, A.G. Potortì, R. Rando, M.T. Russo, G. Dugo, Gas chromatography-tandem mass spectrometry multi-residual analysis of contaminants in Italian honey samples, Food Addit. Contam. Part A. (2017) 1–9. https://doi.org/10.1080/19440049.2017.1292054

[119] F. Barahona, C.L. Bardliving, A. Phifer, J.G. Bruno, C.A. Batt, An aptasensor based on polymer-gold nanoparticle composite microspheres for the detection of malathion using surface-enhanced raman spectroscopy, Ind. Biotechnol. 9 (2013) 42–50. https://doi.org/10.1089/ind.2012.0029

[120] D. Moreno-González, P. Pérez-Ortega, B. Gilbert-López, A. Molina-Díaz, J.F. García-Reyes, A.R. Fernández-Alba, Evaluation of nanoflow liquid chromatography high resolution mass spectrometry for pesticide residue analysis

Materials Research Forum LLC

doi: http://dx.doi.org/10.21741/9781644900130-1

in food, J. Chromatogr. A. 1512 (2017) 78–87.
https://doi.org/10.1016/j.chroma.2017.07.019

[121] A. Goon, Z. Khan, D. Oulkar, R. Shinde, S. Gaikwad, K. Banerjee, A simultaneous screening and quantitative method for the multiresidue analysis of pesticides in spices using ultra-high performance liquid chromatography-high resolution (Orbitrap) mass spectrometry, J. Chromatogr. A. 1532 (2018) 105–111. https://doi.org/10.1016/j.chroma.2017.11.066

[122] L. Jia, M. Su, X. Wu, H. Sun, Rapid selective accelerated solvent extraction and simultaneous determination of herbicide atrazine and its metabolites in fruit by ultra high performance liquid chromatography, J. Sep. Sci. 39 (2016) 4512–4519. https://doi.org/10.1002/jssc.201600883

[123] G. Qian, L. Wang, Y. Wu, Q. Zhang, Q. Sun, Y. Liu, F. Liu, A monoclonal antibody-based sensitive enzyme-linked immunosorbent assay (ELISA) for the analysis of the organophosphorous pesticides chlorpyrifos-methyl in real samples, Food Chem. 117 (2009) 364–370. https://doi.org/10.1016/j.foodchem.2009.03.097

[124] J.K. Lee, K.C. Ahn, O.S. Park, S.Y. Kang, B.D. Hammock, Development of an ELISA for the detection of the residues of the insecticide imidacloprid in agricultural and environmental samples, J. Agric. Food Chem. 49 (2001) 2159–2167. https://doi.org/10.1021/jf001140v

[125] E. Watanabe, S. Miyake, Y. Yogo, Review of enzyme-linked immunosorbent assays (ELISAs) for analyses of neonicotinoid insecticides in agro-environments, J. Agric. Food Chem. 61 (2013) 12459–12472. https://doi.org/10.1021/jf403801h

[126] D.S. Bol'shakova, V.G. Amelin, Determination of pesticides in environmental materials and food products by capillary electrophoresis, J. Anal. Chem. 71 (2016) 965–1013. https://doi.org/10.1134/S1061934816100026

[127] N. Srivastava, S. Kumari, K. Nair, S. Alam, S.K. Raza, Determination of organophosphorous pesticides in environmental water samples using surface-engineered C18 functionalized silica-coated core-shell magnetic nanoparticles–based extraction coupled with GC-MS/MS analysis, J. AOAC Int. 100 (2017) 804–809. https://doi.org/10.5740/jaoacint.16-0312

[128] M. Pirsaheb, M. Rezaei, N. Fattahi, M. Karami, K. Sharafi, H.R. Ghaffari, Optimization of a methodology for the simultaneous determination of deltamethrin, permethrin and malathion in stored wheat samples using dispersive liquid–liquid microextraction with solidification of floating organic drop and

HPLC-UV, J. Environ. Sci. Heal. Part B. 52 (2017) 641–650.
https://doi.org/10.1080/03601234.2017.1330078

[129] E.M. Brun, M. Garcés-García, M.J. Bañuls, J.A. Gabaldón, R. Puchades, Á.
Maquieira, Evaluation of a novel malathion immunoassay for groundwater and
surface water analysis, Environ. Sci. Technol. 39 (2005) 2786–2794.
https://doi.org/10.1021/es048945u

[130] H. Chen, Y. Wu, W. Yang, S. Zhan, S. Qiu, P. Zhou, Ultrasensitive and selective
detection of isocarbophos pesticide based on target and random ssDNA triggered
aggregation of hemin in polar organic solutions, Sensors Actuators B Chem. 243
(2017) 445–453. https://doi.org/10.1016/j.snb.2016.12.014

[131] X. Fang, W. Tan, Aptamers generated from Cell-SELEX for molecular medicine:
A chemical biology approach, Acc. Chem. Res. 43 (2010) 48–57.
https://doi.org/10.1021/ar900101s

[132] R. Sharma, K.V. Ragavan, M.S. Thakur, K.S.M.S. Raghavarao, Recent advances
in nanoparticle based aptasensors for food contaminants, Biosens. Bioelectron. 74
(2015) 612–627. https://doi.org/10.1016/j.bios.2015.07.017

[133] X. Dou, X. Chu, W. Kong, J. Luo, M. Yang, A gold-based nanobeacon probe for
fluorescence sensing of organophosphorus pesticides, Anal. Chim. Acta. 891
(2015) 291–297. https://doi.org/10.1016/j.aca.2015.08.012

[134] Y. Qi, F.-R. Xiu, M. Zheng, B. Li, A simple and rapid chemiluminescence
aptasensor for acetamiprid in contaminated samples: Sensitivity, selectivity and
mechanism, Biosens. Bioelectron. 83 (2016) 243–249.
https://doi.org/10.1016/j.bios.2016.04.074

[135] L. Wang, W. Ma, W. Chen, L. Liu, W. Ma, Y. Zhu, L. Xu, H. Kuang, C. Xu, An
aptamer-based chromatographic strip assay for sensitive toxin semi-quantitative
detection, Biosens. Bioelectron. 26 (2011) 3059–3062.
https://doi.org/10.1016/j.bios.2010.11.040

[136] N. Mohammad Danesh, M. Ramezani, A. Sarreshtehdar Emrani, K. Abnous, S.M.
Taghdisi, A novel electrochemical aptasensor based on arch-shape structure of
aptamer-complimentary strand conjugate and exonuclease I for sensitive detection
of streptomycin, Biosens. Bioelectron. 75 (2016) 123–128.
https://doi.org/10.1016/j.bios.2015.08.017

[137] L. Madianos, G. Tsekenis, E. Skotadis, L. Patsiouras, D. Tsoukalas, A highly
sensitive impedimetric aptasensor for the selective detection of acetamiprid and

Biosensors: Materials and Applications   Materials Research Forum LLC
Materials Research Foundations **47** (2019) 1-50   doi: http://dx.doi.org/10.21741/9781644900130-1

atrazine based on microwires formed by platinum nanoparticles, Biosens. Bioelectron. 101 (2018) 268–274. https://doi.org/10.1016/j.bios.2017.10.034

[138] S.M. Marín A, Martínez Vidal JL, Egea Gonzalez FJ, Garrido Frenich A, Glass CR, Assessment of potential (inhalation and dermal) and actual exposure to acetamiprid by greenhouse applicators using liquid chromatography-tandem mass spectrometry., J Chromatogr B Anal. Technol Biomed Life Sci. 804 (2004) 269–75. https://doi.org/10.1016/j.jchromb.2004.01.022

[139] R.P. Kaur, V. Gupta, A.F. Christopher, P. Bansal, Potential pathways of pesticide action on erectile function – A contributory factor in male infertility, Asian Pacific J. Reprod. 4 (2015) 322–330. https://doi.org/10.1016/j.apjr.2015.07.012

[140] L. Fan, G. Zhao, H. Shi, M. Liu, Z. Li, A highly selective electrochemical impedance spectroscopy-based aptasensor for sensitive detection of acetamiprid, Biosens. Bioelectron. 43 (2013) 12–18. https://doi.org/10.1016/j.bios.2012.11.033

[141] A. Fei, Q. Liu, J. Huan, J. Qian, X. Dong, B. Qiu, H. Mao, K. Wang, Label-free impedimetric aptasensor for detection of femtomole level acetamiprid using gold nanoparticles decorated multiwalled carbon nanotube-reduced graphene oxide nanoribbon composites, Biosens. Bioelectron. 70 (2015) 122–129. https://doi.org/10.1016/j.bios.2015.03.028

[142] D. Jiang, X. Du, Q. Liu, L. Zhou, L. Dai, J. Qian, K. Wang, Silver nanoparticles anchored on nitrogen-doped graphene as a novel electrochemical biosensing platform with enhanced sensitivity for aptamer-based pesticide assay, Analyst. 140 (2015) 6404–6411. https://doi.org/10.1039/C5AN01084E

[143] R. Rapini, A. Cincinelli, G. Marrazza, Acetamiprid multidetection by disposable electrochemical DNA aptasensor, Talanta. 161 (2016) 15–21. https://doi.org/10.1016/j.talanta.2016.08.026

[144] P. Weerathunge, R. Ramanathan, R. Shukla, T.K. Sharma, V. Bansal, Aptamer-controlled reversible inhibition of gold nanozyme activity for pesticide sensing, Anal. Chem. 86 (2014) 11937–11941. https://doi.org/10.1021/ac5028726

[145] P. Hernandez, Y. Ballesteros, F. Galan, L. Hernandez, Determination of carbendazim with a graphite electrode modified with silicone OV-17, Electroanalysis. 8 (1996) 941–946. https://doi.org/10.1002/elan.1140081018

[146] W. Prashantkumar, R.S. Sethi, D. Pathak, S. Rampal, S.P.S. Saini, Testicular damage after chronic exposure to carbendazim in male goats, Toxicol. Environ. Chem. 94 (2012) 1433–1442. https://doi.org/10.1080/02772248.2012.693493

[147] J.M. Goldman, G.L. Rehnberg, R.L. Cooper, L.E. Gray, J.F. Hein, W.K. McElroy, Effects of the benomyl metabolite, carbendazim, on the hypothalamic-pituitary reproductive axis in the male rat, Toxicology. 57 (1989) 173–182. https://doi.org/10.1016/0300-483X(89)90163-7

[148] S. Eissa, M. Zourob, Selection and characterization of dna aptamers for electrochemical biosensing of carbendazim, Anal. Chem. 89 (2017) 3138–3145. https://doi.org/10.1021/acs.analchem.6b04914

[149] L. Karila, R. Zarmdini, A. Petit, G. Lafaye, W. Lowenstein, M. Reynaud, Addiction à la cocaïne : données actuelles pour le clinicien, Presse Med. 43 (2014) 9–17. https://doi.org/10.1016/j.lpm.2013.01.069

[150] Center for behavioral health statistics and quality (CBHSQ) (2015) behavioral health trends in the united states, results from 2014 Natl. Surv. Drug Use Heal. Rockville, MD Subst. Abus. Andm. Heal. Serv. Adm. HHS Publ. No. SMA, 2014. NSDUH Ser. H-50 P. (n.d.) 15–4927

[151] D. Roncancio, H. Yu, X. Xu, S. Wu, R. Liu, J. Debord, X. Lou, Y. Xiao, A label-free aptamer-fluorophore assembly for rapid and specific detection of cocaine in biofluids, Anal. Chem. 86 (2014) 11100–11106. https://doi.org/10.1021/ac503360n

[152] G. Bozokalfa, H. Akbulut, B. Demir, E. Guler, Z.P. Gumus, D. Odaci Demirkol, E. Aldemir, S. Yamada, T. Endo, H. Coskunol, S. Timur, Y. Yagci, Polypeptide functional surface for the aptamer immobilization: electrochemical cocaine biosensing, Anal. Chem. 88 (2016) 4161–4167. https://doi.org/10.1021/acs.analchem.6b00760

[153] L. Asturias-Arribas, M.A. Alonso-Lomillo, O. Domínguez-Renedo, M.J. Arcos-Martínez, Sensitive and selective cocaine electrochemical detection using disposable sensors, Anal. Chim. Acta. 834 (2014) 30–36. https://doi.org/10.1016/j.aca.2014.05.012

[154] S.P. Wren, T.H. Nguyen, P. Gascoine, R. Lacey, T. Sun, K.T.V. Grattan, Preparation of novel optical fibre-based Cocaine sensors using a molecular imprinted polymer approach, Sensors Actuators B Chem. 193 (2014) 35–41. https://doi.org/10.1016/j.snb.2013.11.071

[155] P. Fernández, N. Lafuente, A.M. Bermejo, M. López-Rivadulla, A. Cruz, HPLC Determination of cocaine and benzoylecgonine in plasma and urine from drug abusers, J. Anal. Toxicol. 20 (1996) 224–228. https://doi.org/10.1093/jat/20.4.224

[156] G. Floriani, J.C. Gasparetto, R. Pontarolo, A.G. Gonçalves, Development and validation of an HPLC-DAD method for simultaneous determination of cocaine, benzoic acid, benzoylecgonine and the main adulterants found in products based on cocaine, Forensic Sci. Int. 235 (2014) 32–39. https://doi.org/10.1016/j.forsciint.2013.11.013

[157] K. Kang, A. Sachan, M. Nilsen-Hamilton, P. Shrotriya, Aptamer functionalized microcantilever sensors for cocaine detection, Langmuir. 27 (2011) 14696–14702. https://doi.org/10.1021/la202067y

[158] Y. Lu, R.M. O'Donnell, P.B. Harrington, Detection of cocaine and its metabolites in urine using solid phase extraction-ion mobility spectrometry with alternating least squares, Forensic Sci. Int. 189 (2009) 54–59. https://doi.org/10.1016/j.forsciint.2009.04.007

[159] L. Skender, V. Karačić, I. Brčić, A. Bagarić, Quantitative determination of amphetamines, cocaine, and opiates in human hair by gas chromatography/mass spectrometry, Forensic Sci. Int. 125 (2002) 120–126. https://doi.org/10.1016/S0379-0738(01)00630-2

[160] M. Yonamine, N. Tawil, R. Moreau, O. Alvessilva, Solid-phase micro-extraction–gas chromatography–mass spectrometry and headspace-gas chromatography of tetrahydrocannabinol, amphetamine, methamphetamine, cocaine and ethanol in saliva samples, J. Chromatogr. B. 789 (2003) 73–78. https://doi.org/10.1016/S1570-0232(03)00165-X

[161] J.L. da Costa, F.G. Tonin, L.A. Zanolli, A.A. da Matta Chasin, M.F.M. Tavares, Simple method for determination of cocaine and main metabolites in urine by CE coupled to MS, Electrophoresis. 30 (2009) 2238–2244. https://doi.org/10.1002/elps.200900032

[162] M. Roushani, F. Shahdost-fard, Impedimetric detection of cocaine by using an aptamer attached to a screen printed electrode modified with a dendrimer/silver nanoparticle nanocomposite, Microchim. Acta. 185 (2018) 214. https://doi.org/10.1007/s00604-018-2709-6

[163] P. Hashemi, H. Bagheri, A. Afkhami, Y.H. Ardakani, T. Madrakian, Fabrication of a novel aptasensor based on three-dimensional reduced graphene oxide/polyaniline/gold nanoparticle composite as a novel platform for high sensitive and specific cocaine detection, Anal. Chim. Acta. 996 (2017) 10–19. https://doi.org/10.1016/j.aca.2017.10.035

Biosensors: Materials and Applications   Materials Research Forum LLC
Materials Research Foundations **47** (2019) 1-50   doi: http://dx.doi.org/10.21741/9781644900130-1

[164] B. Shen, J. Li, W. Cheng, Y. Yan, R. Tang, Y. Li, H. Ju, S. Ding, Electrochemical aptasensor for highly sensitive determination of cocaine using a supramolecular aptamer and rolling circle amplification, Microchim. Acta. 182 (2015) 361–367. https://doi.org/10.1007/s00604-014-1333-3

[165] S.M. Taghdisi, N.M. Danesh, A.S. Emrani, M. Ramezani, K. Abnous, A novel electrochemical aptasensor based on single-walled carbon nanotubes, gold electrode and complimentary strand of aptamer for ultrasensitive detection of cocaine, Biosens. Bioelectron. 73 (2015) 245–250. https://doi.org/10.1016/j.bios.2015.05.065

[166] F. Su, S. Zhang, H. Ji, H. Zhao, J.-Y. Tian, C.-S. Liu, Z. Zhang, S. Fang, X. Zhu, M. Du, Two-dimensional zirconium-based metal–organic framework nanosheet composites embedded with au nanoclusters: a highly sensitive electrochemical aptasensor toward detecting cocaine, ACS Sensors. 2 (2017) 998–1005. https://doi.org/10.1021/acssensors.7b00268

[167] Z. Chen, M. Lu, Target-responsive aptamer release from manganese dioxide nanosheets for electrochemical sensing of cocaine with target recycling amplification, Talanta. 160 (2016) 444–448. https://doi.org/10.1016/j.talanta.2016.07.052

[168] Y. Du, B. Li, S. Guo, Z. Zhou, M. Zhou, E. Wang, S. Dong, G-Quadruplex-based DNAzyme for colorimetric detection ofcocaine: Using magnetic nanoparticles as the separation and amplification element, Analyst. 136 (2011) 493–497. https://doi.org/10.1039/C0AN00557F

[169] J.-L. He, Z.-S. Wu, H. Zhou, H.-Q. Wang, J.-H. Jiang, G.-L. Shen, R.-Q. Yu, Fluorescence aptameric sensor for strand displacement amplification detection of cocaine, Anal. Chem. 82 (2010) 1358–1364. https://doi.org/10.1021/ac902416u

[170] Y. Shi, H. Dai, Y. Sun, J. Hu, P. Ni, Z. Li, Fluorescent sensing of cocaine based on a structure switching aptamer, gold nanoparticles and graphene oxide, Analyst. 138 (2013) 7152. https://doi.org/10.1039/c3an00897e

[171] S. Latini, F. Pedata, Adenosine in the central nervous system: release mechanisms and extracellular concentrations, J. Neurochem. 79 (2008) 463–484. https://doi.org/10.1046/j.1471-4159.2001.00607.x

[172] J.. Phillis, Adenosine in the control of the cerebral circulation., Cerebrovasc Brain Metab Rev. 1 (1989) 26–54

[173] J.M. Brundege, T. V. Dunwiddie, Role of adenosine as a modulator of synaptic activity in the central nervous system, in: 1997: pp. 353–391. https://doi.org/10.1016/S1054-3589(08)60076-9

[174] M.R. McMillan, G. Burnstock, S.G. Haworth, Vasodilatation of intrapulmonary arteries to P2-receptor nucleotides in normal and pulmonary hypertensive newborn piglets, Br. J. Pharmacol. 128 (1999) 543–548. https://doi.org/10.1038/sj.bjp.0702815

[175] D.-W. Huang, C.-G. Niu, G.-M. Zeng, M. Ruan, Time-resolved fluorescence biosensor for adenosine detection based on home-made europium complexes, Biosens. Bioelectron. 29 (2011) 178–183. https://doi.org/10.1016/j.bios.2011.08.014

[176] X. Wang, P. Dong, P. He, Y. Fang, A solid-state electrochemiluminescence sensing platform for detection of adenosine based on ferrocene-labeled structure-switching signaling aptamer, Anal. Chim. Acta. 658 (2010) 128–132. https://doi.org/10.1016/j.aca.2009.11.007

[177] S. Giglioni, R. Leoncini, E. Aceto, A. Chessa, S. Civitelli, A. Bernini, G. Tanzini, F. Carraro, A. Pucci, D. Vannoni, Adenosine Kinase Gene Expression in Human Colorectal Cancer, Nucleosides, Nucleotides and Nucleic Acids. 27 (2008) 750–754. https://doi.org/10.1080/15257770802145629

[178] G. Luippold, U. Delabar, D. Kloor, B. Mühlbauer, Simultaneous determination of adenosine, S-adenosylhomocysteine and S-adenosylmethionine in biological samples using solid-phase extraction and high-performance liquid chromatography, J. Chromatogr. B Biomed. Sci. Appl. 724 (1999) 231–238. https://doi.org/10.1016/S0378-4347(98)00580-5

[179] Y. Zhu, P.S.. Wong, Q. Zhou, H. Sotoyama, P.T. Kissinger, Identification and determination of nucleosides in rat brain microdialysates by liquid chromatography/electrospray tandem mass spectrometry, J. Pharm. Biomed. Anal. 26 (2001) 967–973. https://doi.org/10.1016/S0731-7085(01)00450-2

[180] L.-F. Huang, F.-Q. Guo, Y.-Z. Liang, B.-Y. Li, B.-M. Cheng, Simple and rapid determination of adenosine in human synovial fluid with high performance liquid chromatography–mass spectrometry, J. Pharm. Biomed. Anal. 36 (2004) 877–882. https://doi.org/10.1016/j.jpba.2004.07.038

[181] Hua Lin, Dan-Ke Xu, Hong-Yuan Chen, Simultaneous determination of purine bases, ribonucleosides and ribonucleotides by capillary electrophoresis-

electrochemistry with a copper electrode, J. Chromatogr. A. 760 (1997) 227–233. https://doi.org/10.1016/S0021-9673(96)00776-5

[182] H.M. Siragy, J. Linden, Sodium intake markedly alters renal interstitial fluid adenosine, Hypertension. 27 (1996) 404–407. https://doi.org/10.1161/01.HYP.27.3.404

[183] D.B. Northrop, F.B. Simpson, New concepts in bioorganic chemistry beyond enzyme kinetics: Direct determination of mechanisms by stopped-flow mass spectrometry, Bioorg. Med. Chem. 5 (1997) 641–644. https://doi.org/10.1016/S0968-0896(97)00020-5

[184] M.E. Revenis, M.A. Kaliner, Lactoferrin and lysozyme deficiency in airway secretions: Association with the development of bronchopulmonary dysplasia, J. Pediatr. 121 (1992) 262–270. https://doi.org/10.1016/S0022-3476(05)81201-6

[185] S. Lee-Huang, P.L. Huang, Y. Sun, P.L. Huang, H. -f. Kung, D.L. Blithe, H.-C. Chen, Lysozyme and RNases as anti-HIV components in -core preparations of human chorionic gonadotropin, Proc. Natl. Acad. Sci. 96 (1999) 2678–2681. https://doi.org/10.1073/pnas.96.6.2678

[186] J. Ireland, J. Herzog, E.R. Unanue, Cutting edge: unique T cells that recognize citrullinated peptides are a feature of protein immunization, J. Immunol. 177 (2006) 1421–1425. https://doi.org/10.4049/jimmunol.177.3.1421

[187] Z.N. and S.Y. Chunyan Deng, Jinhua Chen, Lihua Nie, Sensitive bifunctional aptamer-based electrochemical biosensor for small molecules and protein, Anal. Chem. 81 (2009) 9972–9978

[188] F. Shahdost-fard, A. Salimi, E. Sharifi, A. Korani, Fabrication of a highly sensitive adenosine aptasensor based on covalent attachment of aptamer onto chitosan-carbon nanotubes-ionic liquid nanocomposite, Biosens. Bioelectron. 48 (2013) 100–107. https://doi.org/10.1016/j.bios.2013.03.060

[189] D. Wu, X. Ren, L. Hu, D. Fan, Y. Zheng, Q. Wei, Electrochemical aptasensor for the detection of adenosine by using PdCu@MWCNTs-supported bienzymes as labels, Biosens. Bioelectron. 74 (2015) 391–397. https://doi.org/10.1016/j.bios.2015.07.003

[190] Y. Wang, J. Feng, Z. Tan, H. Wang, Electrochemical impedance spectroscopy aptasensor for ultrasensitive detection of adenosine with dual backfillers, Biosens. Bioelectron. 60 (2014) 218–223. https://doi.org/10.1016/j.bios.2014.04.022

[191] www.who.int/mediacentre/factsheets/fs310/en/index1.html., (n.d.)

[192] A. Davydova, M. Vorobjeva, D. Pyshnyi, S. Altman, V. Vlassov, A. Venyaminova, Aptamers against pathogenic microorganisms, Crit. Rev. Microbiol. 42 (2016) 847–865. https://doi.org/10.3109/1040841X.2015.1070115

[193] M. Ikanovic, W.E. Rudzinski, J.G. Bruno, A. Allman, M.P. Carrillo, S. Dwarakanath, S. Bhahdigadi, P. Rao, J.L. Kiel, C.J. Andrews, Fluorescence assay based on aptamer-quantum dot binding to bacillus thuringiensis spores, J. Fluoresc. 17 (2007) 193–199. https://doi.org/10.1007/s10895-007-0158-4

[194] W. Wu, M. Li, Y. Wang, H. Ouyang, L. Wang, C. Li, Y. Cao, Q. Meng, J. Lu, Aptasensors for rapid detection of Escherichia coli O157:H7 and Salmonella typhimurium, Nanoscale Res. Lett. 7 (2012) 658. https://doi.org/10.1186/1556-276X-7-658

[195] www.who.int/news-room/fact-sheets/detail/hiv-aids., (n.d.)

[196] www.who.int/news-room/fact-sheets/detail/hepatitis-c., (n.d.)

[197] James Miller, Interference in immunoassays: avoiding erroneous results., Clin. Lab. 28 (2004) 14–17

[198] S. Laperche, N. Le Marrec, N. Simon, F. Bouchardeau, C. Defer, M. Maniez-Montreuil, T. Levayer, J.-P. Zappitelli, J.-J. Lefrere, A new HCV core antigen assay based on disassociation of immune complexes: an alternative to molecular biology in the diagnosis of early HCV infection, Transfusion. 43 (2003) 958–962. https://doi.org/10.1046/j.1537-2995.2003.00430.x

[199] S. Lee, Y.S. Kim, M. Jo, M. Jin, D. Lee, S. Kim, Chip-based detection of hepatitis C virus using RNA aptamers that specifically bind to HCV core antigen, Biochem. Biophys. Res. Commun. 358 (2007) 47–52. https://doi.org/10.1016/j.bbrc.2007.04.057

[200] R. Wang, Y. Li, Hydrogel based QCM aptasensor for detection of avian influenzavirus, Biosens. Bioelectron. 42 (2013) 148–155. https://doi.org/10.1016/j.bios.2012.10.038

[201] R. Yamamoto, M. Katahira, S. Nishikawa, T. Baba, K. Taira, P.K.R. Kumar, A novel RNA motif that binds efficiently and specifically to the Tat protein of HIV and inhibits the trans-activation by Tat of transcription in vitro and in vivo, Genes to Cells. 5 (2000) 371–388. https://doi.org/10.1046/j.1365-2443.2000.00330.x

[202] S. Tombelli, M. Minunni, E. Luzi, M. Mascini, Aptamer-based biosensors for the detection of HIV-1 Tat protein, Bioelectrochemistry. 67 (2005) 135–141. https://doi.org/10.1016/j.bioelechem.2004.04.011

Biosensors: Materials and Applications   Materials Research Forum LLC
Materials Research Foundations **47** (2019) 1-50   doi: http://dx.doi.org/10.21741/9781644900130-1

[203]   P.R. Torgerson, B. Devleesschauwer, N. Praet, N. Speybroeck, A.L. Willingham,
        F. Kasuga, M.B. Rokni, X.-N. Zhou, E.M. Fèvre, B. Sripa, N. Gargouri, T. Fürst,
        C.M. Budke, H. Carabin, M.D. Kirk, F.J. Angulo, A. Havelaar, N. de Silva, World
        Health Organization Estimates of the Global and Regional Disease Burden of 11
        Foodborne Parasitic Diseases, 2010: A Data Synthesis, PLOS Med. 12 (2015)
        e1001920. https://doi.org/10.1371/journal.pmed.1001920

[204]   http://www.who.int/malaria/en/., No Title, (n.d.)

[205]   D. Bell, C. Wongsrichanalai, J.W. Barnwell, Ensuring quality and access for
        malaria diagnosis: how can it be achieved?, Nat. Rev. Microbiol. 4 (2006) 682–
        695. https://doi.org/10.1038/nrmicro1474

[206]   T. Hänscheid, M.P. Grobusch, How useful is PCR in the diagnosis of malaria?,
        Trends Parasitol. 18 (2002) 395–398. https://doi.org/10.1016/S1471-
        4922(02)02348-6

[207]   S. Lee, K.-M. Song, W. Jeon, H. Jo, Y.-B. Shim, C. Ban, A highly sensitive
        aptasensor towards Plasmodium lactate dehydrogenase for the diagnosis of
        malaria, Biosens. Bioelectron. 35 (2012) 291–296.
        https://doi.org/10.1016/j.bios.2012.03.003

[208]   E. Scallan, R.M. Hoekstra, F.J. Angulo, R. V. Tauxe, M.-A. Widdowson, S.L.
        Roy, J.L. Jones, P.M. Griffin, Foodborne illness acquired in the United States—
        major pathogens, Emerg. Infect. Dis. 17 (2011) 7–15.
        https://doi.org/10.3201/eid1701.P11101

[209]   www.cdc.gov/parasites/crypto/index.html., (n.d.)

[210]   U. Ryan, R. Fayer, L. Xiao, Cryptosporidium species in humans and animals:
        current understanding and research needs, Parasitology. 141 (2014) 1667–1685.
        https://doi.org/10.1017/S0031182014001085

[211]   S.P. Buckwalter, L.M. Sloan, S.A. Cunningham, M.J. Espy, J.R. Uhl, M.F. Jones,
        E.A. Vetter, J. Mandrekar, F.R. Cockerill, B.S. Pritt, R. Patel, N.L. Wengenack,
        Inhibition controls for qualitative real-time PCR assays: are they necessary for all
        specimen matrices?, J. Clin. Microbiol. 52 (2014) 2139–2143.
        https://doi.org/10.1128/JCM.03389-13

[212]   M. Labib, M. V. Berezovski, Electrochemical aptasensors for microbial and viral
        pathogens, in: 2013: pp. 155–181. https://doi.org/10.1007/10_2013_229

[213]   A. Iqbal, M. Labib, D. Muharemagic, S. Sattar, B.R. Dixon, M. V. Berezovski,
        Detection of Cryptosporidium parvum Oocysts on fresh produce using DNA

47

aptamers, PLoS One. 10 (2015) e0137455.
https://doi.org/10.1371/journal.pone.0137455

[214] globocan.iarc.fr/Pages/fact_sheets_cancer.aspx, (n.d.)

[215] L. Wu, X. Qu, Cancer biomarker detection: recent achievements and challenges, Chem. Soc. Rev. 44 (2015) 2963–2997. https://doi.org/10.1039/C4CS00370E

[216] B. Sarkar, J. Dosch, D.M. Simeone, Cancer stem cells: A new theory regarding a timeless disease, Chem. Rev. 109 (2009) 3200–3208. https://doi.org/10.1021/cr9000397

[217] B. Davidson, H.P. Dong, A. Holth, A. Berner, B. Risberg, Flow cytometric immunophenotyping of cancer cells in effusion specimens: Diagnostic and research applications, Diagn. Cytopathol. 35 (2007) 568–578. https://doi.org/10.1002/dc.20707

[218] I. Alevizos, M. Mahadevappa, X. Zhang, H. Ohyama, Y. Kohno, M. Posner, G.T. Gallagher, M. Varvares, D. Cohen, D. Kim, R. Kent, R.B. Donoff, R. Todd, C.M. Yung, J.A. Warrington, D.T.W. Wong, Oral cancer in vivo gene expression profiling assisted by laser capture microdissection and microarray analysis, Oncogene. 20 (2001) 6196–6204. https://doi.org/10.1038/sj.onc.1204685

[219] B.S. Ghossein RA, Molecular detection and characterisation of circulating tumour cells and micrometastases in solid tumours., Eur J Cancer. 36 (2000) 1681–94

[220] K. Xiao, J. Liu, H. Chen, S. Zhang, J. Kong, A label-free and high-efficient GO-based aptasensor for cancer cells based on cyclic enzymatic signal amplification, Biosens. Bioelectron. 91 (2017) 76–81. https://doi.org/10.1016/j.bios.2016.11.057

[221] M. Vestergaard, K. Kerman, E. Tamiya, An Overview of Label-free electrochemical protein sensors, Sensors. 7 (2007) 3442–3458. https://doi.org/10.3390/s7123442

[222] C.-H. Pui, M. V. Relling, J.R. Downing, Acute lymphoblastic leukemia, N. Engl. J. Med. 350 (2004) 1535–1548. https://doi.org/10.1056/NEJMra023001

[223] G.S. Zamay, T.N. Zamay, V.A. Kolovskii, A. V. Shabanov, Y.E. Glazyrin, D. V. Veprintsev, A. V. Krat, S.S. Zamay, O.S. Kolovskaya, A. Gargaun, A.E. Sokolov, A.A. Modestov, I.P. Artyukhov, N. V. Chesnokov, M.M. Petrova, M. V. Berezovski, A.S. Zamay, Electrochemical aptasensor for lung cancer-related protein detection in crude blood plasma samples, Sci. Rep. 6 (2016) 34350. https://doi.org/10.1038/srep34350

[224] www.cdc.gov/diabetes/statistics/slides/long_term_trends.pdf., (n.d.)

[225] K. Hashimoto, S. Noguchi, Y. Morimoto, S. Hamada, K. Wasada, S. Imai, Y. Murata, S. Kasayama, M. Koga, A1C but not serum glycated albumin is elevated in late pregnancy owing to iron deficiency, Diabetes Care. 31 (2008) 1945–1948. https://doi.org/10.2337/dc08-0352

[226] C. Apiwat, P. Luksirikul, P. Kankla, P. Pongprayoon, K. Treerattrakoon, K. Paiboonsukwong, S. Fucharoen, T. Dharakul, D. Japrung, Graphene based aptasensor for glycated albumin in diabetes mellitus diagnosis and monitoring, Biosens. Bioelectron. 82 (2016) 140–145. https://doi.org/10.1016/j.bios.2016.04.015

[227] N. Varghese, U. Mogera, A. Govindaraj, A. Das, P.K. Maiti, A.K. Sood, C.N.R. Rao, Binding of DNA nucleobases and nucleosides with graphene, ChemPhysChem. 10 (2009) 206–210. https://doi.org/10.1002/cphc.200800459

[228] B. Liu, S. Salgado, V. Maheshwari, J. Liu, DNA adsorbed on graphene and graphene oxide: Fundamental interactions, desorption and applications, Curr. Opin. Colloid Interface Sci. 26 (2016) 41–49. https://doi.org/10.1016/j.cocis.2016.09.001

[229] S. Ghosh, D. Datta, M. Cheema, M. Dutta, M.A. Stroscio, Aptasensor based optical detection of glycated albumin for diabetes mellitus diagnosis, Nanotechnology. 28 (2017) 435505. https://doi.org/10.1088/1361-6528/aa893a

[230] T.P. Peacock, Z.K. Shihabi, A.J. Bleyer, E.L. Dolbare, J.R. Byers, M.A. Knovich, J. Calles-Escandon, G.B. Russell, B.I. Freedman, Comparison of glycated albumin and hemoglobin A1c levels in diabetic subjects on hemodialysis, Kidney Int. 73 (2008) 1062–1068. https://doi.org/10.1038/ki.2008.25

[231] M. Koga, Glycated albumin; clinical usefulness, Clin. Chim. Acta. 433 (2014) 96–104. https://doi.org/10.1016/j.cca.2014.03.001

[232] T. Kohzuma, T. Yamamoto, Y. Uematsu, Z.K. Shihabi, B.I. Freedman, Basic performance of an enzymatic method for glycated albumin and reference range determination, J. Diabetes Sci. Technol. 5 (2011) 1455–1462. https://doi.org/10.1177/193229681100500619

[233] Y. Kasahara, S. Kitadume, K. Morihiro, M. Kuwahara, H. Ozaki, H. Sawai, T. Imanishi, S. Obika, Effect of 3'-end capping of aptamer with various 2',4'-bridged nucleotides: Enzymatic post-modification toward a practical use of polyclonal aptamers, Bioorg. Med. Chem. Lett. 20 (2010) 1626–1629. https://doi.org/10.1016/j.bmcl.2010.01.028

[234] J. Ruckman, L.S. Green, J. Beeson, S. Waugh, W.L. Gillette, D.D. Henninger, L. Claesson-Welsh, N. Janjic, 2′-Fluoropyrimidine RNA-based aptamers to the 165-amino acid form of vascular endothelial growth factor (VEGF 165), J. Biol. Chem. 273 (1998) 20556–20567. https://doi.org/10.1074/jbc.273.32.20556

[235] B. Wlotzka, S. Leva, B. Eschgfaller, J. Burmeister, F. Kleinjung, C. Kaduk, P. Muhn, H. Hess-Stumpp, S. Klussmann, In vivo properties of an anti-GnRH Spiegelmer: An example of an oligonucleotide-based therapeutic substance class, Proc. Natl. Acad. Sci. 99 (2002) 8898–8902. https://doi.org/10.1073/pnas.132067399

[236] P. Kalra, A. Dhiman, W.C. Cho, J.G. Bruno, T.K. Sharma, Simple methods and rational design for enhancing aptamer sensitivity and specificity, Front. Mol. Biosci. 5 (2018). https://doi.org/10.3389/fmolb.2018.00041

Biosensors: Materials and Applications   Materials Research Forum LLC
Materials Research Foundations **47** (2019) 51-76   doi: http://dx.doi.org/10.21741/9781644900130-2

Chapter 2

Applications of Molecularly Imprinted Polymers to Genobiosensors

Mayank Garg[1a], Satish Pandey[1], Vijay Kumar Meena[1], Amit L. Sharma[1a], Suman Singh[1a*]

[1]CSIR- Central Scientific Instruments Organisation, Sector 30-C, Chandigarh-160030, India

[a]Academy of Scientific and Innovative Research (AcSIR), Ghaziabad-201002, India

*ssingh@csio.res.in

Abstract

The molecularly imprinted polymer (MIP) is a special variety of polymer which acts as a recognition element for the analytes whose imprint is present on the polymer during its synthesis. These can be custom tuned to have strong or weak affinity towards the targeted analyte. Biosensors are the typical devices having a recognition element of biological origin to track down certain analytes specific to the recognition element. This biological origin can be enzyme, antibody or even genetic material. The traditional biosensors have inherent disadvantages of being thermally unstable and not able to work at extreme conditions because of the biological elements used for detection purposes. Use of MIP technology marks the beginning of a new era for use as biosensing elements as "bio-mimetics" having many advantages over traditional ones. This chapter covered some insight into MIPs and their application as a biosensor for the detection of genetic materials i.e DNA biosensors.

Keywords

Molecularly Imprinted Polymers, Bio-mimetics, Biosensors, Diagnostics, DNA Biosensors

Contents

1.   Introduction

1.1   Molecularly Imprinted Polymers

Molecularly imprinted polymers (MIPs) are polymeric matrices in which imprinting technology is used. This technology is based on the formation of a complex between an analyte (template) and a mixture of functional monomer which helps in the creation of artificial recognition sites for the particular target analytes. The synthesis involves two major mechanisms; free radical polymerization and the sol-gel process. The imprinting methods in case of free radical polymerization are bulk, suspension, emulsion, seed and precipitation polymerization. The polymer network is formed due to polymerization after adding a cross-linking agent, initiator and porogen. In the case of sol-gel process, sol-gel imprinting is the sole method [1]. Polymerization can be initiated either by heat or UV treatment, UV treatment is the more commonly used one because of its ease of use and simplicity of the method.

Since the analyte to be sensed uses a template and a cavity corresponding to that analyte is made in the polymer, it is said to be imprinted. This cavity is very sensitive, stable and only binds to the analyte used to create the imprint [2]. Figure 1 shows the synthesis schematic of MIPs. The interaction between the template molecule and functional groups present in the polymer matrix takes place via intermolecular interactions like H-bonds, dipole-dipole and ionic interactions. These interactions derive the molecular recognition

Biosensors: Materials and Applications   Materials Research Forum LLC
Materials Research Foundations **47** (2019) 51-76   doi: http://dx.doi.org/10.21741/9781644900130-2

phenomena. Among different types of polymers used for producing MIPs are conducting polymers like polypyrrole, polyaniline, polythiophene, etc.

Based on their characteristics, the MIPs are successfully being applied for applications like solid phase extraction, chromatographic fractionation and biosensors. The use of MIPs in biosensing application can play a dual role; as an immobilization matrix and as a replacement of bioreceptors. These are often referred as synthetic bioreceptors since these are highly selective towards their templated analyte. Also as compared to biological systems, MIPs have high resistance to elevated temperature and inertness towards metal ions, organic solvents etc., are cheap compared to bioreceptors and have a good shelf life, which qualify them for the onsite applications as well.

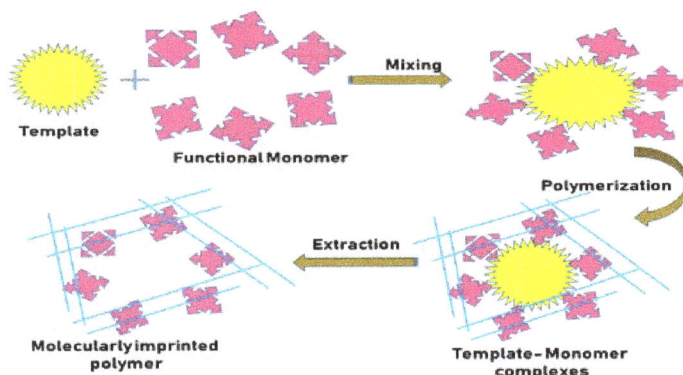

Figure 1: A schematic of the molecularly imprinted polymer synthesis.

1.2   Precursors of MIPs

1. **Functional Monomers:** The functional monomers are an essential component for polymerisation and are responsible for forming binding sites imprinted in the polymer. The choice of functional monomer decides whether non-covalent or covalent imprinting will be done.

2. **Cross-Linkers:** As the name indicates, the function of a cross linker is to provide linkage of functional monomers with the target template. Examples of these are ethylene glycol dimethacrylate and trimethylolpropane trimethacrylate. The selectivity of

Biosensors: Materials and Applications   Materials Research Forum LLC
Materials Research Foundations **47** (2019) 51-76   doi: http://dx.doi.org/10.21741/9781644900130-2

imprinted polymers greatly depends on the kind and amount of crosslinking agent used. Even the polymer morphology can be controlled by the crosslinking, at the same time, this also imparts mechanical stability to the polymer matrix. Generally, to have adequate mechanical stability, a high cross-link ratio is used, which can be due to the fact that high degree of cross-linking enables the microcavities to maintain three-dimensional structure complementary in both shape and chemical functionality to that of the template after removal of the template [3].

3. **Radical Initiators:** The radical initiators start producing radicals required for a polymerization process which starts when either heat or UV light is applied to the reaction mixture. 2,2'-Azobis(isobutyronitrile) (AIBN) and Ammonium persulfate are the most commonly used radical initiators. Compared to the monomer, initiators are used at low concentrations. The light, heat, chemical/electrochemical reactions can trigger the rate and mode of decomposition of an initiator to radicals.

4. **Solvents:** This is the most vital component in this process as it provides the medium for reactions to take place. The polymer morphology, polymer performance, porosity of polymer, strength of interactions, etc. are highly influenced by the nature and level of solvents being used. Solvents with low solubility phase have tendency to form materials with large pores and lower surface areas. The choice of solvents affects the reaction dynamics. A solvent with low polarity is preferred to reduce the interferences during the complex formation. Water, acetone, toluene, methanol and dimethyl sulfoxide are some of the commonly used solvents in MIPs preparation.

1.3   Approaches to MIP synthesis

The covalent, non-covalent and semi-covalent are three strategies commonly used for the synthesis of molecularly imprinted polymers and all these strategies have their advantages and disadvantages. Of course the selection of the strategies is done based on the desired properties and applications.

Covalent approach:

In the covalent approach, there is strong binding between the template and functional monomer and extra energy is required to free the template from the monomer. The covalent approach gives more stability as compared to the non-covalent approach, though it is difficult to design an appropriate template-monomer complex where covalent bond formation and cleavage can take place reversibly [4]. For covalent molecular imprinting, since the number of functional monomers that can be covalently attached is decided by the template, changing the template to functional monomer ratio might not affect the overall reaction.

Biosensors: Materials and Applications   Materials Research Forum LLC
Materials Research Foundations **47** (2019) 51-76   doi: http://dx.doi.org/10.21741/9781644900130-2

Non-covalent approach:

This approach first introduced in 1981 by Mosbach et al. [5] involves the formation of non-covalent interactions between templates and monomers before polymerization. Since the bonds in the non-covalent approach are weak, it is easy to separate the template from the monomers. Also, as compared to the covalent approach, this approach is favored among researchers as the experimental strategy is easy, template removal is easy and a variety of functionality can be introduced using this technique. In this approach, the functional monomer is used in excess relative to the template to favour the formation of template-functional monomer assemblies. But like the covalent approach, this approach also suffers the disadvantage of the requirement of a large amount of monomer to displace the equilibrium to form a monomer-template complex. Moreover, the functionality of the template should match the functionality of functional monomer.

Semi-covalent approach:

This approach is an intermediate approach and combines the advantages of both covalent and non-covalent strategies. The template is bound to a functional monomer covalently but template rebinding is based only on the non-covalent approach.

1.4   Applications of MIPs

Since MIPs can imitate the receptor–ligand, antibody-antigen, or enzyme–substrate biorecognition these can be used in applications relying on selective molecular binding. Their specificity and selectivity towards their target molecule are similar to natural antibodies [6]. Also, in terms of stability, chemical inertness, and ability to withstand extremes of pH and temperature, MIPs are superior. Moreover, natural receptors can only function in aqueous environments, but MIPs are inherently equally active in aqueous and non-aqueous conditions. Numerous analytes having no optical or electrochemical properties can also be detected by exploiting MIPs' capacity of generating optical or electrochemical signals in response to the template binding to the functional groups in the imprinting site. The transducers used with MIPs in recording the binding of the target analyte can be a change in mass in acoustic sensors, change in refractive index in Surface Plasmon Resonance (SPR) sensors, or Electrochemical Impedance Spectroscopy (EIS) in electrochemical sensors. The surface imprinting of MIPs offers an option that can be used in connection to biosensors. Several research groups highlighted the present and future aspects of the molecularly imprinted polymers, stating their use in the areas of purification, pharma and biosensing [7–10]. Figure 2 depicts the different applications of MIPs like as chemosensor, as a catalyst, as a kit, etc.

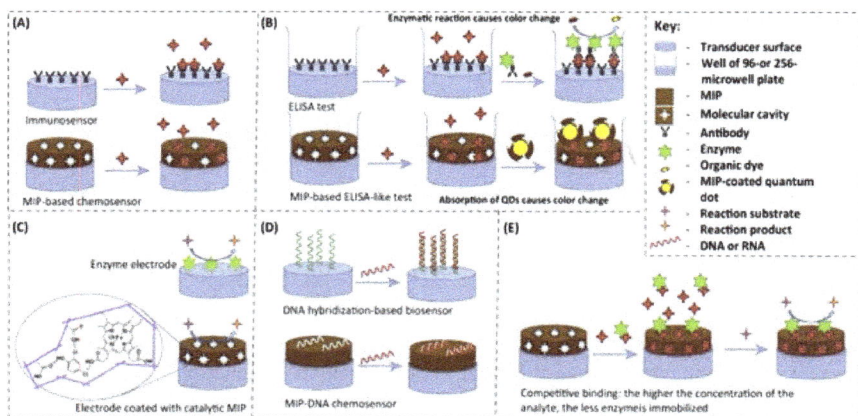

Figure 2: Application of MIPs as (A) as chemosensor, (B) as ELISA like kit, (C) as catalytic activity, and (D) as DNA based sensor. (E) shows the competitive binding (Reprinted from Trends in Biotechnology, 34(11), Cieplak, M. and W. Kutner, Artificial Biosensors: How Can Molecular Imprinting Mimic Biorecognition?, p. 922-941, 2016 [11] with permission from Elsevier).

2. Biosensors

2.1 Introduction to Biosensors

The term 'Biosensor' coined by Cammann [12,13] refers to a sensor which either convert a biological response into a readable signal or uses bioreceptors for catalysing the reaction. The signal generated can be in the form of some colour change, generation of an electrical signal, production of heat or can be changed in the mass of the system [14], depending upon the bio/chemical reaction involved. These biosensors are used for a number of applications like diagnostics, health care, microbiology, biotechnology, food industry, tissue engineering and so on [15–17].

Figure 3 shows the basic components of a biosensor which help in chemical reactions and signal generation and signal to capture [18].

2.2 Components of Biosensor

Broadly the components are divided into

- Bioreceptors
- Transducers

Biosensors: Materials and Applications   Materials Research Forum LLC
Materials Research Foundations **47** (2019) 51-76   doi: http://dx.doi.org/10.21741/9781644900130-2

- Amplifier
- Electronics
- Interface or display

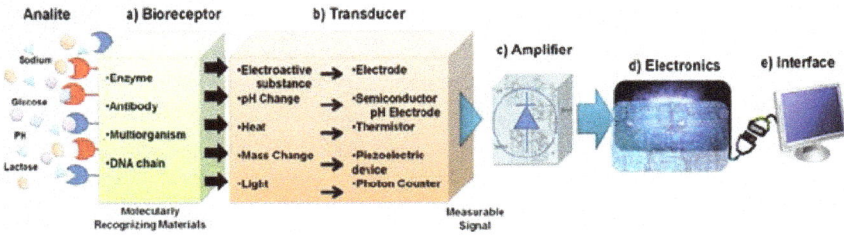

Figure 3: The components of a biosensor in general (Reproduced with permission from ref. [18]).

2.3   Bioreceptors

Bioreceptors are the recognition elements in the biosensor, selection of which governs the sensing element. These are major contributors to specificity. Various bioreceptors are possible in a biosensor, ranging from enzymes, antibodies, microorganisms or even genetic material such as DNA or RNA and can be used against a number of analytes. But broadly these can generally be classified into:

(i) Enzyme-based biosensors

These are the most prominent biosensors which have found their way into the market. The most popular being the glucose sensor in which glucose oxidase enzyme is used to convert the glucose in the sample to peroxide which leads to the generation of a current proportional to the amount of glucose present in the sample [19]. Many other enzymes have also been used for the sensing of the substrates they are specific to. For instance, fructose dehydrogenase enzyme is used for the sensing of fructose [20], alcohol oxidase for sensing of alcohol [21] and amino acid oxidase for sensing of amino acids [22].

(ii) Antibody-based biosensors

Antibodies are the most specific entities which only binds to antigens corresponding to which they are produced. Polyclonal antibodies recognize multiple epitopes of the same antigen whereas monoclonal antibodies recognize only the same epitope of the antigen. The antibodies specifically bind to the antigens which result in the generation of a signal, depending upon the transducer used [23]. Antibodies based biosensors are also used for diverse applications like detection of pathogens in food [24], water [25], detection of disease biomarkers [26]. Currently, wearable biosensors are being developed. One such instance is sensing of Interleukin-6 and Cortisol from human sweat. These are wearable antibodies based biosensors [27].

(iv) Genetic material based biosensors

This kind of biosensor generally uses DNA or RNA as the biorecognition element. These are highly sophisticated biosensors and are much more selective and sensitive than other biosensors. *Neisseria gonorrhoeae*, the causative agent for gonorrhoea can be detected by using DNA based biosensors [28]. DNA based sensors can also be used to determine the damage caused to the DNA by reactive oxygen species and to check the effects of anti-oxidants on them [29]. Similarly, RNA based fluorescent biosensors are also proposed for imaging live cells [30].

2.2.2 Transducers

Transducers are the devices which convert the obtained biological signal into a measurable form. And depending upon the type of property changing, these can be broadly divided into optical, calorimetric, piezoelectric, and electrochemical-based biosensors.

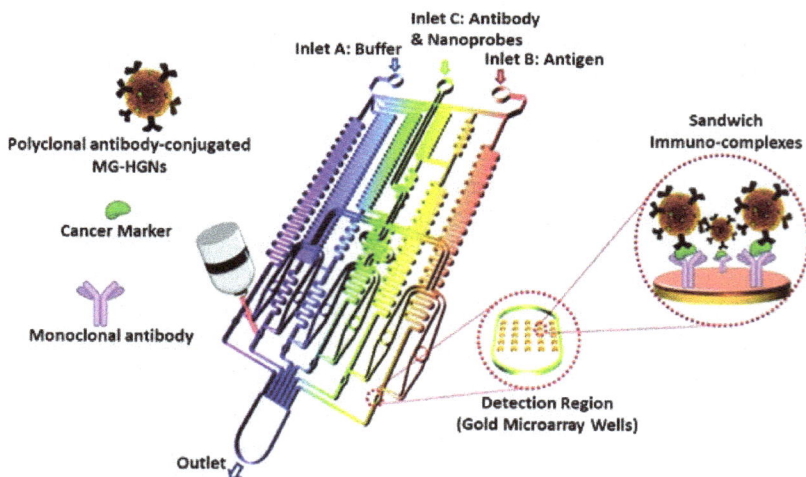

Figure 4: A gold-based SERS platform for optical based biosensing of cancer biomarkers (Reproduced from [33] with permission of The Royal Society of Chemistry).

Biosensors: Materials and Applications   Materials Research Forum LLC
Materials Research Foundations **47** (2019) 51-76   doi: http://dx.doi.org/10.21741/9781644900130-2

Optical-based

When the generated signal is in the form of fluorescence, luminescence, light absorption or involves a change in the refractive index, the biosensor is classified as an optical based biosensor. These biosensors are based on fibre optics, Surface Plasmon Resonance (SPR), planar guide wave, interferometric, colorimetric and Raman signal [31] and can be used for sensing of various analytes such as drugs, enzymes, pesticides or even genetic material [32].

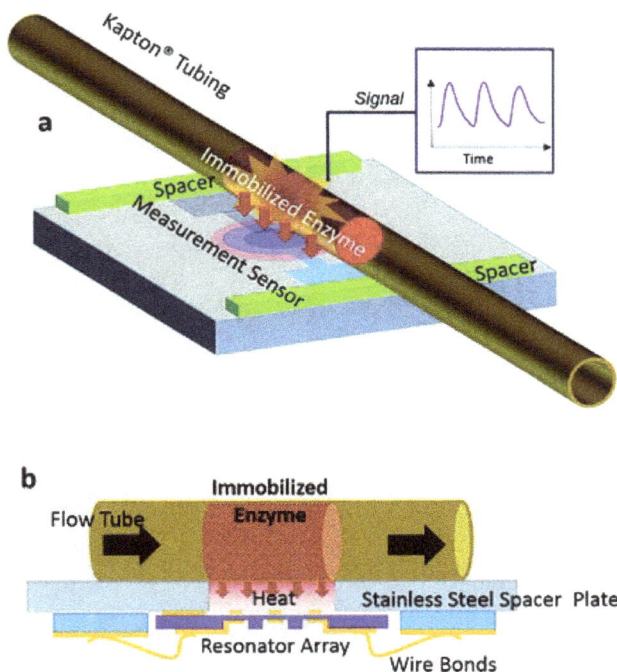

Figure 5: Schematic showing a calorimetric based sensor for urea detection. (a) and (b) shows the sensor in complete 3D mode and the cross-sectional view respectively (Reproduced from [35] with permission of The Royal Society of Chemistry).

Calorimetric based

When two complementary things react together, the reaction sometimes might involve either the production or consumption of heat which is often referred to as exothermic or

Biosensors: Materials and Applications   Materials Research Forum LLC
Materials Research Foundations **47** (2019) 51-76   doi: http://dx.doi.org/10.21741/9781644900130-2

endothermic reactions respectively. This heat change can be used as a signal in the biosensor. Enzyme thermistors are used to detect the changes in the temperatures when the substrate reacts with the enzymes. These are of significant interest given the fact that many enzymes in our body are important from the diagnostic point of view and hence their monitoring is very important [34]. Urea, an important marker signifying the state of health of a person can be detected by thermal based biosensors [35]. There have been efforts to detect cancer cells by using this kind of biosensors.

Piezoelectric based

This biosensor is based on the piezoelectric effect which states that when a piezoelectric material is subjected to an external electric field, a strain is produced in that material and vice-versa. The most common piezoelectric material is quartz crystal [36]. This biosensor is the most sensitive amongst other biosensors and is very difficult to manufacture and is very expensive. It has been used for detection of HIV viruses which is very difficult with the traditional methods [37]. Due to its sensitivity, it has also been used for the sensing of pesticides of organophosphate and carbamate origin [38].

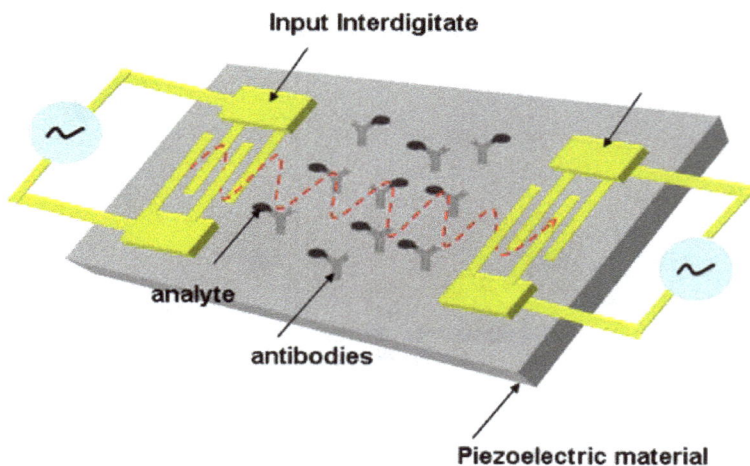

Figure 6: A schematic for a piezoelectric based biosensor for the detection of a particular analyte whose antibody is immobilized on the surface (Reproduced with permission from ref [39]).

Biosensors: Materials and Applications   Materials Research Forum LLC
Materials Research Foundations **47** (2019) 51-76   doi: http://dx.doi.org/10.21741/9781644900130-2

Electrochemical-based

Electrochemical-based biosensors are those in which the analyte and the bioreceptor cause a change in the electrical properties of the system. This electrical change is due to the oxidations or reductions taking place by the interactions of the analyte and receptor. These are highly sensitive, portable and easy to construct [40,41]. The first biosensor invented by Clark and Lyons (1962) to measure glucose in biological samples utilized the strategy of electrochemical detection of oxygen or hydrogen peroxide using immobilized glucose oxidase electrode [42]. Glucose biosensors are widely popular among hospitals or diagnostic clinics as these are essential for diabetic patients for periodic monitoring of blood glucose. In recent times, electrochemical biosensors Wang et al. [43] are typically prepared by modifying the surface of metal and carbon electrodes using biomaterials, such as enzyme, antibody, or DNA. Electrochemical biosensors have different detection techniques. These are amperometric (based on measuring the current), potentiometric (based on measuring the potential) or conductometric (ability to conduct current between the analyte and the receptor) [44].

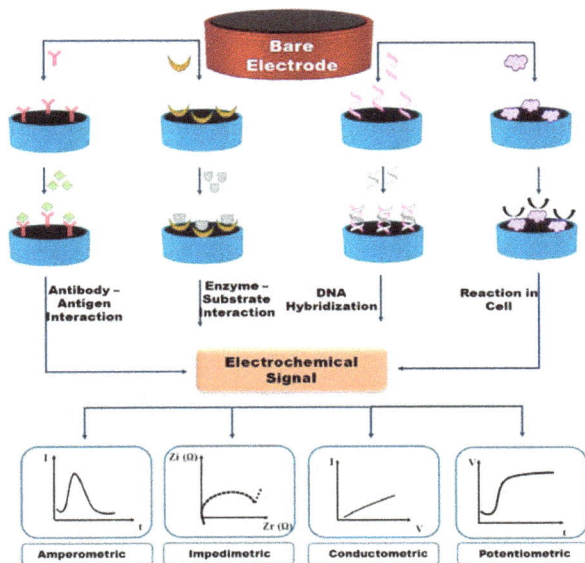

Figure 7: Various types of electrochemical sensors based on the electrochemical signal obtained after analyte bioreceptor binding (Reprinted from Biosensors and Bioelectronics, 68, Bahadır, E.B. and M.K. Sezgintürk ,Electrochemical biosensors for hormone analyses, p. 62-71, 2015 with permission from Elsevier[45]).

Amperometric biosensors

As mentioned above, amperometric based electrochemical sensors are those which measures the current generated from oxidation or reduction of biological entity binding to its specific receptor in a biosensor [46]. Glucose sensing has been the key application for the amperometric based method. Many authors have reported amperometric based glucose sensing having good linearity and good limit of detection [47–49]. The amperometric based biosensors are majorly used for diagnostic applications [50]. They have been used in the immunosensing of diseases like malaria and Chagas disease as highlighted by Ferreira et.al. [51]. Apart from these diseases, there have been attempts to synthesize a point of care device for amperometric sensing of *Mycobacterium tuberculosis,* the causative agent for Tuberculosis. The developed sensor showed a detection limit of 100 CFU/ml which is comparable to other detection methods [52]. Amperometric based detection of cancer has also been reported highlighted by Gómez et.al. [53].

Potentiometric biosensors

Potentiometric based biosensors rely on the variation of potential in the system which is further amplified and recorded. These are mainly used for determination of adulterants in food samples and in environmental samples [54]. Urea, a waste product of many organisms can be detected using the potentiometric method. Urea sensing is important both from a diagnostic as well as from a food quality point of view. Recombinant urease based potentiometric sensors for determination of urea in blood serum and hemodialyzate has shown good working linearity and good response time of 1-2 minutes [55]. Monitoring of urea is essential for patients having malfunctioning kidneys or liver [56]. Apart from urea detection in blood samples, these have been used for detection contamination in milk as well [57]. Aflatoxin B1, a carcinogen has been found to be present in many food samples of sesame, walnut and pea using an enzyme-based potentiometric biosensor [58]. Just like amperometric based sensors, potentiometric biosensors are also used for on-site detection of diseases. Tarasov et al. developed a potentiometric biosensor for detection of Bovine Herpes Virus-1 (BHV-1), which causes Bovine Respiratory Disease (BRD). The developed sensor is significantly faster than the conventional ELISA based platform [59].

Conductometric biosensors

Conductometric based biosensors are of particular interest to the scientific community owing to its simplicity and ease of construction. These are highly sensitive and can be used to determine even trace concentrations of the analytes. This biosensor employs those components which are electroactive in nature and conduct current. The most common are nanoparticles which are metallic in nature. These have a wide variety of applications. Urea sensing using silica nanoparticles is one such common application [60]. From a more diagnostic point of view, conductometric based sensors are used for the detection of methamphetamine in human urine [61], cardiac markers [62] and Johne's disease [63]. The conductometric sensors are also used for the detection of various pollutants in an environment such as heavy metals [64,65], organophosphate pesticides [66], formaldehyde [67] and hydrogen peroxides in water. Various food pathogens such as enterohemorrhagic *Escherichia coli* O157:H7 and *Salmonella* spp are also detected using conductometric based methods within 10 minutes [68].

2.2.3 Amplifier, electronics and interface or display

The signal obtained is quite weak and needs amplification which is generally done with the help of an amplifier. Usually, a CMOS based amplifier is used in a biosensor. The intermediate electronics obtain the amplified signal and converts it into a readable signal which can be interpreted by a user. A monitor or a screen displays the signal as graphs or some images which a user can understand and proceed further.

3.   Immobilization Matrices for Biosensors

One of the most important steps in biosensor fabrication is to immobilize the biomolecules. Immobilization is basically confinement of a bio-receptor to a phase (matrix/support) and it provides the platform for the biological receptor onto which the target analyte binds and creates a bio-signal. A successful matrix should immobilize biomolecules stably at a transducer surface. The functionality of the biomolecules should also be maintained efficiently as well as the matrix should provide accessibility towards the target analyte. Generally, polymers having inertness, stability, regenerability, etc. are used. However, the suitability of support material depends on the bioreceptor being used on application target. Key parameters for selection of suitable material and method are the binding capacity of the material, stability, retention of bioreceptor activity and minimizing leakage of bioreceptor after immobilization on the material.

Polymers are promising candidates as immobilization matrix for biosensing applications [69]. Both natural and artificially occurring polymers along with their composites can be used. Among various polymers, molecularly imprinted polymers (MIPs) are one of the emerging classes of polymers used for biosensing applications.

4. Molecularly Imprinted Polymers based Geno-biosensors:

MIPs have been used as a biosensor for many analytes like cancer biomarkers, viruses, heavy metals, pesticides, etc., where biomolecular receptors (such as antibodies, enzymes, and histones) have been increasingly replaced with these artificial recognition elements; MIPs. MIPs can mimic biological recognition well enough to be sometimes called "plastic antibodies" [70,71].

DNA being part of the Central Dogma, its detection is very essential from the diagnostic point of view. In recent times, there has been a demand for detection of multiple analytes at a time i.e. multiplexing approach. One such example is the detection of folic acid, folate receptors, Hg^{2+} and DNA using the MIP approach wherein DNA based MIP was synthesized (Figure 8). The platform showed detection limits of 30 nM, 0.3 ng.ml^{-1}, 3.45 pM and 40 nM for folic acid, folate receptors, Hg^{2+} and DNA respectively. The developed sensor had good reproducibility, re-generability and stability [72].

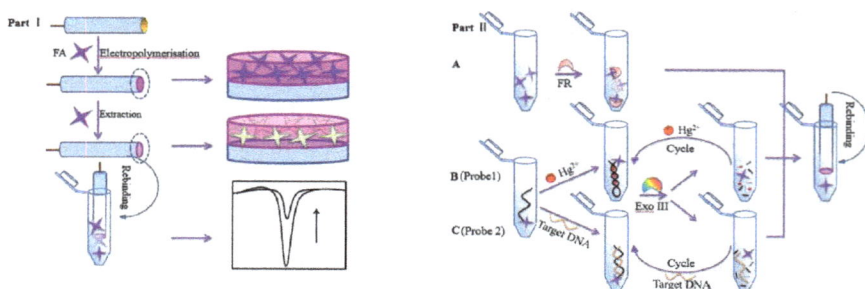

Figure 8: Part I of the schematic shows the fabrication process of the biosensor. Part II shows the sensing of the analytes i.e. folic acid, folate receptors, Hg2+ and DNA (Reproduced from [70] with permission of The Royal Society of Chemistry).

Another approach for DNA detection is the incorporation of a specific dsDNA sequence template into the MIP gel. This gel can be used as a matrix for electrophoresis. When the gel is run, the target DNA sequences will bind to binding sites in the MIP and can be detected. This proof of concept was provided by Ogiso et.al. [73] by using the Ha-ras gene and its point mutants. The developed system showed that mutant dsDNA can be easily differentiated from the target wild-type dsDNA except A·T to T·A base pair substitution.

Biosensors: Materials and Applications   Materials Research Forum LLC
Materials Research Foundations **47** (2019) 51-76   doi: http://dx.doi.org/10.21741/9781644900130-2

Single nucleotide polymorphism (SNP) is responsible for variation in the human genome and is the "look out for" elements for various disease detection states and hence, its detection is vital. Hexameric 2,2′-bithien-5-yl DNA analogues were used as a template for the synthesis of MIP which can differentiate a purine- nucleobase mismatch. This mismatch was detected by using piezo-microgravimetric, Surface Plasmon Resonance, or Capacitive Impedimetry (CI) techniques [74]. The CI based method had a detection limit of 5 nM [75].

At the cellular level, ribosomes are involved in the translation for the generation of proteins and any error in this process can cause a cancer state. Therefore, detection of these mRNA or messenger RNA is the need of the hour. King et al. [76] provided a proof of concept by using polyacrylamide based MIP for monitoring changes in gene expression in a cell by sensing these RNAs. This system can isolate ribosomes and translated mRNAs from merely 1000 cells which are 1000 fold less than the conventional methods.

Adenine is one of the most important nucleic bases present in the body and is responsible for many biological functions [77]. This nucleic base is now been used as a biomarker for the diagnosis of many diseases, especially cancer. Liu et al. [78] used high-performance liquid chromatography for selective adsorption and separation of adenine with MIP-PMAA/SiO_2 particles as the stationary phase. The particles were reliable and showed adsorption of adenine even after multiple cycles and had good regeneration strength. This nucleic base has also been found to be present in urine samples, indicating a cancer state in the human body. This is very alarming and its monitoring can be done by imprinting 2-deoxyadenosine (2-dA) on the molecularly imprinted polymer membrane. The membrane showed a response time of 30 minutes as compared to hours. The current system has a rapid rate of binding of 30 minutes in comparison to the other similar systems which usually takes 150 to 180 minutes to achieve equilibrium. The sensor showed a linear response in 0.5-0.1 mmol.L^{-1} with a R^2 value of 0.9995. The molecularly imprinted membrane was found to be highly selective [79].

Piezomicrogravimetric (acoustic) chemosensor was also developed for the determination of adenine with the help of adenine imprinted MIP. The chemosensor was able to discriminate between adenine and structurally similar molecules. The detection limit for this sensor was 5 nM with 1 ml of the sample having a flow rate of 35 μL.min^{-1} [80].

Hydrogel-based MIPs have been shown as a low-cost platform for sensing of cognate and non-cognate proteins. These proteins are essential templates from a diagnostics point of view. The researchers used bovine haemoglobin (BHb) and trypsin (Tryp) as templates [81].

Metronidazole (MNZ) an antifungal agent used for the treatment of protozoal diseases has been detected using MIPs with the electrochemical method and showed a detection limit of 1.8×10^{-11} mol.L^{-1} (S/N=3) [82]. Serum C-terminal telopeptide of type I collagen (CTxI) detection is very important from bone loss monitoring point of view. Electrochemical methods employing MIPs have been used for this application. The developed sensor showed a good linear range of 0.1 to 2.5 ng.mL^{-1} and detection limit of 0.09 ng.mL^{-1}. The results were well corroborated with the traditional ELISA based methods [83].

Uric acid is an important waste product of the human body and tells us the state of kidneys and liver functioning. High levels of these are deleterious to the body. Chen et al. [84] used conducting polymer imprinted onto the ITO substrate with uric acid as a template for determination of uric acid. The sensor worked fine in the linear range of 0 to 1.125 mM with a detection limit of 0.3 µM having a signal to noise ratio of 3 [82].

The flexibility of MIPs has been used for screening of prostate-specific antigen (PSA), a gold biomarker for prostate cancer using electrochemical impedance spectroscopy [85]. An aptamer having an affinity for PSA was imprinted onto the MIP and was used for detection of this antigen. The sensor showed a detection limit of 1 pg.mL^{-1} with the linear response from 100 pg.mL^{-1} to 100 ng.mL^{-1}. The developed platform was highly selective and the interfering agent such as Human Kallikrein 2 and Human Serum Albumin couldn't generate much response from the sensor. MIP has also been used to track the point mutations occurring in the p53 gene [86]. The point mutations in this gene are responsible for human prostate cancer. The developed biosensor was single-stranded oligodeoxyribonucleotide (ss-ODN) based. The ss-ODN was used as a template and o-phenylenediamine was used as a functional monomer for the synthesis of MIP on indium-tin-oxide (ITO) coated glass substrate. The biosensor showed a linear response in the range of 0.01- 300 fM with a sensitivity of 0.62 µA.fM^{-1} and exhibited response time of fewer than 20 seconds.

Apart from sensing the disease associated biomarkers, MIPs have also been used for the detection of various anticancer drugs. This type of study is very essential to evaluate whether only a sufficient dose is injected into the body, thereby preventing side effects of the chemotherapy. A RNA- based MIP geno-biosensor is reported by Huynh et al. for the detection of 5-fluorouracil (FU), an anti-cancer agent for the treatment of colorectal cancer [87]. The MIP was synthesized from 6-aminopurine (adenine) derivative of bis(2,2'-bithienyl)- methane, vis., 4-[2-(6-amino-9H-purin-9-yl)ethoxy]phenyl-4-[bis(2,2'- bithienyl)methane] or Ade-BTM which selectively recognizes FU. The geno-biosensor showed the limit of detection of 56 nM, 75 nM, and 0.26 mM by different

Biosensors: Materials and Applications   Materials Research Forum LLC
Materials Research Foundations **47** (2019) 51-76   doi: http://dx.doi.org/10.21741/9781644900130-2

transduction platforms such as differential pulse voltammetry (DPV), capacitive impedimetry (CI), and piezoelectric microgravimetry (PM) respectively.

Conclusion

Molecularly imprinted polymers are promising candidates to be used as biomimetic agents because of their versatilities. The ability to accommodate analytes in their cavities makes them potential agents to be used as artificial antibodies and in other biological applications. The current MIPs technology offers several advantages over the traditional methods. The most important being sensitivity and quick detection time. Detection of genetic biomarkers is very important from a erly disease detection point of view. Tracing them in biological fluids gives an important understanding of their levels in the body which certainly can help in the early diagnosis of life-threatening diseases.

Acknowledgement

The authors acknowledge the support of Director, CSIR-CSIO for his constant encouragement. MG acknowledges the SRF-GATE fellowship from Council of Scientific and Industrial Research (CSIR), New Delhi. SP acknowledges the financial support received through National Post Doc research fellowship from DST-SERB, New Delhi.

References

[1]   L. Chen, X. Wang, W. Lu, X. Wu, J. Li, Molecular imprinting: perspectives and applications, Chem. Soc. Rev. 45 (2016) 2137–2211. https://doi.org/10.1039/C6CS00061D

[2]   G. Ertürk, B. Mattiasson, molecular imprinting techniques used for the preparation of biosensors, Sensors. 17 (2017) 288. https://doi.org/10.3390/s17020288

[3]   H. Yan, K. Row, characteristic and synthetic approach of molecularly imprinted polymer, Int. J. Mol. Sci. 7 (2006) 155–178. https://doi.org/10.3390/i7050155

[4]   F. Meier, B. Mizaikoff, Molecularly imprinted polymers as artificial receptors, in: Artif. Recept. Chem. Sensors, Wiley-VCH Verlag GmbH & Co. KGaA, Weinheim, Germany, 2010: pp. 391–437. https://doi.org/10.1002/9783527632480.ch13

[5]   Reza Arshady Klaus Mosbach, Synthesis of substrate-selective polymers by host-guest polymerization, Macromol. Chem. Phys. 182 (1981) 687–692

Biosensors: Materials and Applications   Materials Research Forum LLC
Materials Research Foundations **47** (2019) 51-76   doi: http://dx.doi.org/10.21741/9781644900130-2

[6]   L. Chen, S. Xu, J. Li, Recent advances in molecular imprinting technology: current status, challenges and highlighted applications, Chem. Soc. Rev. 40 (2011) 2922. https://doi.org/10.1039/c0cs00084a

[7]   G. Vasapollo, R. Del Sole, L. Mergola, M.R. Lazzoi, A. Scardino, S. Scorrano, G. Mele, Molecularly imprinted polymers: present and future prospective, Int. J. Mol. Sci. 12 (2011) 5908–5945. https://doi.org/10.3390/ijms12095908

[8]   F. Navarrovilloslada, J. Urraca, M. Morenobondi, G. Orellana, Zearalenone sensing with molecularly imprinted polymers and tailored fluorescent probes, Sensors Actuators B Chem. 121 (2007) 67–73. https://doi.org/10.1016/j.snb.2006.09.042

[9]   F. Puoci, F. Iemma, N. Picci, Stimuli-responsive molecularly imprinted polymers for Drug Delivery: A review, Curr. Drug Deliv. 5 (2008) 85–96. https://doi.org/10.2174/156720108783954888

[10]   W.C. Lee, C.H. Cheng, H.H. Pan, T.H. Chung, C.C. Hwang, Chromatographic characterization of molecularly imprinted polymers, Anal. Bioanal. Chem. 390 (2008) 1101–1109

[11]   M. Cieplak, W. Kutner, Artificial biosensors: How can molecular imprinting mimic biorecognition?, Trends Biotechnol. 34 (2016) 922–941. https://doi.org/10.1016/j.tibtech.2016.05.011

[12]   P. Mehrotra, Biosensors and their applications - A review., J. Oral Biol. Craniofacial Res. 6 (2016) 153–9. https://doi.org/10.1016/j.jobcr.2015.12.002

[13]   K. Cammann, Bio-sensors based on ion-selective electrodes, Fresenius' Zeitschrift Fur Anal. Chemie. 287 (1977) 1–9. https://doi.org/10.1007/BF00539519

[14]   D. Kim, D. Kang, molecular recognition and specific interactions for biosensing applications, Sensors. 8 (2008) 6605–6641. https://doi.org/10.3390/s8106605

[15]   M.S. Thakur, K. V. Ragavan, Biosensors in food processing, J. Food Sci. Technol. 50 (2013) 625–641. https://doi.org/10.1007/s13197-012-0783-z

[16]   and A.A.J. Anwarul Hasan, Md Nurunnabi, Mahboob Morshed, Arghya Paul, Alessandro Polini, Tapas Kuila,Moustafa Al Hariri, Yong-kyu Lee, Recent advances in application of biosensors in tissue engineering, Biomed Res. Int. (2014). https://doi.org/10.1155/2014/307519

[17]   S. Kanchi, M.I. Sabela, P.S. Mdluli, Inamuddin, K. Bisetty, Smartphone based bioanalytical and diagnosis applications: A review, Biosens. Bioelectron. 102 (2018) 136–149. https://doi.org/10.1016/j.bios.2017.11.021

[18] A.I.R.-V. and J.S. Jordi Colomer-Farrarons, Pere Ll. Miribel-Català, Portable Bio-Devices: Design of electrochemical instruments from miniaturized to implantable devices, in: P.A. Serra (Ed.), New Perspect. Biosens. Technol. Appl., 2011: pp. 373–400

[19] E.-H. Yoo, S.-Y. Lee, Glucose biosensors: An overview of use in clinical practice, Sensors. 10 (2010) 4558–4576. https://doi.org/10.3390/s100504558

[20] U.B. Trivedi, D. Lakshminarayana, I.L. Kothari, P.B. Patel, C.J. Panchal, Amperometric fructose biosensor based on fructose dehydrogenase enzyme, Sensors Actuators B Chem. 136 (2009) 45–51. https://doi.org/10.1016/j.snb.2008.10.020

[21] M. Hämmerle, K. Hilgert, M.A. Horn, R. Moos, Analysis of volatile alcohols in apple juices by an electrochemical biosensor measuring in the headspace above the liquid, Sensors Actuators B Chem. 158 (2011) 313–318. https://doi.org/10.1016/j.snb.2011.06.026

[22] S. Lata, B. Batra, N. Singala, C.S. Pundir, Construction of amperometric l-amino acid biosensor based on l-amino acid oxidase immobilized onto ZnONPs/c-MWCNT/PANI/AuE, Sensors Actuators B Chem. 188 (2013) 1080–1088. https://doi.org/10.1016/j.snb.2013.08.025

[23] S. Sharma, H. Byrne, R.J. O'Kennedy, Antibodies and antibody-derived analytical biosensors, Essays Biochem. 60 (2016) 9–18. https://doi.org/10.1042/EBC20150002

[24] M.M. and K.W.C.L. Guangfu Wu, Graphene field-effect transistors-based biosensors for Escherichia coli detection, in: Int. Conf. Nanotechnol. Sendai, Japan, 2016: pp. 22–25

[25] J.W.F. Law, N.S. Ab Mutalib, K.G. Chan, L.H. Lee, Rapid methods for the detection of foodborne bacterial pathogens: principles, applications, advantages and limitations, Front Microbiol. 5 (2014) 770–788

[26] S.T. Sanjay, G. Fu, M. Dou, F. Xu, R. Liu, H. Qi, X. Li, Biomarker detection for disease diagnosis using cost-effective microfluidic platforms, Analyst. 140 (2015) 7062–7081. https://doi.org/10.1039/C5AN00780A

[27] R.D. Munje, S. Muthukumar, B. Jagannath, S. Prasad, A new paradigm in sweat based wearable diagnostics biosensors using room temperature ionic liquids (RTILs), Sci. Rep. 7 (2017) 1950. https://doi.org/10.1038/s41598-017-02133-0

Materials Research Forum LLC

doi: http://dx.doi.org/10.21741/9781644900130-2

[28]   R. Singh, G. Sumana, R. Verma, S. Sood, K.N. Sood, R.K. Gupta, B.D. Malhotra, Fabrication of Neisseria gonorrhoeae biosensor based on chitosan–MWCNT platform, Thin Solid Films. 519 (2010) 1135–1140. https://doi.org/10.1016/j.tsf.2010.08.057

[29]   Ziyatdinova Guzel, Galandova Julia, Labuda Jan, Impedimetric Nanostructured disposable DNA-based biosensors for the detection of deep dna damage and effect of antioxidants, Int. J. Electrochem. Sci. 3 (2008) 223–235

[30]   C.A. Kellenberger, C. Chen, A.T. Whiteley, D.A. Portnoy, M.C. Hammond, RNA-Based Fluorescent biosensors for live cell imaging of second messenger cyclic di-AMP, J. Am. Chem. Soc. 137 (2015) 6432–6435. https://doi.org/10.1021/jacs.5b00275

[31]   F. Long, A. Zhu, H. Shi, Recent advances in optical biosensors for environmental monitoring and early warning, Sensors. 13 (2013) 13928–13948. https://doi.org/10.3390/s131013928

[32]   S.M. Borisov, O.S. Wolfbeis, Optical biosensors, Chem. Rev. 108 (2008) 423–461. https://doi.org/10.1021/cr068105t

[33]   M. Lee, K. Lee, K.H. Kim, K.W. Oh, J. Choo, SERS-based immunoassay using a gold array-embedded gradient microfluidic chip, Lab Chip. 12 (2012) 3720. https://doi.org/10.1039/c2lc40353f

[34]   B.X. B.Danielsson, U.Hedberg, M.Rank, Recent investigations on calorimetric biosensors, Sensors Actuators B Chem. 6 (1992) 138–142

[35]   D.E. Gaddes, M.C. Demirel, W.B. Reeves, S. Tadigadapa, Remote calorimetric detection of urea via flow injection analysis, Analyst. 140 (2015) 8033–8040. https://doi.org/10.1039/C5AN01306B

[36]   P. Skládal, Piezoelectric biosensors, TrAC Trends Anal. Chem. 79 (2016) 127–133. https://doi.org/10.1016/j.trac.2015.12.009

[37]   M. Bisoffi, V. Severns, D.W. Branch, T.L. Edwards, R.S. Larson, Rapid detection of human immunodeficiency virus types 1 and 2 by use of an improved piezoelectric biosensor, J. Clin. Microbiol. 51 (2013) 1685–1691. https://doi.org/10.1128/JCM.03041-12

[38]   G. Marrazza, Piezoelectric biosensors for organophosphate and carbamate pesticides: A review, Biosensors. 4 (2014) 301–317. https://doi.org/10.3390/bios4030301

[39] Y. Zhou, C.-W. Chiu, H. Liang, interfacial structures and properties of organic materials for biosensors: An overview, Sensors. 12 (2012) 15036–15062. https://doi.org/10.3390/s121115036

[40] C. Zhu, G. Yang, H. Li, D. Du, Y. Lin, Electrochemical sensors and biosensors based on nanomaterials and nanostructures, Anal. Chem. 87 (2015) 230–249. https://doi.org/10.1021/ac5039863

[41] N.J. Ronkainen, H.B. Halsall, W.R. Heineman, Electrochemical biosensors, Chem. Soc. Rev. 39 (2010) 1747. https://doi.org/10.1039/b714449k

[42] A.P.F. Turner, Biosensors: sense and sensibility, Chem. Soc. Rev. 42 (2013) 3184. https://doi.org/10.1039/c3cs35528d

[43] B. Wang, S. Takahashi, X. Du, J. Anzai, electrochemical biosensors based on ferroceneboronic acid and its derivatives: A review, Biosensors. 4 (2014) 243–256. https://doi.org/10.3390/bios4030243

[44] D. Grieshaber, R. MacKenzie, J. Vörös, E. Reimhult, Electrochemical biosensors - sensor principles and architectures, Sensors. 8 (2008) 1400–1458. https://doi.org/10.3390/s80314000

[45] E.B. Bahadır, M.K. Sezgintürk, Electrochemical biosensors for hormone analyses, Biosens. Bioelectron. 68 (2015) 62–71. https://doi.org/10.1016/j.bios.2014.12.054

[46] S.J. Sadeghi, Amperometric Biosensors, in: Encycl. Biophys., Springer Berlin Heidelberg, Berlin, Heidelberg, 2013: pp. 61–67. https://doi.org/10.1007/978-3-642-16712-6_713

[47] T.B. Goriushkina, A.P. Soldatkin, S. V. Dzyadevych, Application of amperometric biosensors for analysis of ethanol, glucose, and lactate in wine, J. Agric. Food Chem. 57 (2009) 6528–6535. https://doi.org/10.1021/jf9009087

[48] Z.-D. Gao, Y. Qu, T. Li, N.K. Shrestha, Y.-Y. Song, Development of amperometric glucose biosensor based on prussian blue functionlized TiO_2 nanotube arrays, Sci. Rep. 4 (2015) 6891. https://doi.org/10.1038/srep06891

[49] Marianna Portaccio and Maria Lepore, Determination of different saccharides concentration by means of a multienzymes amperometric biosensor, J. Sensors. (2017). https://doi.org/10.1155/2017/7498945

[50] M. Belluzo, M. Ribone, C. Lagier, Assembling amperometric biosensors for clinical diagnostics, Sensors. 8 (2008) 1366–1399. https://doi.org/10.3390/s8031366

[51]   A.V.B. and H.Y. Antonio Aparecido Pupim Ferreira, Carolina Venturini Uliana, Michelle de Souza Castilho, Naira Canaverolo Pesquero, Marcos Vinicius Foguel, Glauco Pilon dos Santos, Cecílio Sadao Fugivara, Amperometric biosensor for diagnosis of disease, in: State Art Biosens., 2013

[52]   M. Hiraiwa, J.H. Kim, H.B. Lee, S. Inoue, A.L. Becker, K.M. Weigel, G.A. Cangelosi, K.-H. Lee, J.-H. Chung, Amperometric immunosensor for rapid detection of Mycobacterium tuberculosis, J. Micromechanics Microengineering. 25 (2015) 055013. https://doi.org/10.1088/0960-1317/25/5/055013

[53]   M.H.H.G. and R.G.G. Luis Jesús Villarreal Gómez, Irma Esthela Soria Mercado, Detection of molecular markers of cancer through the use of biosensors, Biol. Med. (2015). https://doi.org/10.4172/0974-8369.S2-005

[54]   A.M. Pisoschi, Potentiometric Biosensors: Concept and analytical applications-An editorial, Biochem. Anal. Biochem. 5 (2016). https://doi.org/10.4172/2161-1009.1000e164

[55]   S.V. Marchenko, I.S. Kucherenko, A.N. Hereshko, I.V. Panasiuk, O.O. Soldatkin, A.V. El'skaya, A.P. Soldatkin, Application of potentiometric biosensor based on recombinant urease for urea determination in blood serum and hemodialyzate, Sensors Actuators B Chem. 207 (2015) 981–986. https://doi.org/10.1016/j.snb.2014.06.136

[56]   C.-Y. Lai, P. Foot, J. Brown, P. Spearman, A urea potentiometric biosensor based on a thiophene copolymer, Biosensors. 7 (2017) 13. https://doi.org/10.3390/bios7010013

[57]   U.B. Trivedi, D. Lakshminarayana, I.L. Kothari, N.G. Patel, H.N. Kapse, K.K. Makhija, P.B. Patel, C.J. Panchal, Potentiometric biosensor for urea determination in milk, Sensors Actuators B Chem. 140 (2009) 260–266. https://doi.org/10.1016/j.snb.2009.04.022

[58]   K.V. Stepurska, O.O. Soldatkin, V.M. Arkhypova, A.P. Soldatkin, F. Lagarde, N. Jaffrezic-Renault, S.V. Dzyadevych, Development of novel enzyme potentiometric biosensor based on pH-sensitive field-effect transistors for aflatoxin B1 analysis in real samples, Talanta. 144 (2015) 1079–1084. https://doi.org/10.1016/j.talanta.2015.07.068

[59]   A. Tarasov, D.W. Gray, M.-Y. Tsai, N. Shields, A. Montrose, N. Creedon, P. Lovera, A. O'Riordan, M.H. Mooney, E.M. Vogel, A potentiometric biosensor for rapid on-site disease diagnostics, Biosens. Bioelectron. 79 (2016) 669–678. https://doi.org/10.1016/j.bios.2015.12.086

Biosensors: Materials and Applications   Materials Research Forum LLC
Materials Research Foundations **47** (2019) 51-76   doi: http://dx.doi.org/10.21741/9781644900130-2

[60]   T.P. Velychko, O.O. Soldatkin, V.G. Melnyk, S. V. Marchenko, S.K. Kirdeciler,
       B. Akata, A.P. Soldatkin, A. V. El'skaya, S. V. Dzyadevych, A Novel
       Conductometric Urea Biosensor with improved analytical characteristic based on
       recombinant urease adsorbed on nanoparticle of silicalite, Nanoscale Res. Lett. 11
       (2016) 106. https://doi.org/10.1186/s11671-016-1310-3

[61]   K. Yagiuda, A. Hemmi, S. Ito, Y. Asano, Y. Fushinuki, C.-Y. Chen, I. Karube,
       Development of a conductivity-based immunosensor for sensitive detection of
       methamphetamine (stimulant drug) in human urine, Biosens. Bioelectron. 11
       (1996) 703–707. https://doi.org/10.1016/0956-5663(96)85920-3

[62]   I. Lee, X. Luo, J. Huang, X.T. Cui, M. Yun, Detection of cardiac biomarkers using
       single polyaniline nanowire-based conductometric biosensors, Biosensors. 2
       (2012) 205–220. https://doi.org/10.3390/bios2020205

[63]   C. Okafor, D. Grooms, E. Alocilja, S. Bolin, comparison between a
       conductometric biosensor and ELISA in the evaluation of Johne's disease,
       Sensors. 14 (2014) 19128–19137. https://doi.org/10.3390/s141019128

[64]   S.A. Soldatkin OO, Kucherenko IS, Pyeshkova VM, Kukla AL, Jaffrezic-Renault
       N, El'skaya AV, Dzyadevych SV, Novel conductometric biosensor based on
       three-enzyme system for selective determination of heavy metal ions.,
       Bioelectrochemistry. 83 (2012) 25–30

[65]   A.L.B.; C.D.; J.-M.C.; S.V.D.; C. Tran-Minh, Whole-cell conductometric
       biosensor for determination heavy-metals in water, in: TRANSDUCERS 2007 -
       2007 Int. Solid-State Sensors, Actuators Microsystems Conf., n.d

[66]   Ani Mulyasuryani and Sasangka Prasetyawan, Organophosphate hydrolase in
       conductometric biosensor for the detection of organophosphate pesticides, Anal
       Chem Insights. 10 (2015) 23–27

[67]   T.-T. Nguyen-Boisse, J. Saulnier, N. Jaffrezic-Renault, F. Lagarde, Miniaturised
       enzymatic conductometric biosensor with Nafion membrane for the direct
       determination of formaldehyde in water samples, Anal. Bioanal. Chem. 406 (2014)
       1039–1048. https://doi.org/10.1007/s00216-013-7197-2

[68]   Z. Muhammad-Tahir, E.C. Alocilja, A conductometric biosensor for biosecurity,
       Biosens. Bioelectron. 18 (2003) 813–819. https://doi.org/10.1016/S0956-
       5663(03)00020-4

Materials Research Forum LLC
doi: http://dx.doi.org/10.21741/9781644900130-2

[69]   T. Wang, M. Farajollahi, Y.S. Choi, I.-T. Lin, J.E. Marshall, N.M. Thompson, S. Kar-Narayan, J.D.W. Madden, S.K. Smoukov, Electroactive polymers for sensing, Interface Focus. 6 (2016) 20160026. https://doi.org/10.1098/rsfs.2016.0026

[70]   Y. Wang, Z. Zhang, V. Jain, J. Yi, S. Mueller, J. Sokolov, Z. Liu, K. Levon, B. Rigas, M.H. Rafailovich, Potentiometric sensors based on surface molecular imprinting: Detection of cancer biomarkers and viruses, Sensors Actuators B Chem. 146 (2010) 381–387. https://doi.org/10.1016/j.snb.2010.02.032

[71]   G. Selvolini, G. Marrazza, MIP-based sensors: Promising new tools for cancer biomarker determination, Sensors. 17 (2017) 718. https://doi.org/10.3390/s17040718

[72]   C. Wang, Z. Guo, L. Zhang, N. Zhang, K. Zhang, J. Xu, H. Wang, H. Shi, M. Qin, L. Ren, DNA based signal amplified molecularly imprinted polymer electrochemical sensor for multiplex detection, RSC Adv. 6 (2016) 49597–49603. https://doi.org/10.1039/C6RA05797G

[73]   M. Ogiso, N. Minoura, T. Shinbo, T. Shimizu, DNA detection system using molecularly imprinted polymer as the gel matrix in electrophoresis, Biosens. Bioelectron. 22 (2007) 1974–1981. https://doi.org/10.1016/j.bios.2006.08.026

[74]   A. Seidel, S. Brunner, P. Seidel, G.I. Fritz, O. Herbarth, Modified nucleosides: an accurate tumour marker for clinical diagnosis of cancer, early detection and therapy control, Br. J. Cancer. 94 (2006) 1726–1733. https://doi.org/10.1038/sj.bjc.6603164

[75]   K. Bartold, A. Pietrzyk-Le, T.-P. Huynh, Z. Iskierko, M. Sosnowska, K. Noworyta, W. Lisowski, F. Sannicolò, S. Cauteruccio, E. Licandro, F. D'Souza, W. Kutner, programmed transfer of sequence information into a molecularly imprinted polymer for hexakis(2,2′-bithien-5-yl) DNA analogue formation toward single-nucleotide-polymorphism detection, ACS Appl. Mater. Interfaces. 9 (2017) 3948–3958. https://doi.org/10.1021/acsami.6b14340

[76]   H.A. King, H.F. El-Sharif, A.M. Matia-González, V. Iadevaia, A. Fowotade, S.M. Reddy, A.P. Gerber, Generation of ribosome imprinted polymers for sensitive detection of translational responses, Sci. Rep. 7 (2017) 6542. https://doi.org/10.1038/s41598-017-06970-x

[77]   R. Malathi, I.M. Johnson, From RNA world to Protein: An eagle's eye view of the role of guanosine in tracing the antiquity of the intron, J. Biomol. Struct. Dyn. 18 (2001) 709–712. https://doi.org/10.1080/07391102.2001.10506701

[78] Y.C. Liu, C. Tian, H.L. Cong, Q.H. Peng, S.H. Xu, B. Yu, Selective adsorption and separation of adenine by molecularly imprinted polymethacrylic acid on surface of silica particles, Integr. Ferroelectr. 178 (2017) 11–22. https://doi.org/10.1080/10584587.2017.1321464

[79] S. Scorrano, L. Mergola, M. Di Bello, M. Lazzoi, G. Vasapollo, R. Del Sole, Molecularly imprinted composite membranes for selective detection of 2-deoxyadenosine in urine samples, Int. J. Mol. Sci. 16 (2015) 13746–13759. https://doi.org/10.3390/ijms160613746

[80] A. Pietrzyk, S. Suriyanarayanan, W. Kutner, R. Chitta, M.E. Zandler, F. D'Souza, Molecularly imprinted polymer (MIP) based piezoelectric microgravimetry chemosensor for selective determination of adenine, Biosens. Bioelectron. 25 (2010) 2522–2529. https://doi.org/10.1016/j.bios.2010.04.015

[81] S.M. Reddy, Q.T. Phan, H. El-Sharif, L. Govada, D. Stevenson, N.E. Chayen, Protein crystallization and biosensor applications of hydrogel-based molecularly imprinted polymers, Biomacromolecules. 13 (2012) 3959–3965. https://doi.org/10.1021/bm301189f

[82] Y. Li, Y. Liu, J. Liu, J. Liu, H. Tang, C. Cao, D. Zhao, Y. Ding, Molecularly imprinted polymer decorated nanoporous gold for highly selective and sensitive electrochemical sensors, Sci. Rep. 5 (2015) 7699. https://doi.org/10.1038/srep07699

[83] K.M. Afsarimanesh N, Mukhopadhyay SC, molecularly imprinted polymer-based electrochemical biosensor for bone loss detection., IEEE Trans Biomed Eng. 65 (2018) 1264–1271

[84] P.-Y. Chen, R. Vittal, P.-C. Nien, G.-S. Liou, K.-C. Ho, A novel molecularly imprinted polymer thin film as biosensor for uric acid, Talanta. 80 (2010) 1145–1151. https://doi.org/10.1016/j.talanta.2009.08.041

[85] P. Jolly, V. Tamboli, R.L. Harniman, P. Estrela, C.J. Allender, J.L. Bowen, Aptamer–MIP hybrid receptor for highly sensitive electrochemical detection of prostate specific antigen, Biosens. Bioelectron. 75 (2016) 188–195. https://doi.org/10.1016/j.bios.2015.08.043

[86] A. Tiwari, S.R. Deshpande, H. Kobayashi, A.P.F. Turner, Detection of p53 gene point mutation using sequence-specific molecularly imprinted PoPD electrode, Biosens. Bioelectron. 35 (2012) 224–229. https://doi.org/10.1016/j.bios.2012.02.053

Biosensors: Materials and Applications
Materials Research Foundations **47** (2019) 51-76

Materials Research Forum LLC
doi: http://dx.doi.org/10.21741/9781644900130-2

[87] T.-P. Huynh, P. Pieta, F. D'Souza, W. Kutner, Molecularly imprinted polymer for recognition of 5-Fluorouracil by RNA-type nucleobase pairing, Anal. Chem. 85 (2013) 8304–8312. https://doi.org/10.1021/ac401598k

Biosensors: Materials and Applications   Materials Research Forum LLC
Materials Research Foundations **47** (2019) 77-130   doi: http://dx.doi.org/10.21741/9781644900130-3

Chapter 3

Application of Functional Metal Nanoparticles for Biomarker Detection

Goutam Ghosh

UGC-DAE Consortium for Scientific Research, Mumbai Centre, Trombay, Mumbai 400085, India

Abstract

Recent advances in metal nanoparticles (mNPs), such as gold and silver nanoparticles, based biosensor technology for biomarkers detection have been reviewed. The localized surface plasmon resonance signal appearing from the surface of mNPs upon irradiation of light provides the immense scope of improving the sensitivity and lowering of the detection limit of biosensors. Moreover, mNPs have advantages such as biocompatibility, functional flexibility and large surface-to-volume ratio. The interaction of functional nanoparticles with the cell membrane and their subsequent internalization into the cell play an important role in the imaging of diseased cells/tissues. Several factors such as surface functionalization, size and shape of nanoparticles influence these processes. Recent reports indicate that non-spherical nanoparticles such as nanorods have a better yield than spherical ones for cellular uptake, longer blood circulation time, and higher catalytic activity. Potential toxicity of mNPs in an *in vivo* application has also been reviewed.

Keywords

Metal Nanoparticles, Biosensors, Biomarkers, Cancer Detection, Toxicity

Contents

Materials Research Forum LLC
doi: http://dx.doi.org/10.21741/9781644900130-3

Abbreviations

Ab	Antibody
ADC	Antibody-drug conjugation
AFP	α-Fetoprotein
AgNPs	Silver nanoparticles
AuNPs	Gold nanoparticles
AuNRs	Gold nanorods
BBB	Blood-brain barrier
BRCA	BReast CAncer
CA125	Cancer antigen 125
CT	Computed tomography

Biosensors: Materials and Applications   Materials Research Forum LLC
Materials Research Foundations **47** (2019) 77-130   doi: http://dx.doi.org/10.21741/9781644900130-3

DCS	Differential centrifugal sedimentation
DLS	Dynamic light scattering
DOS	Density of state
EGFR	Epidermal growth factor receptor
ELISA	Enzyme−linked immunosorbent assay
FA	Folic acid
FR	Folate receptor
FRET	Fluorescence resonance energy transfer
FTIR	Fourier-transform infrared
GPI	Glycosylphosphatidylinositol, or, glycophosphatidylinositol
HER	Human epidermal receptor
LSPR	Localized surface plasmon reference
NTA	Nanoparticle tracking analysis
mNPs	Metal nanoparticles
mNRs	Metal nanorods
MRI	Magnetic resonance imaging
MW	Molecular weight
MWNT	Multi-walled nanotube
NIR	Near infra-red
PAT	Phtoacoutic tomography
PET	Positron emission tomography
PSA	Prostate specific antigen
PVP	Polyvinyl-pyrrolidone
ROS	Reactive oxygen species
SAED	Selected area electron diffraction
SDS	Sodium dodecyl sulphate
SEIRAS	Surface-enhanced infrared absorption spectroscopy
SEM	Scanning electron microscopy

Biosensors: Materials and Applications
Materials Research Foundations **47** (2019) 77-130

Materials Research Forum LLC
doi: http://dx.doi.org/10.21741/9781644900130-3

SERS   Surface enhanced Raman scattering

SERRS   Surface enhance resonance Raman scattering

TEM   Transmission electron microscopy

UV   Ultra-violet

XPS   X-ray photoelectron spectroscopy

XRD   X-ray diffraction

1. Background

A biomarker (or simply marker) is a biochemical or molecular change which takes place in cells, tissues or fluids and is evaluated to assess normal biological processes or the presence of disease/abnormality in any organism or responses to therapeutic intervention. Biomarkers are specific cells, molecules, genes, proteins, enzymes or hormone which are detected in order to identify certain severity in the body, like heart disease, multiple sclerosis, cancer, etc. At present, there are a large number of biomolecules under investigation in the direction of their role as a cancer biomarker for diagnostic, prognostic and therapeutic purposes. Depending on their clinical applications, biomarkers can be classified as molecular biomarkers, cellular biomarkers or imaging biomarkers [1,2]. A molecular biomarker is a molecule that can be used to detect a specific disease; for example, glucose levels are used as a biomarker in managing diabetes. There are four types of molecular biomarkers: genomic, transcriptomic, proteomic and metabolic. A genomic biomarker is a measurable DNA and/or RNA characteristic that indicates normal biological processes, pathogenic processes, and/or response to therapeutic or other interventions. Transcriptomic biomarkers can distinguish the gene expression between a healthy cell and a diseased cell. Proteomic biomarkers are proteins that are produced or over expressed during specific diseases. Metabolic biomarkers are used to identify diseases, such as diabetes mellitus, arising from metabolic disorders. Cellular biomarkers, on the other hand, help to identify or isolating diseased cells by characterizing their morphology and physiology. Cellular biomarkers such as normal human fibroblast-like cells are studied for identifying aging, epilepsy, dry eye disease and so on. An imaging biomarker is a biological feature, or biomarker detectable in an image. For example, a number of biomarkers are frequently used to determine the risk of lung cancer. Imaging biomarkers allow earlier detection of disease compared to molecular biomarkers.

Protein receptor located either on the surface or within the cytoplasm or nucleus of a cell binds to a specific ligand, initiating signal transduction and a change in cellular activity.

Biosensors: Materials and Applications   Materials Research Forum LLC
Materials Research Foundations **47** (2019) 77-130   doi: http://dx.doi.org/10.21741/9781644900130-3

The receptors have demonstrated important roles as diagnostic, prognostic, and predictive biomarkers in cancer research and therapy. Cancer cells over express certain proteins which are used as markers to identify the type and the progress of cancer in the patient's body. To prevent cancer from spreading, early detection is necessary which is also beneficial for treatment. Early diagnosis of cancer requires detection of a very low-level presence of markers in the bloodstream or body fluid. Detection of the marker also helps in risk assessment, screening, prognosis and treatment predictions, monitoring treatment response and recurrence of the diseases [3]. Different markers have been reported in recognition of different cancers such as AFP for liver cancer, BRCA1/BRCA2 or HER-2 for breast cancer, CA-125 for ovarian cancer, PSA for prostate cancer, etc [4–13].

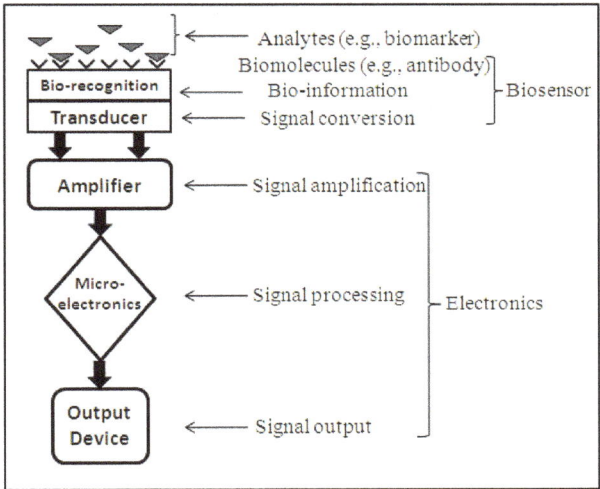

Fig. 1. Schematic diagram of the biosensor device.

Biosensors are used for *in vitro* detection of biomarkers in a fluid sample collected from the patient, and only a right type of biosensor can detect a specific biomarker associated with a specific severity [14,15]. An efficient biosensor gives an unambiguous output signal upon interaction of a biomarker with its transducing surface, eventually to use for either diagnostic or therapeutic purpose. To understand the structure and function of a biosensor, a schematic diagram of the usual biosensor device is shown in Fig.1. It has two main intimately coupled parts: a biorecognition layer and a transducing layer, acting together convert a biochemical signal to an electronic, electrochemical, electrochemiluminescent, magnetic, gravimetric, or optical signal. Based on biorecognition element or transducing method biosensors are classified into several types, such as an electrochemical biosensor, amperometric biosensor, potentiometric biosensor, conduct metric biosensor, thermometric biosensor, and piezoelectric biosensor. For example, the oligonucleotide is used as a biorecognition element in DNA sensor, the antibody is used in immunosensor and so on. While the biorecognition element determines the degree of selectivity or specificity of the biosensor, the biosensor's ability to detect low concentrations is mainly influenced by the transducer, as it transforms the biological or biochemical response into a quantifiable output signal [16]. The biorecognition layer typically contains an enzyme or a binding protein such as an antibody. It may also contain oligonucleotide sequences, sub-cellular fragments such as organelles (e.g. mitochondria) and receptor carrying fragments (e.g. cell wall), single whole cells, small numbers of cells on synthetic scaffolds, or thin slices of animal or plant tissues. As a result of the presence and biochemical action of the analyte (i.e., the target of interest), a physicochemical change is produced within the biorecognition layer that is measured by the physicochemical transducer producing a signal that is proportional to the concentration of the analyte in the sample collected from the patient's body. This signal is then processed further to get an observable output. By analyzing the change in the output signal compared to the original signal (i.e., before binding of biomarker) information about the disease condition is collected.

Use of nanomaterials can increase sensitivities and lower detection limits down to molecular level due to their capacity of immobilizing an enhanced quantity of analytes at reduced volumes. Many nanomaterials as metal NPs (mNPs), oxide NPs, and semiconductor NPs have been studied widely. For example, gold (Au) and silver (Ag) NPs, or silver–silica hybrid NPs [17–19], quantum dots such as CdS NPs [20] have been used as biosensor substrates. Metal NPs with large diameter (>30 nm) exhibit strong light scattering in the visible region and could be used directly for light scattering labels in a biochemical assay. However, small mNPs, which apparently do not feature light-scattering, can also be used to sense chemical interactions (e.g., antigen-antibody, avidin-biotin, DNA hybridization, and electrostatic attraction), since enhanced light scattering signals would be produced if these NPs were to aggregate during the interactions. Because the enhanced light scattering signals from the aggregated species are sufficiently sensitive to monitor NP aggregation, in a simple procedure, a biochemical assay based on such light scattering signals has been widely used in the determination of DNA, proteins, and drugs [21]. Many signalling techniques can be used for biomarker detection, such as localized surface plasmon resonance (LSPR), surface enhanced Raman spectroscopy

Biosensors: Materials and Applications   Materials Research Forum LLC
Materials Research Foundations **47** (2019) 77-130   doi: http://dx.doi.org/10.21741/9781644900130-3

(SERS), colourimetry, fluorescence, photoacoustic, electrochemistry, dynamic light scattering (DLS) [22]. At present, organic fluorescence molecules are used at the biorecognition layer (Fig. 1) for clinical detection of biomarkers [23,24], but the signal resolution remains poor due to large widths of their excitation and emission bands. In order to improve the sensitivity and resolution of the signal, functionalized mNPs are being used in developing sensing devices and further research towards this direction is on [25–28]. Since the first report [29] on the application of gold nanoparticles, several nanomaterials have been reported to use in biosensing applications with high sensitivity based on their unique optical [30], electrical [31], magnetic [32], electrochemical [33], and thermal [34] properties.

2.   Metal nanomaterials

Metal nanomaterials are nanosized metals with 1−, 2−, or 3− dimension(s) in the range of 1–100 nm, and are accordingly known as 1D, 2D or 3D nanomaterials, respectively. 1D nanomaterials are thin films or nano-sheets (like graphene), 2D nanomaterials are nanorods (NRs), and 3D nanomaterials are nanoparticles. The interaction of nanomaterials with biological molecules or membranes depends on many factors such as size, shape and surface composition. Metal NPs, particularly gold (Au) and silver (Ag) NPs have been widely studied for several biomedical applications including molecular diagnosis, imaging, drug delivery and therapeutics due to their unique properties such as localized surface plasmon resonance (LSPR), biocompatibility, stability and surface tunability. LSPR signal arises due to resonant oscillation (known as plasmonic oscillation) of free electrons at the surface of mNPs upon interaction with light [35–39] and is extremely sensitive to the changes at the interface between the mNPs surface and a surrounding medium such as binding of molecules (analytes or biomarkers) [40]. A tentative change in the signal upon binding of analytes with the ligands on the mNPs surface is shown by the cartoon in Fig.2. The change in signal wavelength is indicated by the change in colour of the signal. This change occurs more strongly with the change in the aspect ratio of rod-shaped mNRs rather than the change in the diameter of spherical mNPs [41–43]. For example, the longitudinal LSPR signal of gold nanorods (Au NRs) showed more than 300 nm red−shift for an increment of the aspect ratio by 3 [41], in contrast, to a red−shift of less than 100 nm for an increment of the diameter of gold nanoparticles (AuNPs) by 40 nm [44]. A relation between the LSPR band position (λ_{max}) and the aspect ratio (R) was given by [45],

$$\lambda_{max} = 95R + 420 \tag{1}$$

Biosensors: Materials and Applications   Materials Research Forum LLC
Materials Research Foundations **47** (2019) 77-130   doi: http://dx.doi.org/10.21741/9781644900130-3

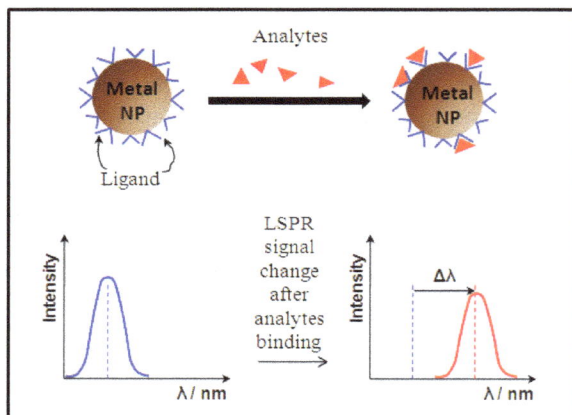

Fig.2. Cartoon shows the basic concept ofbiosensingby the mNPs, such as LSPR band shift upon binding with analytes.

For spherical particles, $R = 1$ and the LSPR band should appear at (λ_{max}) around 515 nm, which is indeed in agreement with an earlier report [44]. Similarly, λ_{max} of the longitudinal mode of LSPR signal for a given aspect ratio of AuNRs [41] has to be consistent with Eq. (1). Nanorods (NRs) have many advantages over nanoparticles (NPs) in *in vivo* applications such as a higher cellular internalization rate, a longer blood circulation time, and higher catalytic activity [46]. Therefore, the synthesis of metal nanomaterials with well defined size and shape is very important for smart biosensor fabrication, and a brief over view of different synthesis methods has been given below.

2.1   Synthesis

Nanomaterials can be synthesized by either top-down or bottom-up method. The top-down method involves reducing the size from a macroscopic body to nanometric particles. Afterwards, nanometric particles are stabilized against agglomeration by coating with surfactants [47]. Many techniques are used in a top-down procedure such as ball−mill [48] for solid NPs and photolithography and electron beam lithography [49] for thin film preparation. Bottom-up methods involve growing of nano-dimensional entities from an atomistic level using either physical or chemical processes (e.g., chemical reduction of metal ions). Bottom-up methods are most widely used as they give better control over the size and shape of nanoparticles. Few bottom−up methods are listed in

Biosensors: Materials and Applications   Materials Research Forum LLC
Materials Research Foundations **47** (2019) 77-130   doi: http://dx.doi.org/10.21741/9781644900130-3

Table 1, and some of them have been briefly reviewed below. Detail methods may be seen elsewhere [50].

Physical methods—There are several methods to grow nanoparticles from vapour phase of solid, such as chemical vapour deposition, pulsed laser ablation, sputtering and so on. In vapour synthesis, the material in the vapour phase is brought in a hot wall reactor and by controlling different factors like pressure and temperature nanoparticles are nucleated [51].

Table 1. List of few physical and chemical methods of the synthesis of metal nanoparticles

Physical methods	Chemical methods
• Chemical vapour deposition • Microwave irradiation • Pulsed laser ablation • Supercritical fluids • Wire electrical explosion • Gamma radiation	• Chemical reduction • Microemulsion • Thermal decomposition • Electrochemical synthesis

Chemical methods—The most ancient and widely used chemical method for synthesis of nanoparticles is the reduction of metal ions in solution. Metal ions (M^+) which belong to a specific precursor are reduced to metal atoms by providing some extra energy using different types of chemical reductants. The provided energy such as photo energy, electrical energy or thermal energy, is used to decompose the precursor. The reduced metal atoms subsequently self-assemble to nucleate metal clusters and grow into nanostructures with time [52]. This procedure is shown by the cartoons in Fig.3. The Turkevich method has been used to synthesize gold nanoparticles (AuNPs) by reducing gold ions (Au^+) in gold hydrochlorate ($HAuCl_4$) in solution by tri-sodium citrate ($Na_3C_6H_5O_7$) [53]. For synthesizing silver nanoparticles (AgNPs), silver nitrate ($AgNO_3$) is taken as a precursor and sodium citrate [54] or sodium borohydride ($NaBH_4$) [55,56] is used to reduce silver ions (Ag^+) in the solution to silver atoms (Ag^o) which then self-assemble to nucleate as AgNPs. Many other mNPs are also synthesized using the chemical reduction method [54–56]. Mesityl derivative of respective metal ion has been used as one of the several precursors for synthesizing Au, Ag and Cu NPs [57]. Au, Ag, Pt and Pd NPs have been reported to synthesize using commonly available sugars as one of the several reducing agents [58]. Ethylene glycol and ethanol have also been used as reducing agents for synthesizing AgNPs [59]. The size and shape of NPs are controlled

Biosensors: Materials and Applications   Materials Research Forum LLC
Materials Research Foundations **47** (2019) 77-130   doi: http://dx.doi.org/10.21741/9781644900130-3

by adding stabilizer to the solution in a controlled manner [60]. For example, 2D superlattice of Au and Ag nanoclusters were stabilized using sodium citrate and sodium dodecyl sulfate [61]. Chemical reduction is also done via a sonochemical method for synthesis of various NPs such as FeNPs [62], CdSNPs [63], TiO_2NPs [64], and so on.

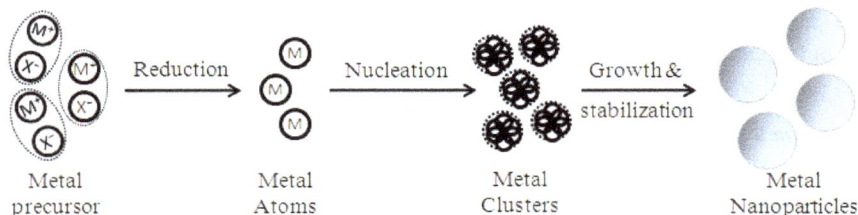

| Metal precursor | Metal Atoms | Metal Clusters | Metal Nanoparticles |

Fig.3. Cartoon shows the basic concept of synthesis of metal nanoparticles (mNPs) using a chemical reduction method.

Microemulsion method has been successfully utilized to synthesize nanoparticles with controlled size, size distribution and shape [65,66]. Capek [67] has reviewed the microemulsion method for the synthesis of many mNPs such as Ag, Au, Cd, Co, Cu, Fe, Ni, Pd, and Pt NPs. This method allows synthesis of monodisperse NPs of size in a range of 5–50 nm. Briefly, synthesis takes place upon fusing of two microemulsions, or microemulsion and aqueous solution, carrying the appropriate reactants. Upon collision of microemulsions, interchanging of reactants takes place very fast inside the droplets where nanoparticles are synthesized with size, shape and distribution that are pre–controlled size, shape and size distribution of the microemulsions. This method is shown by the cartoons in Fig. 4. Direct reduction of silver ions by the chloride counter ions to synthesize silver chloride nanoparticles in the water phase of single microemulsion (reverse micelle) of dioctyldimethyl ammonium chloride surfactants dispersed in n–decanol/isooctane (oil) phase was reported earlier [68]. In another report [69], size-controlled octahedron nanocrystals of cuprous oxide (Cu_2O) were synthesized in Triton X–100 water–in–oil (w/o) microemulsion by reducing copper nitrate using γ–irradiation.

Thermal decomposition is one of the best ways of synthesizing monodisperse mNPs. In this method, nucleation of mNPs takes place when the metal precursor is added into a heated solution in presence of surfactants, while growth takes place at a higher reaction temperature. Salavati–Niasari and coworkers [70] used this method to synthesize Cu NPs from copper oxalate (CuO_4C_2) precursor in presence of oleylamine ($C_{18}H_{35}NH_2$).

Biosensors: Materials and Applications   Materials Research Forum LLC
Materials Research Foundations **47** (2019) 77-130   doi: http://dx.doi.org/10.21741/9781644900130-3

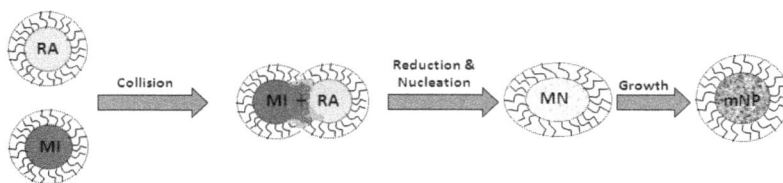

Fig. 4. Cartoon shows the basic concept of synthesis of metal nanoparticles (mNPs) using microemulsion method. MI: metal ions, RA: reducing agent, MN: metal nuclei, mNP: metal nanoparticle.

Green synthesis—Plants, algae, microbes contain several organic compounds such as phenolic and glycoside compounds which work as reducing agents in mNPs synthesis. The major benefit in using a green synthesis method over chemical reduction methods includes no toxic by-products during mNPs synthesis. The rate of reduction of metal ions by the plant extracts is faster than that with microorganisms. Moreover, using plant extracts are of low−cost and eco-friendly; therefore, the most preferred for the green synthesis of mNPs. Gold and silver nanoparticles were the first to be synthesized inside living plants [71,72].

2.2   Characterization methods

Characterization of functional nanomaterials is very important for technological applications and many techniques are used for that. Transmission electron microscopy (TEM) and scanning electron microscopy (SEM) is used for characterizing size, morphology and crystallinity of nanoparticles. Powder X-ray Diffractometer (XRD) is used for the characterizationof chemical and structural phase, crystallinity and grain size of the nanoparticle core, as well as the surface coating of nanoparticles. X-ray photoelectron spectroscopy (XPS) is used to characterize the chemical composition of the core and surface coating of nanoparticles. Dynamite light scattering (DLS) technique is also used to measure the size and the surface charge (ζ−potential) of nanoparticles. However, in the case of agglomeration and hydrophilicity, DLS provides inaccurate measurement. In such cases, differential centrifugal sedimentation (DCS) method is used [73]. Nanoparticle tracking analysis (NTA) is used to measure the size of nanoparticles in biological systems such as proteins and DNA. This technique allows finding the size distribution of NPs in the range of 10—1000 nm in a liquid medium [74], and has been found to be very precise for sizing monodisperse as well as polydispersed samples, with substantially better peak resolution compared to DLS [75]. As for examples, gold (Au), palladium (Pd), silver (Ag) and zinc (Zn) NPs were characterized using XRD, UV spectroscopy and Fourier transformed infrared spectroscopy (FTIR) techniques [76–81]. TEM, a selected area diffraction (SAED) pattern, high-resolution transmission electron microscopy (HRTEM) have been used for characterization to AgNPs [78]. Au, Ag, and ZnNPs have been reported to characterize also by SEM [79−81].

2.3   Biomedical applications

Biomedical applications of mNPs are involved mainly in diagnosis and therapy. For diagnosis and prognosis of disease inside the body both *in vitro* sensing and *in vivo* imaging methods are used, and mNPs such as Au and AgNPs are well studied for such applications. For therapeutic applications, drug delivery and hyperthermia have been investigated using mNPs. In this chapter, sensing and imaging applications of mNPs and their toxicity have been reviewed. For biomedical applications, mNPs are functionalized (coated) with appropriate chemical or biological functional molecules and thus are called functional metal nanoparticles.

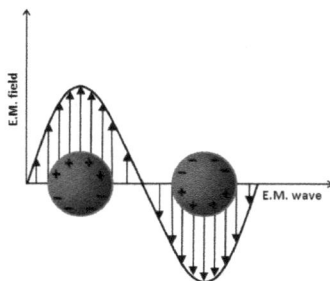

Fig. 5. Schematic representation of Localized Surface Plasmon Resonance (LSPR) of metal nanoparticles (mNPs) due to the influence of the electromagnetic field (E.M. field) vector upon application of electromagnetic wave (E.M. wave).

3.   Functional metal nanoparticles

Metal nanoparticles (mNPs) have been of great research interest due to their unique physicochemical and plasmonic properties and various potential applications. Noble mNPs such as Au and AgNPs produce localized surface plasmon resonance (LSPR) absorbance signal upon irradiation with light [35–39] and are used in imaging/labelling of biological systems [82]. The LSPR absorbance appears due to the collective and coherent oscillation of the polarized conduction electrons at the surface of mNPs upon irradiation

Biosensors: Materials and Applications   Materials Research Forum LLC
Materials Research Foundations **47** (2019) 77-130   doi: http://dx.doi.org/10.21741/9781644900130-3

by an electromagnetic wave, e.g. light, of matching frequency. This phenomenon has been demonstrated schematically in Fig. 5.

Functionalization of NPs means the introduction of the coating of organic/inorganic molecules, polymers on the surface of NPs giving them stability against the agglomeration, and some functional properties, such as targeting of specific cell or tissue, delivering drug at the targeted site, imaging of targeted cells and so on. The coating modifies many of their physical and chemical properties, as well as the targetability of functional mNPs in the medium such as blood or body fluid. Functional mNPs have a 'core-shell' structure where the core contains nanoparticles and the shell contains a protective coating and functional elements, as shown in Fig. 6. The shell may be composed of polyethylene glycol (PEG) and/or linker to conjugate the cell targeting elements such as an antibody or folic acid, and drugs, proteins, vaccines, biological macromolecules, imaging dye, drugs, etc depending on required applications. Functions of various elements have been discussed briefly in the following sections.

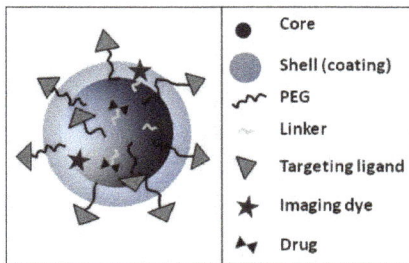

Fig. 6. The cartoon shows the functional nanoparticle with 'core-shell' structure. 'Core' contains the nanomaterial; 'shell' contains coating elements (for dispersion), polyethylene glycol (PEG), with targeting ligand (antibody or folic acid), a linker with drug or imaging dye.

Functionalization of NPs means the introduction of the coating of organic/inorganic molecules, polymers on the surface of NPs giving them stability against the agglomeration, and some functional properties, such as targeting of specific cell or tissue, delivering drug at the targeted site, imaging of targeted cells and so on. The coating modifies many of their physical and chemical properties, as well as the targetability of functional mNPs in the medium such as blood or body fluid. Functional mNPs have a 'core-shell' structure where the core contains nanoparticles and the shell contains a protective coating and functional elements, as shown in Fig. 6. The shell may be composed of polyethylene glycol (PEG) and/or linker to conjugate the cell targeting elements such as an antibody or folic acid, and drugs, proteins, vaccines, biological macromolecules, imaging dye, drugs, etc depending on required applications. Functions of various elements have been discussed briefly in the following sections.

PEGylation—PEG coating (also called PEGylation) on NPs reduces their uptake by the reticulo endothelial system (RES) and, in turn, increases their circulation time in the liver, spleen, etc. compared to non-PEGylated NPs [83]. PEGylation of NPs has many other advantages such as reduction of their association with non−targeted serum and tissue proteins, reduction of the charge−based contact with proteins and small−molecules, modification of the interface between NPs and the circulation medium and increment of the circulation time [84]. A longer circulation time assures functional NPs reaching the target and remaining there long enough for imaging of tissues. Circulation halftime (t½) describes blood pool residence and is the period over which the concentration of circulating nanoparticles remains above 50% of the injected dose, analogous to a drug's half-life [85]. PEG also works as a linker for various imaging dyes, macromolecules, drugs [86]. Thus, functionalization of NPs by PEG, called PEGylation, is useful [87–89].

Linkers—Linking of dye, drug and other functional elements with NPs could be done either via covalent or via non-covalent bonding [90]. The covalent bonding or conjugation gives stronger binding of functional molecules at the nanoparticle surface. On the other hand, non-covalent bonding via physical association of ligands with the nanoparticle surface is weaker but has an advantage due to avoiding of rigorous, destructive reaction agents. The non-covalent conjugation may also cause conformational disorder to the functional molecules [91]. To avoid this problem binding of functional elements with NPs is done through linkers. In a single protein, multiple domains are kept separated using short amino acid sequences as linker or spacer [92]. Nickels et al. [93] reported the synthesis and properties of a versatile linker consisting of both epoxide and amine terminals for the surface modification of NPs. This linker has been shown to mediate binding of both folic acid (a targeting ligand) and fluorescein isothiocyanate (an imaging dye) with the dextran-coated iron oxide nanoparticles (IONPs). Lu et al. [94] reviewed the role of linkers as antibody-drug conjugates (ADCs). As mentioned in their paper [95], an appropriate linker between the antibody and the functional element provides a specific bridge, and thus helps the antibody to selectively deliver the functional elements such as a cytotoxic drug to tumour cells. In addition to conjugation, the linkers maintain ADCs' stability during the preparation, storage stages and during the systemic circulation period. A specific antibody (Ab) targets a specific antigen with high tumour expression. Linker binds with Ab typically through cystine or lysine residues.

Biosensors: Materials and Applications   Materials Research Forum LLC
Materials Research Foundations **47** (2019) 77-130   doi: http://dx.doi.org/10.21741/9781644900130-3

Targeting ligands— Targeting ligands are molecules such as an antibody (Ab, also called Immunoglobulin, Ig) or folic acid (FA) which bind with biomolecules/cell membrane to serve the desired application towards diagnosis or therapy. In the case of protein binding, the ligand binds to a specific site of the targeted protein and produces a signal. The binding of a ligand to the target may take place via ionic or hydrogen bonding or van der Waals attraction. Cell-membrane contains few proteins which spans the plasma membrane and performs signal transduction, converting an extracellular signal into an intracellular signal. These proteins act as receptors and their cognate ligands have designated them as especially useful clinical targets. Cancer cells over express such receptors, considered as biomarkers, at the cell-membrane site which can be used for *in vivo* diagnosis and staging of the disease [95,96]. Ligand, receptor or marker may have many forms, but they come in closely matched pairs. One (or few) specific receptors recognizes only one (or few) specific ligands. Selectivity of interaction between surface ligands of NPs and cell receptors plays an important role for the internalization and implementation of desired functions of nanobiomedicine [97]. Nanoparticles can be functionalized with antibodies, aptamers, or ligands for specific interaction and targeting of cells [98–100]. Such specific ligand–receptor interaction does not differentiate between healthy and diseased cells [101,102]. To achieve targeting, ligands-conjugated NPs have to recognize the highly densed distribution of the receptor at the membrane of targeted diseased cells, avoiding other low receptor–density healthy cells [103,104]. The ligand–receptor binding helps in internalizing nanoparticles into the cell via endocytosis process. This process has been demonstrated by the cartoons in Fig. 7.

Imaging dye—An imaging dye is used to improve the contrast of the image of a tissue or cell compared to its peripheral tissues or cells, and also called contrast dye. In the case of fluorescence imaging, the dye needs to be excited by light of fixed wavelength (λ_{ex}) to obtain a secondary emission at a longer wavelength (λ_{em}) to be used for imaging applications in biology [105]. Other imaging methods such as magnetic resonance imaging (MRI), positron emission tomography (PET), and x-ray computed tomography (CT) involve the use of appropriate imaging dyes to improve the image quality and obtain more accurate information about the internal organs which can be used for diagnosis of disease as well as monitoring the effects of ongoing therapy [106,107]. Imaging techniques, other than MRI, use electromagnetic radiation in a range of wavelengths between gamma and infrared as the imaging probe. High energy gamma or x-ray radiation can penetrate tissues and get images from deeper tissues or the whole body, but they could be potentially harmful to healthy tissues [108]. On the other hand, low energy radiations in the visible to infrared are harmless as they do not penetrate tissues and, therefore, frequently used for imaging purpose in clinical diagnosis. Nowadays

Biosensors: Materials and Applications   Materials Research Forum LLC
Materials Research Foundations **47** (2019) 77-130   doi: http://dx.doi.org/10.21741/9781644900130-3

fluorescence imaging using near-infrared (NIR) radiation is used routinely for whole-body imaging of small animals in preclinical studies [109,110]. Infrared (IR) imaging dyes such as indocyanine green (ICG), methylene blue and omocyanine [106] have been used clinically; IR780 [111] and IR820 [112] have been used in laboratory research. As mentioned, these dyes are conjugated with coated–NPs via a linker.

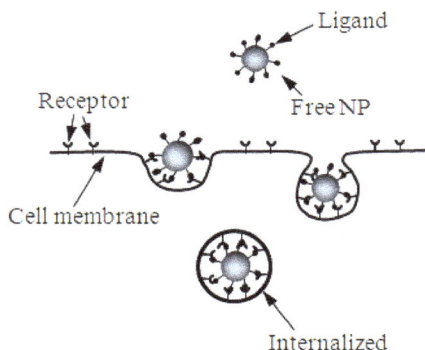

Fig. 7. The cartoon shows the binding of nanoparticle (NP) functionalized with the ligands for the receptors of the cell membrane, and internalization via endocytosis.

4.   Tumour markers and targeting of nanoparticles

Cell receptors such as folate receptor (FR), transferrin receptor (TR), epidermal growth factor receptor (EGFR), and various antigens are usually expressed in high density or muted at the membrane of cancer cells [113–118]. Thus these over-expressed receptors are considered as tumour (cancerous) markers. Folate and transferrin receptors are the most common among all tumour markers [119–122]. Thus, using the appropriate coating on NPs such as folate, transferrin or antibody targeting of specific cancer cell scans be achieved. For example, Goldberg et al. [123] reported CD8$^+$ T cell-specific targeting of nanoparticles by conjugating anti-CD8a F(ab')$_2$ fragments to the particle surface.

Dysregulated folate metabolism has been associated with many diseases [124]. Folate receptors (FR) are proteins which are attached to the cell membrane by a GPI anchor. There are four types of FR: FRα, FRβ, FRγ and FRδ, whose molecular weight varies in the range from 38 to 45 kDa [125]. The tumour-associated antigen FRα (folate receptor alpha) is a membrane protein which is over expressed in tumours such as ovarian, breast and lung cancers, and low and restricted distribution in normal tissues of kidney, lung and

Biosensors: Materials and Applications   Materials Research Forum LLC
Materials Research Foundations **47** (2019) 77-130   doi: http://dx.doi.org/10.21741/9781644900130-3

choroid plexus [126]. Folic acid (FA) or folate is a small molecule ($C_{19}H_{19}N_7O_6$, M.W. ~441 Da), stable over a broad range of temperatures and pH values, inexpensive and non-immunogenic which can bind with the FR [127] of cancer cells and, thus, used in functionalizing NPs for cancer targeting [128]. Transferrin receptor (TR) is a carrier protein expressed at the cell membrane for the import of iron into the cell and is regulated in response to intracellular iron concentration. Low iron concentration causes an increased level of transferrin receptor at the cell membrane to increase iron intake into the cell. The epidermal growth factor receptor (EGFR) is a transmembrane protein at the cell membrane which is the receptor epidermal growth factor family of extracellular protein ligands. When there are too many expression of EGFR due to a mutation, the cancer cells continue to grow and divide. Erythropoietin receptor (Epo−R) is a type I transmembrane protein (MW ~59−70 kDa) which belongs to the cytokine receptor super family. In bone marrow, Epo−R is expressed in erythroid progenitors, and in megakaryocytes, myeloid cells, and endothelial progenitors. Recently, Epo−R was identified in non-erythroid cells in the heart, lung, brain, and the immune system as well as in several cancer types [129] and in the central nervous system (CNS) [130]. Erythropoietin (Epo) binds to Epo−R for production of matured red cells [131]. Epo is also a most used drug for anaemia treatment. Epo may be used as Epo−R binding ligands for functional NPs. A review has been published earlier [132] discussing different receptors/markers that are associated with different cancer cells and their targeting for therapeutic applications.

5.   Sensing and imaging applications of metal nanoparticles

As described earlier, the sensing mechanism is based on the conversion of analytes−antibody binding interaction into output signals which can be amplified and detected. Metal NPs based biosensor shave been widely investigated for *in vitro* detection of molecular markers due to the presence of cancer, neurodegenerative diseases or infectious diseases [133,134] in patients' body fluids. The challenge in biosensing technique lies in the efficient signal capturing of biorecognition event such as interaction of analytes (receptor, biomarker or antigens) with the biological elements (e.g., antibody) in biorecognition layer of the biosensor. Transducer translates this signal into an appropriate outcome (see Fig. 1) such as electrical, optical or electrochemical signal. Recent reports revealed that the sensitivity of transducer may be enhanced by using noble mNPs and that can also reduce the detection limits down to even individual molecule [135]. This is because noble mNPs have a large surface−to−volume ratio allowing them to immobilize an enhanced quantity of analytes, localized surface plasmon resonance (LSPR), excellent conductivity and anomalous transmission and reflection of light. Nanoparticle itself can act as transducing agent, i.e., transfers the biorecognition event into output signal depending on the type of NPs used. For example, SPR-based biosensors utilize an optical method to measure the refractive index near the biorecognition layer. Upon binding of analytes or biomarkers with the biorecognition molecules at the sensor surface the refractive index increases. This change of refractive index is measured in real time, and the result is plotted as a response or resonance units (RUs) versus time, called sensor gram. Because of their unique optical properties, such as LSPR (see Fig. 5), mNPs, especially Au and AgNPs, show high potentiality in 'label-free' optical image transducing application. But the immobilization strategy of analytes on NPs remains challenging; and thus in most cases, the enzyme immobilization technique is used [136]. In general, non−covalent interactions (such as electrostatic interaction, π−π stacking, entrapment in polymers, or van der Waals forces) between nanomaterials and analytes have an advantage over covalent interaction as these interactions do not change the specificities of either nanomaterial or analytes [135]. By functionalizing the NPs with appropriate ligands (see Fig. 6) a reproducible non−covalent strategy for immobilizing the analytes can be achieved. The strategy for the transduction of biorecognition event is another concern for biosensor development. Secondary bio−components (second antibody or DNA strands) modified with labels for optical or electrochemical transduction have to be used for appropriate signal conversion [135]. Metal NPs also contribute to signal amplifications when used as labels. There are three generations of biosensors: enzyme−dependent biosensor, mediator−based biosensor, and catalytic electrode−dependent biosensor, among which the third generation biosensors attract the most. The third−generation sensors rely on co-immobilization of the metallic enzyme and mediator on an electrode in making the biorecognition component of the transducer. Metal NPs introduced into a catalytic structure (e.g., enzyme) give control over both the catalytic fine-tuning and cost−saving. Binary nanostructured materials exhibit different physical and chemical properties compared to their individual components, which confer greater catalytic activity and can be further improved and tailored specifically [133]. For example, Ghodselahi et al. [137] reported more response sensitivity LSPR sensor chip of Ag@Au BNPs with respect to Ag NPs and Au NPs. Rick et al. [138] published a review of electrochemical bio-sensing for target analytes based on the use of electrocatalytic bimetallic nanoparticles (NPs), which can improve both the sensitivity and selectivity of biosensors.

Metal NPs based biosensors have also been used for *in vivo* sensing, cell tracking and monitoring disease pathogenesis or therapy monitoring [139,140] in combination with an imaging modality such as MRI, PET, and CT. Zhang [141] reported a study on PEGylated AuNPs as a contrast agent for *in vivo* tumour imaging with photo acoustic tomography (PAT). PAT is a rapidly emerging non-invasive imaging technology that

Biosensors: Materials and Applications   Materials Research Forum LLC
Materials Research Foundations **47** (2019) 77-130   doi: http://dx.doi.org/10.21741/9781644900130-3

integrates the merits of high optical contrast with high ultrasound resolution. Metal NPs have recently emerged as ubiquitous surface-enhanced Raman scattering (SERS) agents also for nanoimaging and nanoanalysis [142]. These applications make use of the unique optical properties of mNPs to enhance the efficiency of Raman scattering, and in addition, NPs can localize the SERS signal at the nanoscale due to their small size which enhances the resolution of image or spectrum. Actually, photons slow down at the surface of mNPs and get more chances to interact with surrounding molecules that cause an enhancement of the optical response of the latter. Due to small size mNPs can enter into areas that are rather difficult to reach (e.g., lymph nodes) and can produce high resolution and brightness in the image [143,144]. There are many other types of optical biosensors based on the SPR property of mNPs [145]. Metal NPs can be made sensitive to the medium in cells through surface functionalization with molecules sensitive to the change in chemical conditions of the surrounding medium and, thus, can be used in a medium sensitive biosensor. For example, 4−mercaptobenzoic acids (4−MBA) coated mNPs allows measurement of the pH of surrounding medium by measuring the intensity of the SERS band at 1430 cm^{-1} originating from 4−MBA [146], and Kneipp et al. [147,148] obtained the pH map of the NIH/3T3 cell by using 4−MBA functionalized mNPs. In the following sections, more detail of few mNPs in biosensor application has been discussed.

5.1   Gold nanoparticles

Gold nanoparticles (AuNPs) are most widely studied for biosensing and imaging applications due to their ease of synthesis, stable surface chemistry, biocompatibility, low short-term toxicity, high atomic number in association with unique optical and electronic properties, and high X-ray absorption coefficient [149,150]. AuNPs upon exposure to light provides very strong LSPR signal, the position of which depends on the size and shape of nanoparticles, as well as the dielectric strength of the surrounding medium [151]. Typically, colloidal AuNPs of size around 40 nm shows the LSPR signal at around 520 nm [152]. The interparticle distance of NPs also strongly influences the position of the LSPR band. That is why aggregation leads to a distinct colour change as a consequence of the plasmon coupling between AuNPs, and a concomitant redshift of the band [153]. Using this agglomeration effect, DNA−AuNPs based biosensors have been developed [154,155]. Almost all *in vivo* applications require AuNPs to possess a LSPR peak in near-infrared (NIR) region where the haemoglobin (blood) and water absorption spectra are minimal. Since the LSPR frequency and cross-section of scattering and absorption can be tuned by the size and shape of nanoparticles, varieties of gold nanomaterials such as AuNRs, nanoshells, hollow AuNPs and nanocages have been developed and explored for tuning LSPR signal into NIR region [156]. The shifts in the

LSPR band have also been explored for antibody (Ab) conjugated AuNRs and AuNPs to monitor the real-time dynamic interactions with a specific secondary antibody [157,158].

The general method of using AuNPs in biosensors is to attach or deposit them onto the electrode. This AuNPs modified electrode can then host the biorecognition layer in the biosensor (Fig. 1). AuNPs introduce many advantages to these sensors, encompassing their ability to provide an efficient loading platform for immobilizing analyte molecules (enzymes, proteins, DNA etc.) that further improve electron transfer between the active site and electrode.The immobilization of analytes can take place through several binding procedures such as electrostatic binding (physisorption) and covalent binding (chemisorption) [136]. Yang et al. [159] developed Au−coated polystyrene NP-based sensors which could effectively detect variation in refractive index of the surrounding medium. The medium dependency represents a great advantage for analytics since the biorecognition event can result in a change of the LSPR wavelength (see Fig. 2), and therefore, in an observable colour change of the dispersion. Use of gold as thin films and NPs for transducing action of biosensor has been reported earlier [160,161]. AuNPs can also serve as a label when attached to secondary antibodies (or DNA strands) of the transducer. AuNPs conjugated with anti−CA15−3−HRP have been used to improve the optical signal of a traditional enzyme-linked immunosorbent assay (ELISA) for detecting the breast cancer biomarker (i.e., CA15−3 antigen) in blood [162]. Functional AuNPs have also been successfully used in SERS technique. By introducing AuNPs into intestinal epithelial cells, Kneipp et al. [162] have reported an enhancement of the Raman scattering from intracellular molecules. Fijita et al. [143] published SERS images of macrophage cells cultured with 50 nm AuNPs. Functional AuNPs of size 1.9 nm have been demonstrated as the X-ray CT contrast agent for detecting tumors in mice with their significant accumulations in kidney (10.62% injected dose (ID)/g), tumor (4.2% ID/g), liver (3.6% ID/g) and muscle (1.2% ID/g) just after 15 min. of injection [163]. Gum Arabic (also known as acacia gum) coated AuNPs has also been reported to be a potential X-ray CT contrast agent with high stability in blood plasma medium [164]. PEG−coated AuNPs injected intravenously into rats showed much longer blood circulation time (> 4 h) than the iodine contrast agent iopromide (< 10 min) [165]. Antibody UM−A9 conjugated AuNRs targeted to squamous cell carcinoma showed an increased attenuation coefficient (ΔHU; 168−170) compared with non−targeted NRs, non−cancerous cells (normal fibroblast cells) and other cancerous cells (melanoma) (ΔHU; 28−32) [95]. Gold nanostructures (rods, shells etc.) have also been used for surface−enhanced infrared absorption spectroscopy (SEIRAS)−mediated biosensing [166].

AuNP−based biosensors can be classified into optical biosensors, electrochemical biosensors and piezo-electric biosensors [167]. Application of AuNPs in biosensors has been reported to improve the sensitivity and lower the limit of detection of analyte molecules such as proteins biomarkers and antigens. For example, use of AuNPs has improved the sensitivity of DNA−based optical biosensors more than 1000 times of those without it [168], and in the piezoelectric biosensor, it has reduced the detection limit down to as low as 10^{-16} mol/L [169]. Lowering the detection limit of biomarkers is advantageous for early detection of cancer [170,171]. For specific cancer, specific biomarkers have been identified [172]. Thus, detection of specific cancer in a patient requires specific molecular/Ab functionalization on NPs in order to use as biorecognition layer in a biosensor. Upon binding of a specific biomarker, the LSPR band of AuNPs will show a shift (Fig. 2) from its actual position indicating the presence of a specific disease. The LSPR excitation of electrons results in selective photon absorption and surface enhanced optical phenomena including Raman scattering, fluorescence and various non-linear effects [173]. For most of the AuNPs array biosensors, sensing is generally based on the dependence of LSPR behaviours on either the refractive index of the surrounding media or the adlayer thickness or both [174]. Applications of different types of Au nanomaterials such as nanospheres, nanorods, nanoshells and nanocages in detection and therapy of cancer have been briefly reviewed earlier [175]. AuNPs have been used for developing various imaging techniques of biological relevance [176].

5.2 Silver nanoparticles

Silver NPs, due to their antibacterial action, are commonly used in textiles [177], food packaging [178], painting [179], water purification [180] or air treatment [181] etc. For biomedical applications also AgNPs are recently getting much attention. Plasmonic and metallic properties of AgNPs are being considered for diagnostic (e.g., biosensors) as well as therapeutic applications (e.g., photothermal treatment). AgNPs dispersion appears yellow with the LSPR signal around 420 nm [152]. Of late, we have reported [182] that alkaline metalions (e.g., Li^+, K^+) conjugated negatively charged AgNPs show stronger LSPR signal than uncoated AgNPs and the intensity of the signal increases with increasing atomic number (Z) of metalions. Geagea et al. [183] reported enhancement of the electrochemical biosensor signal by the use of zwitter ionic polymer coated AgNPs. Coated charged AgNPs can be used as an excellent agent for surface enhanced resonance Raman scattering (SERRS) [184]. Controlled surface chemistry and the use of appropriate protocols also enable them for effective sensing of biomolecules. Enzyme biosensor using AgNPs has been reported for detecting penicillin [185]. In biosensors, AgNPs are used to facilitate the electron exchange between biosensor elements. As silver is known to be the best electrical conductor, AgNPs are the best among mNPs to use as electron exchanging agents in the biosensor. Triangular−shaped AgNPs deposited substrates have been used to monitor interactions between biomolecules, such as biotin-streptavidin [186] and the two biomolecules are related to Alzheimer's disease [187]. Cubic [188] and rhombic [189] AgNPs have also been employed in biosensing of protein interactions. A label-free biosensor based on silver nanoparticles array has been used for clinical detection of serum p53 in head and neck squamous cell carcinoma [190]. AgNPs based plasmonic biosensors are also promising for cancer detection [191]. AgNPs–decorated multi-walled carbon nanotube (MWNT) modified glassy carbon electrode has been used to develop glucose biosensor [192]. Like AuNPs, AgNPs have been tested for imaging applications also. For example, Kravets et al. [193] reported imaging of neural stem and rat basophilic leukaemiacellusing the fluorescent glycine dimer coated AgNPs.

Both gold and silver NPs have been used to form multilayer films for developing SPR biosensor [194]. Though AgNPs are superior to AuNPs as far as LSPR is a concern, but they pose more toxicity to living bodies [195], and therefore, are limited for biomedical applications. Other mNPs (e.g., Pt, Fe, Zn, Cu, Pd, Ir and alloys) have been incorporated into glucose biosensors as modifiers, labelling factors or immobilizer agents [196–199].

5.3   Platinum nanoparticles

Platinum NPs (PtNPs) can be synthesized by reducing hexa chloroplatinate using hydrogen gas as the reducing agent. Spherical platinum nanoparticles can be made with sizes between about 2 and 100 nm, depending on reaction conditions [200,201]. PtNPs with controlled size and shape were synthesized using a modified polyol method [202]. Several other methods of synthesis of PtNPs of different sizes and shapes have been reviewed earlier [203]. PtNPs show LSPR signal around 215 nm which is in the ultra-violet (UV) regime [204–209], unlike the other noble mNPs providing SPR signal in the visible regime. The Pt NPs synthesized via citrate reduction do not show SPR signalling the UV regime [210].

There are few reports on applications of PtNPs in biosensors. For example, the amperometric biosensor (e.g., glucose biosensor) was prepared based on electro-deposition of PtNPs onto carbon nanotubes and immobilizing enzyme with chitosan–SiO_2 sol-gel which showed high sensitivity and stability [211]. In a similar way, PtNPs electro-deposited carbon nanotube has been used to prepare electrochemical sensors [212,213]. Leteba and Lang [214] reported a method to synthesize bimetallic Pt nanoparticles, such as Pt-Co, Pt-Ni and Pt-Fe, to use in biosensors. Platinum nanocluster conjugated with protein-A and antibody was used to image the HeLa cells using the confocal microscopy technique [215]. Application of Pt nanoclusters in bioimaging has been reported by many others as well [216].

5.4   Palladium nanoparticles

Palladium (Pd) is a noble metal like Au, Ag and Pt, and has unique optical properties as others. PdNPs have the potential to be used as a surface-enhanced Raman scattering (SERS) substrate [217–219]. Seed-mediated synthesis of Pd NPs has been reported by Chen et al. [220]. Chemical and electrochemical methods of synthesis of PdNPs and their functionalization with various ligands have been reported by Cookson [221]. PVP stabilized Pd NPs of different size and shape is synthesized using the polyol method [222]. Pd NPs have been successfully used in glucose biosensors [223,224] as well as DNA biosensors [225].

5.5   Other metal nanoparticles

Copper [226] and iridium NPs [227,228] also have been used in biosensing applications. Glucose biosensors are one of few commercially available sensing devices today [229]. Several glucose [230] and DNA biosensors [231] prepared using different mNPs have been reviewed in the literature.

6.   Techniques used for biosensing and imaging applications

Without the aid of targeting ligands, nanoparticles have been observed to target tumours passively through the enhanced permeability and retention effect (EPR) [232]. By coating with tumour-specific ligands (e.g., peptides, small organic molecules, antibodies, etc.), NPs can be actively targeted to specific tumours [233]. Development of nanotechnologies along with imaging modalities has assisted in the detection of disease. Molecular imaging requires molecular probes for visualization of cellular function and for characterization of molecular processes in living organisms [234]. At present, a variety of nanomaterials are being examined for molecular imaging applications in order for diagnosis or treatment of cancer [235]. The optical properties of noble mNPs, such as AuNPs and AgNPs, offer an alternative to the fluorophore-based staining and labelling of biological samples and have potential use in a wide range of biological and physical applications. Among a large choice of set-ups dedicated to *in vivo* measurements in rodents, fluorescence imaging techniques are becoming essential tools in preclinical studies. Human clinical uses of the fluorescence techniques for diagnostic and image-guided surgery [236] are also emerging. In comparison to low-molecular-weight organic dyes, application of fluorescent nanoprobes can improve both the signal sensitivity (better *in vivo* optical properties) and the fluorescence biodistribution (passive "nano" uptake in tumours for instance) [237].

At present, clinical detection of cancer relies on *in vitro* detection of biomarkers in patient's sample or morphological analysis of cells or tissues that are suspected to be diseased, or on *in vivo* imaging of the diseased site. Biosensors are in routine use for detection of biomarkers in *vitro* test. Specific biosensors are used to detect specific biomarkers. A review article [238] has been published with detail on various biosensors and their lower limit of detection. There are many techniques by which biomarkers levels are detected. Some of these techniques are fluorescence spectroscopy, enzyme-linked immunosorbent assay (ELISA), surface enhanced Raman spectroscopy (SERS) which will be discussed briefly.

6.1   Fluorescence sensing techniques

In spectrofluorometry technique, usually an ultraviolet (UV) light is irradiated on the sample to excite atoms (called fluorophores) at a fixed wavelength. The excited atoms then emit a secondary light at a higher wavelength which is called the fluorescence emission. The wavelength or the intensity of the original emission band changes if the dielectric condition of the surrounding medium of the fluorophore changes. Thus, the binding of the biomarker with the fluorophore attached to the biorecognition layer of a biosensor (Fig. 1) causes the original emission band to modify in recognition of the bound biomarker [239]. The bio-element (fluorophore which is attached to the biorecognition layer) interacts with the analyte (biomarker) being tested and the biological response is converted into an electrical signal by the transducer. The fluorescence detection includes fluorophores such as molecular fluorophores, quantum dots, and metallic and silica nanoparticles. A review by Tagit and Hildebrandt [240] presents a concise overview of recent advances in fluorescence sensing techniques for the detection of circulating disease biomarkers.

In addition to single-fluorophore fluorescence technique, a large number of biosensor designs are based on the two-fluorophores fluorescence resonance energy transfer (FRET) technique [241]. In such devices, a donor fluorophore, initially in its electronic excited state, transfers energy to an acceptor fluorophore through nonradiative dipole-dipole coupling [242]. The efficiency of this energy transfer is inversely proportional to the sixth power of the distance between donor and acceptor fluorophores, making FRET extremely sensitive to small changes in their separation [243]. Thus, when an analyte binds with one fluorophore it is easily detected [243,244]. One common pair fluorophores for biological use is a cyan fluorescent protein (CFP)–yellow fluorescent protein (YFP) pair [245]. Another fluorescence technique known as fluorescence correlation spectroscopy (FCS) is also used successfully for the detection of biomarkers. FCS is a technique basically used for spatial and temporal analysis of molecular interactions of extremely low-concentration biomolecules in solution. Shahzad et al. [246] used FCS technique for detecting serum biomarkers. The sensitivity and lower limit of biomarker detection of a biosensor are further improved by using appropriate nanomaterials such as metal nanoparticles. For example, Kurdekar et al. [247] reported early detection of HIV-1 p24 antigen in clinical specimens using fluorescent silver NP-based highly sensitive sandwich immunoassay.

6.2   ELISA technique

ELISA is a plate-based assay technique which is used for detecting and quantifying substances such as peptides, proteins, antibodies and hormones (as biomarkers) in a given sample. An enzyme conjugated with an antigen or antibody reacts with a colourless substrate (assay plate well) to generate a coloured reaction product that directly related to the targeted biomarker concentration. The substrate is called a chromogenic substrate. A number of enzymes have been employed for ELISA, including alkaline phosphatase, horseradish peroxidase, and B-galactosidase enzyme-conjugated antigen or antibodyis immobilized to a solid surfaceeither directly or via the use of a capture antibody itself immobilized on the surface. The antigen or antibody is then complexed to a detection antibody or antigen conjugated with a molecule amenable for detection such as an enzyme or a fluorophore. Depending on the antigen-antibody combination, the ELISA assay is called a direct, indirect, sandwich, competitive etc. Use of metal nanomaterials has improved the sensitivity and detection limit of the ELISA technique [247–251].

6.3   SERS technique

Surface-enhanced Raman spectroscopy (SERS) is an extension of Raman spectroscopy, where metallic nanostructures are used to enhance the intensity of Raman scattering [252].The amplification of the signals in SERS comes (mainly) through the electromagnetic interaction of light with metals (nanoparticles), which produces large amplification of the laser field through plasmonic resonance (LSPR). To profit from these, the molecules (protein, antibody, antigens, DNA etc.) must typically be adsorbed on the metal surface, or at least be very close to it (typically $\approx$ 10 nm maximum). Enhancement factors can be as high as 10^{14}–10^{15}, which are sufficient to allow even single molecule detection [253,254]. Graham et al. [255] reviewed the latest development of *in vivo* biosensing using SERS technique. SPR-based biosensors are very powerful tools for the study of biomolecular interactions, chemical detection and immunoassays. Wijaya et al. [256] reviewed the development of various SPR-based biosensors and their detection schemes.

6.4   *In vivo* imaging

In addition to *in vitro* detection of tumour biomarkers using biosensors, mNPs have been successfully used for *in vivo* detection of tumours using fluorescence imaging techniques. In *in vivo* imaging, mNPs are actively targeted at tumour sites via suitable functionalization on their surfaces. For fluorescence imaging, NPs will be co-functionalized with fluorescence dyes and upon UV/NIR irradiation the distribution of NPs in the cell can be obtained. Targeting of the tumour can also be achieved via EPR effect which is particularly critical for *in vivo* imaging of cancer and atherosclerosis [257] diseases in which neovasculature is prominent, and is by far the most popular mechanism by which to target nanomaterials. The *in vivo* imaging method is useful for locating tumour cells and their stage of progression. The advantage of mNPs in *vivo* imaging is that they improve the contrast of the image. There are several *in vivo* imaging techniques such as PET and CT in which mNPs, especially gold NPs [258–260], have been used for better contrast of the image.

7.   Safety issues of metal nanoparticles

Potential toxic effects are a concern with *in vivo* use of NPs but not with *in vitro* diagnostics, which is the usual procedure of clinical diagnostics [139]. The fate of NPs inside the living body, which may vary considerably due to a huge diversity of materials used and wide variation in their size, is not yet fully understood. NPs are more toxic to human health in comparison to the bulks of the same chemical compositions, and the toxicities are inversely proportional to the size of the NPs [261,262]. Bahadar et al. [263] have reviewed the toxicity of nanoparticles and the current experimental methods for detecting the toxicity of nanoparticles.

Gold NPs are considered to be relatively safe, as its core is inert and non-toxic [264]; though the toxicity of AuNPs has been reported to be dependent on the dose and the detection methods used [265,266]. Toxicity of AuNPs has also been found different in different cell lines [267]. Genotoxicity and cell toxicity of AuNPs have been asset to be not significant toxic [268]. The surface chemistry though plays a crucial role in determining the toxicity of AuNPs [269]. Kim et al. [270] indeed reported significant toxicity of cationic ligand functionalized 1.3 nm AuNPs in an embryonic zebrafish model. A review article has been dedicated to human toxicology of AuNPs [271]. In a brief review, Jia et al. [272] concluded that the toxicity of AuNPs arises through the production of reactive oxygen species (ROS) in the cell.

Being anti-bacterial substance silver NPs are used in the form of wound dressings, a coating of surgical instruments and prostheses [273]. But AgNPs have shown more toxicity, in comparison to other mNPs, in terms of cell viability, ROS production, and lactate dehydrogenase (LDH) leakage [274]. Again, surface chemistry and coatings of AgNPs play a crucial role in cytotoxicity [273]. For example, polyvinyl-pyrrolidone (PVP)-coated AgNPs (6-20 nm) has been reported to have dose-dependent toxicity to human lung cancer cell line and cellular DNA adducts formation [275]. Based on the study on the human leukaemia cell line, Haase et al. [276] reported that peptide-coated AgNPs (20 nm) are more cytotoxic than citrate-coated AgNPs of the same size. Tang et al. [277] reported that AgNPs were transferred to the blood and distributed throughout the brain after subcutaneous injection in rats causing swollen astrocytes, blood-brain barrier (BBB) destruction, and neuronal degeneration in the same test rats.

Feng et al. [278] reviewed the toxicity of metallic NPs to the central nervous system and also discussed the possible mechanisms involved in the toxicity such as generation of ROS [279–282], the interaction of NPs with mitochondria [283–285], and so on. Different factors, such as chemical composition, size and size distribution, shape, surface chemistry and surface charge responsible for the toxicity of metal NPs have been discussed elsewhere [286–288]. Thus, the main limitation for the approval of *in vivo* application of nanomaterials for clinical diagnostics is related to their level of toxicity which needs to be studied thoroughly.

Conclusions

Application of metal nanoparticles (mNPs) in biomarkers detection via either *in vitro* or *in vivo* processes and their possible toxicity effects in *in vivo* administration have been reviewed in this chapter. Gold nanoparticles among all are the most investigated mNPs for biomedical applications. Due to very unique optical properties such as surface plasmon resonance (SPR), mNPs are used in designing biosensor. It is reported that the use of mNPs in biosensor improves the sensitivity and lower the limit of detection of biomarkers present in the sample collected from patients. Other than biosensor application, mNPs have been used as high contrast agents for *in vivo* imaging of targeted tissues. Several detections and imaging techniques have been reported for diagnosis of diseases either via *in vitro* or via *in vivo* processes. Some of these techniques have been discussed briefly in this chapter.

References

[1]   R. Mayeux, Biomarkers: Potential uses and limitations, NeuroRx 1 (2004) 182-188. https://doi.org/10.1602/neurorx.1.2.182

[2]   R. Frank, R. Hargreaves, Clinical biomarkers in drug discovery and development, Nat. Rev. Drug Discov. 2 (2003) 566-580. https://doi.org/10.1038/nrd1130

[3]   N.L. Henry, D.F. Hayes, Cancer biomarkers, Molecular Oncology 6 (2012) 140-146. https://doi.org/10.1016/j.molonc.2012.01.010

[4]   J. Rhea, R.J.Molinaro, Cancer biomarkers: surviving the journey from bench to bedside, Med. Lab. Obs. 43 (2011)10-2, 16, 18; quiz 20, 22; PMID:21446576.

[5]   T. Behne, M.S. Copur, Biomarkers for hepatocellular carcinoma, Int. J. Hepatol. 2012 (2012) 1-7;doi:10.1155/2012/859076. https://doi.org/10.1155/2012/859076

[6]   A. Musolino, M.A. Bella, B. Bortesi, M. Michiara, N. Naldi, P. Zanelli, M. Capelletti, D. Pezzuolo, R. Camisa, M. Savi, T.M. Neri, A. Ardizzoni, BRCA mutations, molecular markers and clinical variables in early-onset breast cancer: a population-based study, Breast16(2007) 280-292; doi:10.1016/j.breast.2006.12.003. PMID 17257844. https://doi.org/10.1016/j.breast.2006.12.003

[7]   R. Dienstmann, J. Tabernero, BRAF as a target for cancer therapy, Anti. Canc. Agents Med. Chem. 11(2011) 285-295;doi:10.2174/187152011795347469. PMID 21426297. https://doi.org/10.2174/187152011795347469

[8]   N. Lamparella, A. Barochia, S. Almokadem, Impact of genetic markers on treatment of non-small cell lung cancer, Adv. Exp. Med. Biol. 779(2013)145-164;doi:10.1007/978-1-4614-6176-0_6. PMID 23288638. https://doi.org/10.1007/978-1-4614-6176-0_6

[9]   G. Orphanos, P. Kountourakis, Targeting the HER2 receptor in metastatic breast cancer, Hematol. Oncol. Stem Cell Ther. 5(2012) 127-137. https://doi.org/10.5144/1658-3876.2012.127

[10]   S.E. DePrimo, X. Huang, M.E. Blackstein, C.R. Garrett, C.S. Harmon, P. Schoffski, M.H. Shah, J. Verweij, C.M. Baum, G.D. Demetri, Circulating levels of soluble KIT serve as a biomarker for clinical outcome in gastrointestinal stromal tumorpatients receiving sunitinib following imatinib failure, Clinic. Can. Res.15 (2009) 5869-5877;doi:10.1158/1078-0432.CCR-08-2480. https://doi.org/10.1158/1078-0432.CCR-08-2480

[11]   A. Bantis, P. Grammaticos, Prostatic specific antigen and bone scan in the diagnosis and follow-up of prostate cancer. Can diagnostic significance of PSA be increased?. Hell. J. Nucl. Med.15 (2012) 241-246; PMID 23227460.

[12]   S. Kruijff, H.J. Hoekstra, The current status of S-100B as a biomarker in melanoma, Eur. J. Surg. Oncol. 38(2012) 281-285;

doi:10.1016/j.ejso.2011.12.005. PMID 22240030.
https://doi.org/10.1016/j.ejso.2011.12.005

[13] J.A. Ludwig, J.N. Weinstein, Biomarkers in cancer staging, prognosis and treatment selection,Nat. Rev. Can. 5 (2005) 845-856;doi:10.1038/nrc1739. PMID 16239904. https://doi.org/10.1038/nrc1739

[14] A. Turner, G. Wilson, I. Kaube, Biosensors: fundamentals and applications, Oxford University Press, Oxford, UK, 1987; ISBN 0198547242.

[15] F.G. Banica, Chemical sensors and biosensors: fundamentals and applications, John Wiley & Sons, Chichester, UK, 2012; ISBN 9781118354230. https://doi.org/10.1002/9781118354162

[16] B. Leca-Bouvier, L.J. Blum, Biosensors for protein detection: a review, Anal. Lett. 38 (2005) 1491-1517. https://doi.org/10.1081/AL-200065780

[17] Y. Xiao, F. Patolsky, E. Katz, J.F. Hainfeld, I. Willner, Plugging into enzymes: nanowiring of redox enzymes by a gold nanoparticle, Science 299 (2003) 1877-1881. https://doi.org/10.1126/science.1080664

[18] M. Schierhorn, S.J. Lee, S.W. Boettcher, G.D. Stucky, M. Moskovits, Metal-silica hybrid nanostructures for surface-enhanced Raman spectroscopy, Adv. Mater., 18 (2006) 2829-2832. https://doi.org/10.1002/adma.200601254

[19] H. Cai, Y. Xu, N. Zhu, P. He, Y. Fang, An electrochemical DNA hybridization detection assay based on a silver nanoparticle label, Analyst 127 (2002) 803-808. https://doi.org/10.1039/b200555g

[20] J. Wang, G. Liu, R. Polsky, A. Merkoçi, Electrochemical stripping detection of DNA hybridization based on cadmium sulfide nanoparticle tags, Electrochem. Comm.4 (2002) 722-726. https://doi.org/10.1016/S1388-2481(02)00434-4

[21] J. Ling, C.Z.Huang, Y.F.Li, L. Zhang, L.Q. Chen, S.J. Zhen, Light scattering signals from nanoparticles in biochemical assay, pharmaceutical analysis and biological imaging, Trends Anal. Chem. 28 (2009) 447-453. https://doi.org/10.1016/j.trac.2009.01.003

[22] X. Huang, R.O. Connor, E.A. Kwizera, Gold nanoparticle based platforms for circulating cancer marker detection, Nanotheranotics 1 (2017) 80-102. https://doi.org/10.7150/ntno.18216

[23] D.L. Nida, M.S. Rahman, K.D. Carlson, R.R. Kortum, M. Follen, Fluorescent nanocrystals for use in early cervical cancer detection,Gynecol. Oncol. 99 (2005) S89-S94. https://doi.org/10.1016/j.ygyno.2005.07.050

[24] D. Karleya, D. Gupta, A. Tewari, Biomarker for cancer: a great promise for future. World J. Oncol. 2 (2011) 151-157.

[25] A. Agah, A. Hassibi, J.D. Plummer, P.B. Griffin, Design requirements for integrated biosensor arrays, European conference on biomedical optics, Proceedings of the SPIE, 5699 (2005) 403-413. https://doi.org/10.1117/12.591300

[26] F. Patolsky, G. Zheng, C.M. Lieber, Nanowire sensors for medicine and the life sciences, Nanomedicine 1(2006) 51-65. https://doi.org/10.2217/17435889.1.1.51

[27] S. Carrara, S. Ghoreishizadeh, J. Olivo, I. Taurino, C. Baj-Rossi, A. Cavallini, M.O. de Beeck, C. Dehollain, W. Burleson, F.G. Moussy, A. Guiseppi-Elie, G. De Micheli, Fully integrated biochip platforms for advanced healthcare, Sensors 12(2012) 11013-11060. https://doi.org/10.3390/s120811013

[28] S. Hou, A. Zhang, M. Su, Nanomaterials for biosensing applications, Nanomaterials (Basel) 6 (2016) 58-62. https://doi.org/10.3390/nano6040058

[29] R. Elghanian, J.J. Storhoff, R.C. Mucic, R.L. Letsinger, C.A. Mirkin, Selective colorimetric detection of polynucleotides based on the distance-dependent optical properties of gold nanoparticles,Science277 (1997) 1078-1081. https://doi.org/10.1126/science.277.5329.1078

[30] W.C. Chan, D.J. Maxwell, X. Gao, R.E. Bailey, M. Han, S. Nie, Luminescent quantum dots for multiplexed biological detection and imaging, Curr. Opin. Biotechnol.13 (2002) 40-46. https://doi.org/10.1016/S0958-1669(02)00282-3

[31] S.J. Park, T.A. Taton, C.A. Mirkin, Array-based electrical detection of DNA with nanoparticle probes,Science295 (2002) 1503-1506. https://doi.org/10.1126/science.1066348

[32] L. Josephson, J.M. Perez, R. Weissleder, Magnetic nanosensors for the detection of oligonucleotide sequences,Angew. Chem.113 (2001) 3304-3306. https://doi.org/10.1002/1521-3757(20010903)113:17<3304::AID-ANGE3304>3.0.CO;2-D

[33] J.A. Hansen, R. Mukhopadhyay, J.Ø.Hansen, K.V. Gothelf, Femtomolar electrochemical detection of DNA targets using metal sulfide nanoparticles, J. Am. Chem. Soc.128 (2006) 3860-3861. https://doi.org/10.1021/ja0574116

[34] C. Wang, Z. Sun, L. Ma, M. Su, Simultaneous detection of multiple biomarkers with over three orders of concentration difference using phase change nanoparticles, Anal. Chem.83 (2011) 2215-2219. https://doi.org/10.1021/ac103102h

Biosensors: Materials and Applications
Materials Research Foundations **47** (2019) 77-130

Materials Research Forum LLC
doi: http://dx.doi.org/10.21741/9781644900130-3

[35] V.M. Shalaev, Transforming light, Science 322 (2008) 384-386. https://doi.org/10.1126/science.1166079

[36] M.L. Brongersma, V.M. Shalaev, The case for plasmonics, Science 328 (2010) 440-441. https://doi.org/10.1126/science.1186905

[37] D.K. Gramotnev, S.I. Bozhevolnyi, Plasmonics beyond the diffraction limit, Nat. Photon. 4 (2010) 83-91. https://doi.org/10.1038/nphoton.2009.282

[38] S. Lal, S. Link, N.J. Halas, Nano-optics from sensing to waveguiding, Nat. Photon.1 (2007) 641-648. https://doi.org/10.1038/nphoton.2007.223

[39] J.A. Schuller, E.S. Barnard, W. Cai, Y.C. Jun, J.S. White, M.L. Brongersma, Plasmonics for extreme light concentration and manipulation, Nat. Mater.9(2010) 193-204. https://doi.org/10.1038/nmat2630

[40] S. Zeng, B. Dominique, H. Ho-Pui, Y. Ken-Tye, Nanomaterials enhanced surface plasmon resonance for biological and chemical sensing applications, Chem. Soc. Rev. 43 (2014) 3426-3452. https://doi.org/10.1039/c3cs60479a

[41] D. Xu, J. Mao, Y. He, E. Yeung, Size-tunable synthesis of high-quality gold nanorods under basic conditions by using H2O2 as the reducing agent, J. Mater. Chem. C 2 (2014) 4989-4996. https://doi.org/10.1039/c4tc00483c

[42] C. Sonnichsen, A.P. Alivisatos, Gold nanorods as novel nonbleaching plasmon-based orientation sensors for polarized single-particle microscopy, Nano Lett. 5 (2005) 301-304. https://doi.org/10.1021/nl048089k

[43] X. Wen, H. Shuai, L. Min, Precise modulation of gold nanorods aspect ratio based on localized surface plasmon resonance, Opt. Mater. 60 (2016) 324-330. https://doi.org/10.1016/j.optmat.2016.08.008

[44] S. Link, M.A. El-Sayed,Size and temperature dependence of the plasmon absorption of colloidal gold nanoparticles, J. Phys. Chem. B 103 (1999) 4212-4217. https://doi.org/10.1021/jp984796o

[45] S. Link, M.B. Mohamed, M.A. El-Sayed,Simulation of the optical absorption spectra of gold nanorods as a function of their aspect ratio and the effect of the medium dielectric constant,J. Phys. Chem. B 103 (1999) 3073-3077. https://doi.org/10.1021/jp990183f

[46] Y. Zhao, Y. Wang, F. Ran, Y. Cui, C. Liu, Q. Zhao, Y. Gao, D. Wang, S. Wang, A comparison between sphere and rod nanoparticles regarding their in vivo biological behavior and pharmacokinetics,Sci. Rep. 7 (2017) 4131(1-11).

[47] R. Richards, H. Bonnemann, Synthetic approaches to metallic nanomaterials,in: C.S.S.R. Kumar, J. Hormes, C. Leuschner (Eds.), Nanofabrication towards

biomedical applications: techniques, tools, applications, and impact, Wiley-VCH Verlag GmbH & Co. KGaA, Weinheim, 2005, pp. 1-32. https://doi.org/10.1002/3527603476.ch1

[48]   H.R. Ghorbani, A review of methods for synthesis of Al nanoparticles, Oriental J. Chem. 30 (2014) 1941-1949. https://doi.org/10.13005/ojc/300456

[49]   C.J. Murphy, T.K. Sau, A.M. Gole, C.J. Orendorff, J. Gao, L. Gou, S.E. Hunyadi, T. Li, Anisotropic metal nanoparticles: synthesis, assembly and optical applications, J. Phys. Chem. B 109 (2005) 13857-13870. https://doi.org/10.1021/jp0516846

[50]   U.Y. Qazi, R. Javaid, A review on metal nanostructures: preparation methods and their potential applications, Advances in Nanoparticles 5 (2016) 27-43. https://doi.org/10.4236/anp.2016.51004

[51]   M.T. Swihart, Vapor-phase synthesis of nanoparticles, Curr. Opin. Colloid Interface Sci. 8 (2003) 127-133. https://doi.org/10.1016/S1359-0294(03)00007-4

[52]   H. Bonnemann, R.M. Richards, Nanoscopic metal particles -synthetic methods and potential applications, Eur. J. Inorg. Chem. 2001 (2001) 2455-2480. https://doi.org/10.1002/1099-0682(200109)2001:10<2455::AID-EJIC2455>3.0.CO;2-Z

[53]   J. Kimling, M. Maier, B. Okenve, V. Kotaidis, H. Ballot, A. Plech, Turkevich method for gold nanoparticle synthesis revisited, J. Phys. Chem. B 110 (2006) 15700-15707. https://doi.org/10.1021/jp061667w

[54]   P.C. Lee, D. Meisel, Adsorption and surface-enhanced Raman of dyes on silver and gold sols. J. Phys. Chem. 86(1982) 3391-3395. https://doi.org/10.1021/j100214a025

[55]   J.A. Creighton, C.G. Blatchford, M.G. Albrecht, Plasma resonance enhancement of Raman scattering bypyridine adsorbed on silver or gold sol particles of size comparable to the excitation wavelength, J. Chem. Soc.,Faraday Trans. 2: Mol. Chem. Phys. 75(1979) 790-798.

[56]   S. Ayyappan, G.R. Srinivasa, G.N. Subbanna, C.N.R. Rao, Nanoparticles of Ag, Au, Pd, and Cu produced by alcohol reduction of the salts, J. Mater. Res. 12(1997) 398-401. https://doi.org/10.1557/JMR.1997.0057

[57]   S.D. Bunge, T.J. Boyle, T.J. Headley, Synthesis of coinage-metal nanoparticles from mesityl precursors, Nanoletters 3(2003) 901-905. https://doi.org/10.1021/nl034200v

[58]   S. Panigrahi, S. Kundu, S.K. Ghosh, S. Nath, T. Pal, General method of synthesis of metal nanoparticles, J. Nanopart. Res. 6 (2004) 411-414. https://doi.org/10.1007/s11051-004-6575-2

[59]   S.M. Landage, A.I. Wasif, P. Dhuppe, Synthesis of nanosilver using chemical reduction method, Int. J. Adv. Res. Engg. Appl. Sci. 3 (2014) 14-22.

[60]   Z.S. Pillai, P.V. Kamat, What factors control the size and shape of silver nanoparticles in the citrate ion reduction method? J. Phys. Chem. B 108 (2004) 945-951. https://doi.org/10.1021/jp037018r

[61]   R.P. Andres, J.D. Bielefeld, J.I. Henderson, D.B. Janes, V.R. Kolagunta, C.P. Kubiak, W.J. Mahoney, R.G. Osifchin, Self-assembly of a two-dimensional superlattice of molecularly linked metal clusters, Science 273 (1996) 1690-1693. https://doi.org/10.1126/science.273.5282.1690

[62]   K.S. Suslick, M. Fang, T. Hyeon, Sonochemical synthesis of iron colloids, J. Am. Chem. Soc. 118 (1996) 11960-11961. https://doi.org/10.1021/ja961807n

[63]   R.A. Hobson, P. Mulvaney, F. Grieser, Formation of Q-state CdS colloids using ultrasound, J. Chem. Soc., Chem. Commun.7 (1994) 823-824. https://doi.org/10.1039/c39940000823

[64]   W. Huang, X. Tang, Y. Wang, Y. Koltypin, A. Gedanken, Selective synthesis of anatase and rutile via ultra-sound irradiation, Chem. Commun.15 (2000) 1415-1416. https://doi.org/10.1039/b003349i

[65]   M.A. Lopez-Quintela, Synthesis of nanomaterials in microemulsions: formation mechanism and growth control, Curr. Opin. Coll. Int. Sci. 8 (2003) 137-144. https://doi.org/10.1016/S1359-0294(03)00019-0

[66]   M.A. Malik, M.Y. Wani, M.A. Hashim, Microemulsion method: a novel route to synthesize organic and inorganic nanomaterials, Arb. J. Chem. 5 (2012) 397-417. https://doi.org/10.1016/j.arabjc.2010.09.027

[67]   I. Capek, Preparation of metal nanoparticles in water-in-oil (w/o) microemulsions, Adv. Colloid Interface Sci. 110 (2004) 49-74. https://doi.org/10.1016/j.cis.2004.02.003

[68]   M. Husein, E. Rodil, J. Vera, Formation of silver chloride nanoparticles in microemulsions by direct precipitation with the surfactant counterions, Langmuir 9 (2003) 846-8474. https://doi.org/10.1021/la0342159

[69]   P. He, X. Shen, H. Gao, Size-controlled preparation of Cu2O octahedron nanocrystals and studies on their optical absorption, J. Colloid Interface Sci. 284 (2005) 510-515. https://doi.org/10.1016/j.jcis.2004.10.060

[70]   M. Salavati-Niasari, F. Davar, N. Mir, Synthesis and characterization of metallic copper nanoparticles via thermal decomposition, Polyhedron 27 (2008) 3514-3518. https://doi.org/10.1016/j.poly.2008.08.020

[71]   J.L. Gardea-Torresdy, J.G. Parsons, E. Gomez, P. Videa, H.E. Troiani, P. Santiago, M.J. Yacaman,Formation and growth of Au nanoparticles inside live alfalfa plants, Nano Lett. 2 (2002) 397-401. https://doi.org/10.1021/nl015673+

[72]   J.L. Gardea-Torresdy, E. Gomez, J.R.P. Videa, J.G. Parsons, H. Troiani, M.J. Yacaman, Alfalfa sprouts: a natural source for the synthesis of silver nanoparticles, Langmuir 19 (2003) 1357-1361. https://doi.org/10.1021/la020835i

[73]   A. Sikora, A.G. Shard, C. Minelli, Size and ζ-potential measurement of silica nanoparticles in serum using tunable resistive pulse sensing, Langmuir 32 (2016) 2216-2224. https://doi.org/10.1021/acs.langmuir.5b04160

[74]   V. Filipe, A. Hawe, W. Jiskoot, Critical evaluation of nanoparticle tracking analysis (NTA) by nanosight for the measurement of nanoparticles and protein aggregates, Pharm. Res. 27 (2010) 796−810. https://doi.org/10.1007/s11095-010-0073-2

[75]   J. Gross, S. Sayle, A.R. Karow, U. Bakowsky, P. Garidel, Nanoparticle tracking analysis of particle size and concentration detection in suspensions of polymer and protein samples: influence of experimental and data evaluation parameters, Eur. J. Pharm. Biopharm, 104 (2016) 30−41. https://doi.org/10.1016/j.ejpb.2016.04.013

[76]   S.K. Srivastava, Y. Yamada, C. Ogino, A. Kondo, Biogenic synthesis and characterization of gold nanoparticles by Escherichia coli K12 and its heterogeneous catalysis in degradation of 4-nitrophenol, Nanoscale res. lett. 8(2013) 1-9. https://doi.org/10.1186/1556-276X-8-70

[77]   U.K. Parida, B.K. Bindhani, P. Nayak, Green synthesis and characterization of gold nanoparticles using onion (Allium cepa) extract, World J. Nano Sci. Engg. 1 (2011) 93-98. https://doi.org/10.4236/wjnse.2011.14015

[78]   R.K. Petla, S. Vivekanandhan, M. Misra, A.K. Mohanty, N. Satyanarayana, Soybean (Glycine max) leaf extract based green synthesis of palladium nanoparticles, J. Biomater. Nanobiotechnol. 3 (2012) 14-19. https://doi.org/10.4236/jbnb.2012.31003

[79]   J.H. Lee, K. Ahn, S.M. Kim, K.S. Jeon, J.S. Lee, I.J. Yu, Continuous 3-day exposure assessment of workplace manufacturing silver nanoparticles, J. Nanopart. Res. 14 (2012) 1-10. https://doi.org/10.1007/s11051-012-1134-8

[80]   H. Yang, Y. Wang, H. Huang, L. Gell, L. Lehtovaara, S. Malola, H. Hakkinen, N. Zheng, All-thiol-stabilized Ag_{44} and $Au_{12}Ag_{32}$ nanoparticles with single-crystal structures, Nat. commun.4 (2013) 1-8. https://doi.org/10.1038/ncomms3422

[81]   A.M. Awwad, N.M. Salem, A green and facile approach for synthesis of magnetite nanoparticles, Nanosci. Nanotechnol. 2 (2012) 208−213. https://doi.org/10.5923/j.nn.20120206.09

[82]   V.P. Zharov, K.E. Mercer, E.N. Galitovskaya, M.S. Smeltzer, Photothermal nanotherapeutics and nanodiagnostics for selective killing of bacteria targeted with gold nanoparticles, Biophys. J. 15 (2006)619-627. https://doi.org/10.1529/biophysj.105.061895

[83]   L.E. van Vlerken, T.K. Vyas, M.M. Amiji,Poly(ethylene glycol)-modified nanocarriers for tumor-targeted and intracellular delivery, Pharm. Res. 24 (2007) 1405-1414. https://doi.org/10.1007/s11095-007-9284-6

[84]   J.V. Jokerst, T. Lobovkina, R.N. Zare, S.S. Gambhir, Nanoparticles PEGylation for imaging and therapy, Nanomedicine 6 (2011) 715-728. https://doi.org/10.2217/nnm.11.19

[85]   G. Prencipe, S.M. Tabakman, K. Welsher, Z. Liu, A.P. Goodwin, L. Zhang, J. Henry, H. Dai, PEG branched polymer for functionalization of nanomaterials with ultralong blood circulation, J. Am. Chem. Soc. 131 (2009) 4783-4787. https://doi.org/10.1021/ja809086q

[86]   V. Wycisk, K. Achazi, O. Hirsch, C. Kuehne, J. Dernedde, R. Haag, K. Licha, Heterobifunctional dye: Highly fluorescent linkers based on cyanine, Chem. Open 6 (2017) 437-446. https://doi.org/10.1002/open.201700013

[87]   D.M. Collard, M.A. Fox, Use of electroactive thiols to study the formation and exchange of alkanethiol monolayers on gold, Langmuir7(1991) 1192-1197.

[88]   Z. Liu, C. Davis, W. Cai, L. He, X. Chen, H. Dai, Circulation and long-term fate of functionalized, biocompatible single-walled carbon nanotubes in mice probed by Raman spectroscopy, Proc. Natl. Acad. Sci., USA105(2008) 1410-1415. https://doi.org/10.1073/pnas.0707654105

[89]   Y. Hong, D. Shin, S. Cho, H. Uhm, Surface transformation of carbon nanotube powder into super-hydrophobic and measurement of wettability, Chem. Phys. Lett. 427 (2006) 390-393. https://doi.org/10.1016/j.cplett.2006.06.033

[90]   G.H. Hermanson, Bioconjugate techniques, second edition, Academic Press, San Diego, Calif, USA, 2008.

[91]   O.M. Koo, I. Rubinstein, H. Onyuksel, Role of nanotechnology in targeted drug delivery and imaging: a concise review, Nanomedicine 1(2005) 193-212. https://doi.org/10.1016/j.nano.2005.06.004

[92]   V.P.R. Chichili, V. Kumar, J. Sivaraman, Linkers in the structural biology of protein-protein interaction, Protein Sci. 22 (2013) 153-167. https://doi.org/10.1002/pro.2206

[93]   M. Nickels, J. Xie, J. Cobb, J.C. Gore, W. Pham, Functionalization of iron oxide nanoparticles with a versatile epoxy amine linker, J. Mater. Chem. 20 (2010) 4776-4780. https://doi.org/10.1039/c0jm00808g

[94]   J. Lu, F. Jiang, A. Lu, G. Zhang, Linkers having a crucial role in antibody-drug conjugates, Int. J. Mol. Sci. 17 (2016) 561-582. https://doi.org/10.3390/ijms17040561

[95]   R. Popovtzer, A. Agrawal, N.A. Kotov, A. Popovtzer, J. Balter, T.E. Carey, R. Kopelman, Targeted gold nanoparticles enable molecular CT imaging of cancer,Nano Lett. 8 (2008) 4593-4596. https://doi.org/10.1021/nl8029114

[96]   J.R. McCarthy, R. Weissleder, Multifunctional magnetic nanoparticles for targeted imaging and therapy, Adv. Drug Deliv. Rev. 60 (2008) 1241-1251. https://doi.org/10.1016/j.addr.2008.03.014

[97]   S. Wang, E.E. Dormidontova, Selectivity of ligand-receptor interactions between nanoparticle and cell surfaces, Phys. Rev. Lett. 109 (2012) 238102 (1-5).

[98]   D. Peer, J.M. Karp, S. Hong, O.C. Farokhzad, R. Margalit, R. Langer, Nanocarriers as an emerging platform for cancertherapy, Nat. Nanotechnol. 2 (2007) 751-760. https://doi.org/10.1038/nnano.2007.387

[99]   K. Loomis, K. McNeeley, R.V. Bellamkonda, Nanoparticles with targeting, triggered release and imaging functionality for cancer applications, Soft Matter 7 (2011) 839-856. https://doi.org/10.1039/C0SM00534G

[100] J.D. Byrne, T. Betancourt, L. Brannon-Peppas, Active targeting schemes for nanoparticle systems in cancer therapeutics, Adv. Drug Delivery Rev.60 (2008) 1615-1626. https://doi.org/10.1016/j.addr.2008.08.005

[101] A.C. Prost, F. Menegaux, P. Langlois, J.M. Vidal, M. Koulibaly, J.L. Jost, J.J. Duron, J.P. Chigot, P. Vayre, A. Aurengo, J.C. Legrand, G. Rosselin, C. Gespach, Differential transferrin receptor density in human colorectal cancer: a potential probe for diagnosis and therapy, Int. J. Oncol. 13 (1998) 871-876. https://doi.org/10.3892/ijo.13.4.871

Biosensors: Materials and Applications   Materials Research Forum LLC
Materials Research Foundations **47** (2019) 77-130   doi: http://dx.doi.org/10.21741/9781644900130-3

[102] W. Cai, S.S. Gambhir, X. Chen, Multimodality tumor imaging targeting integrin αvβ3, Bio.Techniques 39 (2005) S14-S25. https://doi.org/10.2144/000112091

[103] C.B. Carlson, P. Mowery, R.M. Owen, E.C. Dykhuizen, L.L. Kiessling, Selective tumor cell targeting using low-affinity, multivalent interactions, ACS Chem. Biol. 2 (2007) 119-127. https://doi.org/10.1021/cb6003788

[104] F.J. Martinez-Veracoechea, D. Frenkel, Designing super selectivity in multivalent nano-particle binding, Proc. Natl. Acad. Sci., U.S.A 108 (2011) 10963-10968. https://doi.org/10.1073/pnas.1105351108

[105] X. He, K. Wang, Z. Cheng, In vivo near-infrared fluorescence imaging of cancer with nanoparticle-based probes, Wiley Interdiscip. Rev. Nanomed. Nanobiotechnol. 2 (2010) 349-366. https://doi.org/10.1002/wnan.85

[106] T.F. Massoud, S.S. Gambhir, Integrating noninvasive molecular imaging into molecular medicine: an evolving paradigm, Trends Mol. Med. 13 (2007) 183-191. https://doi.org/10.1016/j.molmed.2007.03.003

[107] J.V. Frangioni, New technologies for human cancer imaging, J. Clin. Oncol. 26(2008) 4012-4021. https://doi.org/10.1200/JCO.2007.14.3065

[108] J. Merian, J. Gravier, F. Navarro, I. Texier, Fluorescent nanoprobe dedicated to in vivo imaging: from preclinical validations to clinical translation, Molecules 17 (2012) 5564-5591. https://doi.org/10.3390/molecules17055564

[109] F. Leblond, S.C. Davis, P.A. Valde, B.W. Pogue, Pre-clinical whole-body fluorescences, imaging: Review of instruments, methods and applications, J. Photochem. Photobiol. B: Biol. 98 (2010) 77-94. https://doi.org/10.1016/j.jphotobiol.2009.11.007

[110] K. Licha, C. Olbrich, Optical imaging in drug discovery and diagnostic applications, Adv. Drug. Deliv. Rev. 57 (2005) 1087-1108. https://doi.org/10.1016/j.addr.2005.01.021

[111] S. Li, J. Johnson, A. Peck, Q. Xie, Near infrared fluorescent imaging of brain tumor with IR780 dye incorporated phospholipid nanoparticles, J. Transl. Med. 15 (2017) 18-29. https://doi.org/10.1186/s12967-016-1115-2

[112] T. Lei, A. Fernandez-Fernandez, R. Manchanda, T.C. Huang, A.J. McGoron, Near-infrared dye loaded polymeric nanoparticles for cancer imaging and therapy and cellular response after laser induced heating, Belstein J. Nanotechnol. 5 (2014) 313-322. https://doi.org/10.3762/bjnano.5.35

[113] M. Wang, F. Xie, X. Wen, H. Chen, H. Zhang, J. Liu, H. Zhang, H. Zou, Y. Yu, Y. Chen, Z. Sun, X. Wang, G. Zhang, C. Yin, D. Sun, J. Gao, B. Jiang, Y. Zhong,

Y. Lu, Therapeutic PEG-ceramide nanomicelles synergize with salinomycin to target both liver cancer cells and cancer stem cells, Nanomedicine 12 (2017) 1025-1042. https://doi.org/10.2217/nnm-2016-0408

[114] E. Dickreuter, N. Cordes, The cancer cell adhesion resistome: mechanisms, targeting and translational approaches, Biol. Chem. 398 (2017) 721-735. https://doi.org/10.1515/hsz-2016-0326

[115] J.R. Nedrow, A. Josefsson, S. Park, T. Back, R.F. Hobbs, C. Brayton, F. Bruchertseifer, A. Morgenstern, G. Sgouros, Pharmacokinetics, microscale distribution and dosimetry of alpha-emitter-labeled anti-PD-L1 antibodies in an immune competent transgenic breast cancer model, EJNMMI Res. 7 (2017) 57-72. https://doi.org/10.1186/s13550-017-0303-2

[116] S.S. Dhule, P. Penfornis, J. He, M.R. Harris, T. Terry, V. John, R. Pochampally, The combined effect of encapsulating curcumin and C6 ceramide in liposomal nanoparticles against osteosarcoma, Mol. Pharm. 11 (2014) 417-427. https://doi.org/10.1021/mp400366r

[117] Y.Y. Li, S.K. Lam, C.Y. Zheng, J.C. Ho, The effect of tumor microenvironment on autophagy and sensitivity to targeted therapy in EGFR-mutated lung adenocarcinoma, J. Cancer 6 (2015) 382-386. https://doi.org/10.7150/jca.11187

[118] Y. Li, K. Atkinson, T. Zhang, Combination of chemotherapy and cancer stem cell targeting agents: Preclinical and clinical studies, Cancer Lett. 396 (2017) 103-109. https://doi.org/10.1016/j.canlet.2017.03.008

[119] W. Gao, B. Xiang, T.T. Meng, F. Liu, X.R. Qi, Chemotherapeutic drug delivery to cancer cells using a combination of folate targeting and tumor microenvironment-sensitive polypeptides, Biomaterials 34(2013) 4137-4149. https://doi.org/10.1016/j.biomaterials.2013.02.014

[120] S. Tortorella, T.C. Karagiannis, Transferrin receptor-mediated endocytosis: a useful target for cancer therapy, J. Membr. Biol. 247 (2014) 291-307. https://doi.org/10.1007/s00232-014-9637-0

[121] Y. Liu, J. Sun, W. Cao, J. Yang, H. Lian, X. Li, Y. Sun, Y. Wang, S. Wang, Z. He, Dual targeting folate-conjugated hyaluronic acid polymeric micelles for paclitaxel delivery, Int. J. Pharm. 421 (2011) 160-169. https://doi.org/10.1016/j.ijpharm.2011.09.006

[122] S.K. Sriraman, G. Salzano, C. Sarisozen, V. Torchilin, Anti-cancer activity of doxorubicin-loaded liposomes co-modified with transferrin and folic acid, Eur. J. Pharm. Biopharm. 105(2016) 40-49. https://doi.org/10.1016/j.ejpb.2016.05.023

[123] D. Schmid, C.G. Park, C.A. Hartl, N. Subedi, A.N. Cartwright, R.B. Puerto, Y. Zheng, J. Maiarana, G.J. Freeman, K.W. Wucherpfennig, D.J. Irvine, M.S. Goldberg, T cell-targeting nanoparticles focus delivery of immunotherapy to improve antitumor immunity, Nat. Commun. 8 (2017) 1747-58. https://doi.org/10.1038/s41467-017-01830-8

[124] J.L. Gueant, F. Namour, R.M. Gueant-Rodriguez, J.L. Daval, Folate and fetal programming: a play in epigenomics? Trends Endocrinol Metab. 24 (2013) 279-289. https://doi.org/10.1016/j.tem.2013.01.010

[125] J.A. Ledermann, S. Canevari, T. Thigpen, Targeting the folate receptor: diagnostic and therapeutic approaches to personalize cancer treatments, Ann. Oncol. 26 (2015) 2034-2043. https://doi.org/10.1093/annonc/mdv250

[126] A. Cheung, H.J. Bax, D.H. Josephs, K.M. Ilieva, G. Pellizzari, J. Opzoomer, J. Bloomfield, M. Fittall, A. Grigoriadis, M. Figini, S. Canevari, J.F. Spicer, A.N. Tutt, S.N. Karagiannis, Targeting folate receptor alpha for cancer treatment, Oncotarget 7 (2016) 52553-52574. https://doi.org/10.18632/oncotarget.9651

[127] C. Muller, R. Schibli, Folic acid conjugates for nuclear imaging of folate receptor–positive cancer, J. Nucl. Med. 52 (2011) 1-4. https://doi.org/10.2967/jnumed.110.076018

[128] G.L. Zwicke, G.A. Mansoori, C.J. Jeffery, Utilizing the folate receptor for active targeting of cancer nanotherapeutics, Nano Rev. 3 (2012) 18496(1-11).

[129] Y.H. Ohana, T. Liron, S. Prutchi-Sagiv, M. Mittelman, M.C. Souroujon, D. Neumann, Erythropoietin, in: Abba Kastin (Ed.), Handbook of biologically active peptides, second Ed., Academic Press, Cambridge, Massachusetts, USA, 2013, pp. 1619-1626. https://doi.org/10.1016/B978-0-12-385095-9.00221-9

[130] F. Farrell, A. Lee, The erythropoietin receptor and its expression in tumor cells and other tissues, Oncologist 9 (2004) 18-30. https://doi.org/10.1634/theoncologist.9-90005-18

[131] S.N. Constantinescu, T. Keren, M. Sokolovsky, H.S. Nam, Y.I. Henis, H.F. Lodish, Ligand-independent oligomerization of cell-surface erythropoietin receptor is mediated by the transmembrane domain, Proc. Natl. Acad. Sci., USA 98(2001) 4379-4384. https://doi.org/10.1073/pnas.081069198

[132] M. Richter, H. Zhang, Receptor-targeted cancer therapy, DNA Cell Biol. 24 (2005) 271-282. https://doi.org/10.1089/dna.2005.24.271

[133] S. Hou, A. Zhang, M. Su, Nanomaterials for biosensing applications, Nanomaterials 6 (2016) 58(1-4).

[134] M. Yang, J. Wang, F. Zhou, Biomarker detections using functional noble metal nanoparticles, in: M. Hepel, C.J. Zhong (Eds.), Functional nanoparticles for bioanalysis, nanomedicine and bioelectronic devices, vol. 1, ACS, USA, 2012, pp. 177-205. https://doi.org/10.1021/bk-2012-1112.ch007

[135] M. Holzinger, A. Le Goff, S. Cosnier, Nanomaterials for biosensing applications: a review, Frontiers Chem. 2 (2014) 1-10. https://doi.org/10.3389/fchem.2014.00063

[136] W. Putzbach, N.J. Ronkainen, Immobilization techniques in the fabrication of nanomaterial-based electrochemical biosensors: a review,Sensors 13 (2013) 4811-4840. https://doi.org/10.3390/s130404811

[137] T.Ghodselahi, S.Arsalani, T.Neishaboorynejad, Synthesis and biosensor application of Ag_2Au bimetallic nanoparticles based on localized surface plasmon resonance, Appl. Surf. Sci. 301 (2014) 230-234. https://doi.org/10.1016/j.apsusc.2014.02.050

[138] J. Rick, M.C. Tsai, B.J. Hwang, Biosensors incorporating bimetallic nanoparticles, Nanomaterial (Basal) 6 (2016) 5(1-30).

[139] K.K. Jain, Applications of nanobiotechnology in clinical diagnostics, Clin. Chem. 53 (2007) 2002-2009. https://doi.org/10.1373/clinchem.2007.090795

[140] W. Zhao, J.M. Karp, M. Ferrari, R. Serda, Bioengineering nanotechnology: towards the clinic, Nanotechnology 22 (2011) 490201(1-2). https://doi.org/10.1088/0957-4484/22/49/490201

[141] Q. Zhang, N. Iwakuma, P. Sharma, B.M. Moudgil, C. Wu, J. McNeill, H. Jiang, S.R. Grobmyer, Gold nanoparticles as a contrast agent for in vivo tumor imaging with photoacoustic tomography, Nanotechnology 20 (2009) 395102(1-9).

[142] J. Ando, T. Aki Yano, K. Fujita, S. Kawata, Metal nanoparticles for nano-imaging and nano-analysis, Phys. Chem. Chem. Phys. 15 (2013) 13713-13722. https://doi.org/10.1039/c3cp51806j

[143] K. Fujita, S. Ishitobi, K. Hamada, N.I. Smith, A. Taguchi, Y. Inouye, S. Kawata, Time-resolved observation of surface enhanced Raman scattering from gold nanoparticles during transport through a living cell, J. Biomed. Opt. 14 (2009) 024038(1-7).

[144] J. Ando, K. Fujita, N.I. Smith, S. Kawata, Dynamic SERS imaging of cellular transport pathways with endocytosed gold nanoparticles. Nano Lett. 11 (2011) 5344-5348. https://doi.org/10.1021/nl202877r

[145] P. Damborsky, J. Svitel, J. Katrlik, Optical biosensors, Essay. Biochem. 60 (2016) 91-100. https://doi.org/10.1042/EBC20150010

[146] C.E. Talley, L. Jusinski, C.W. Hollars, S.M. Lane, T. Huser, Intracellular pH sensors based on surface enhanced Raman scattering, Anal. Chem. 76 (2004) 7064-7068. https://doi.org/10.1021/ac049093j

[147] J. Kneipp, H. Kneipp, B. Witting, K. Kneipp, One- and two-photon excited optical pH probing for cells using surface enhanced Raman and hyper Raman nanosensors, Nano Lett. 7 (2007) 2819-2823. https://doi.org/10.1021/nl071418z

[148] J. Kneipp, H. Kneipp, B. Witting, K. Kneipp, Following the dynamics of pH in endosomes of live cells with SERS nanosensors, J. Phys. Chem. C 114(2010) 7421-7426. https://doi.org/10.1021/jp910034z

[149] V. Biju, Chemical modifications and bioconjugate reactions of nanomaterials for sensing, imaging, drug delivery and therapy, Chem. Soc. Rev. 43 (2014) 744-764. https://doi.org/10.1039/C3CS60273G

[150] Y. Li, H. Schluesener, S. Xu, Gold nanoparticle-based biosensors, Gold Bull. 43(2010) 29-41. https://doi.org/10.1007/BF03214964

[151] K.L. Kelly, E. Coronado, L.L. Zhao, G.C. Schatz, The optical properties of metal nanoparticles:? The influence of size, shape, and dielectric environment, J. Phys. Chem. B 107(2002) 668-677. https://doi.org/10.1021/jp026731y

[152] J. Wilcoxon, Optical absorption properties of dispersed gold and silver alloy nanoparticles, J. Phys. Chem. B 113 (2009) 2647-2656. https://doi.org/10.1021/jp806930t

[153] P.K. Jain, M.A. El-Sayed, Universal scaling of plasmon coupling in metal nanostructures: extension from particle pairs to nanoshells, Nano Lett. 7 (2007) 2854-2858. https://doi.org/10.1021/nl071496m

[154] R. Elghanian, J.J. Storhoff, R.C. Mucic, R.L. Letsinger, C.A. Mirkin, Selective colorimetric detection of polynucleotides based on the distance-dependent optical properties of gold nanoparticles, Science 277 (1997) 1078-1081. https://doi.org/10.1126/science.277.5329.1078

[155] C.A. Mirkin, Programming the assembly of two- and three-dimensional architectures with DNA and nanoscale inorganic building blocks, Inorg. Chem. 39 (2000) 2258-2272. https://doi.org/10.1021/ic991123r

[156] S. Moeendarbari, A. Mulgaonkar, A.S. Hande, W. Silvers, C. Zhang,Y. Liu, A.K. Pillai, X. Sun,Y. Hao, Gold nanoparticles in current biomedical applications, Rev. Nanosci. Nanotechnol. 5 (2016) 28-78. https://doi.org/10.1166/rnn.2016.1070

Biosensors: Materials and Applications   Materials Research Forum LLC
Materials Research Foundations **47** (2019) 77-130   doi: http://dx.doi.org/10.21741/9781644900130-3

[157] K.M. Mayer, S. Lee, H. Liao, B.C. Rostro, A. Fuentes, P.T. Scully, C.L. Nehl, J.H. Hafner, A label-free immunoassay based upon localized surface plasmon resonance of gold nanorods, ACS Nano 2 (2008) 687-692. https://doi.org/10.1021/nn7003734

[158] K.M. Mayer, F. Hao, S. Lee, P. Nordlander, J.H. Hafner, A single molecule immunoassay by localized surface plasmon resonance, Nanotechnology 21 (2010) 255503 (1-8).

[159] S. Yang, T. Wu, X. Zhao, X. Li, W. Tan, The optical property of core-shell nanosensors and detection of atrazine based on localized surface plasmon resonance (LSPR) sensing, Sensors 14 (2014) 13273-13284. https://doi.org/10.3390/s140713273

[160] G. Doria, J. Conde, B. Veigas, L. Giestas, C. Almeida, M. Assuncao, J. Rosa, P.V. Baptista, Noble metal nanoparticles for biosensing applications, Sensors 12 (2012) 1657-1687. https://doi.org/10.3390/s120201657

[161] A. Ambrosi, F. Airo, A. Merkoçi, Enhanced gold nanoparticle based ELISA for a breast cancer biomarker, Anal. Chem. 82 (2010) 1151-1156. https://doi.org/10.1021/ac902492c

[162] K. Kneipp, A.S. Haka, H. Kneipp, K. Badizadegan, N. Yoshizawa, C. Boone, K.E. Shafer-Peltier, J.T. Motz, R.R. Dasari, M.S. Feld, Surface-enhanced Raman spectroscopy in single living cells using gold nanoparticles, Appl. Spectrosc. 56 (2002) 150-154. https://doi.org/10.1366/0003702021954557

[163] J.F. Hainfeld, D.N. Slatkin, T.M. Focella, H.M. Smilowitz, Gold nanoparticles: a new X-ray contrast agent, Br. J. Radiol. 79 (2006) 248-253. https://doi.org/10.1259/bjr/13169882

[164] V. Kattumuri, K. Katti, S. Bhaskaran, E.J. Boote, S.W. Casteel, G.M. Fent, D.J. Robertson, M. Chandrasekhar, R. Kannan, K.V. Katti, Gum arabic as a phytochemical construct for the stabilization of gold nanoparticles: in vivo pharmacokinetics and X-ray-contrast-imaging studies, Small 3 (2007) 333-341. https://doi.org/10.1002/smll.200600427

[165] D. Kim, S. Park, J.H. Lee, Y.Y. Jeong, S. Jon, Antibiofouling polymer-coated gold nanoparticles as a contrast agent for in vivo X-ray computed tomography imaging, J. Am. Chem. Soc. 129 (2007) 7661-7665. https://doi.org/10.1021/ja071471p

[166] D. Enders, S. Ruppa, A. Kullerc, A. Puccia, Surface enhanced infrared absorption on Aunanoparticle films deposited on SiO2/Si for optical biosensing: detection of the antibody-antigen reaction, Surf. Sci. 600 (2006) L305-L308. https://doi.org/10.1016/j.susc.2006.09.019

[167] Y. Li, H.J. Schluesener, S. Xu, Gold nanoparticle-based biosensors, Gold Bull. 43 (2010) 29-41. https://doi.org/10.1007/BF03214964

[168] L. He, M.D. Musick, S.R. Nicewarner, F.G. Salinas, S.J. Benkovic, M.J. Natan, C.D. Keating, J. Am. Chem. Soc. 122 (2000) 9071-9077. https://doi.org/10.1021/ja001215b

[169] T. Liu, J. Tang, L. Jiang, The enhancement effect of gold nanoparticles as a surface modifier on DNA sensor sensitivity, Biochem. Biophys. Res. Commun. 313 (2004) 3-7. https://doi.org/10.1016/j.bbrc.2003.11.098

[170] R.V. Devi, M. Doble, R.S. Verma, Nanomaterials for early detection of cancer biomarker with special emphasis on gold nanoparticles in immunoassays/sensors, Biosensors Bioeletron. 68 (2015) 688-698. https://doi.org/10.1016/j.bios.2015.01.066

[171] Y.E. Choi, J.W. Kwak, J.W. Park, Nanotechnology for early cancer detection, Sensors 10 (2010) 428-455. https://doi.org/10.3390/s100100428

[172] Y. Tauran, A. Brioude, A.W. Coleman, M. Rhimi, B. Kim, Molecular recognition by gold, silver and copper nanoparticles, World J. Biol. Chem. 4 (2013) 35. https://doi.org/10.4331/wjbc.v4.i3.35

[173] N. Nath, A. Chilkoti, Label-free biosensing by surface plasmon resonance of nanoparticles on glass: optimization of nanoparticle size, Anal. Chem. 76 (2004) 5370-5378. https://doi.org/10.1021/ac049741z

[174] T. Okamoto, I. Yamaguchi, T. Kobayashi, Local plasmon sensor with gold colloid monolayers deposited upon glass substrates, Opt. Lett. 25 (2000) 372-374. https://doi.org/10.1364/OL.25.000372

[175] W. Cai, T. Gao, H. Hong, J. Sun, Applications of gold nanoparticles in cancer nanotechnology, Nanotechnol. Sci. Appl. 1 (2008) 17-32. https://doi.org/10.2147/NSA.S3788

[176] E. Hutter, D. Maysinger, Gold nanoparticles and quantum dots for bioimaging, Microscopy Res. Tech. 74 (2011) 592-604. https://doi.org/10.1002/jemt.20928

[177] B. Tang, J.F. Wang, S.P. Xu, T. Afrin, W.Q. Xu, L. Sun, X.G. Wang, Application of anisotropic silver nanoparticles: multifunctionalization of wool fabric, J. Colloid. Interface Sci. 356 (2011) 513-518. https://doi.org/10.1016/j.jcis.2011.01.054

[178] S. Loher, O.D. Schneider, T. Maienfisch, S. Bokorny, W.J. Stark, Micro-organism triggered release of silver nanoparticles from biodegradable oxide carriers allows

preparation of self-sterilizing polymer surfaces, Small 4 (2008) 824-832. https://doi.org/10.1002/smll.200800047

[179] A. Kumar, P.K. Vemula, P.M. Ajayan, G. John, Silver-nanoparticle embedded antimicrobial paints based on vegetable oil, Nat. Mater. 7 (2008) 236-241. https://doi.org/10.1038/nmat2099

[180] P. Jain, T. Pradeep, Potential of silver nanoparticle-coated polyurethane foam as an antibacterial water filter, Biotechnol. Bioeng. 90 (2005) 59-63. https://doi.org/10.1002/bit.20368

[181] K.Y. Yoon, J.H. Byeon, C.W. Park, J. Hwang, Antimicrobial effect of silver particles on bacterial contamination of activated carbon fibers, Environ. Sci. Technol. 42 (2008) 1251-1255. https://doi.org/10.1021/es0720199

[182] G. Ghosh, L. Panicker, N.N. Kumar, V. Mallick, Surface plasmon resonance of counterions coated charged silver nanoparticles and application in bio-interaction, Mater. Res. Exp. 5 (2018) 055005(1-9).

[183] R. Geagea, P.H. Aubert, P. Banet, N. Sanson, Signal enhancement of electrochemical biosensors via direct electrochemical oxidation of silver nanoparticle labels coated with zwitterionic polymers, Chem. Comm. 51 (2015) 402-405. https://doi.org/10.1039/C4CC07474B

[184] D. Graham, K. Faulds, W.E. Smith, Biosensing using silver nanoparticles and surface enhanced resonance Raman scattering, Chem. Commun.42 (2006) 4363-4371. https://doi.org/10.1039/b607904k

[185] P. Sistani, L. Sofimaryo, Z.R. Masoudi, A. Sayad, R. Rahimzadeh, B. Salehi, A penicillin biosensor by using silver nanoparticles, Int. J. Electrochem. Sci. 9 (2014) 6201-6212.

[186] A.J. Haes, R.P. van Duyne, A nanoscale optical biosensor: sensitivity and selectivity of an approach based on the localized surface plasmon resonance spectroscopy of triangular silver nanoparticles, J. Am. Ceram. Soc. 124 (2002) 10596-10604.

[187] A.J. Haes, W.P. Hall, L. Chang, W.L. Klein, R.P. van Duyne, A localized surface plasmon resonance biosensor: First steps toward an assay for Alzheimer's disease, Nano Lett. 4 (2004) 1029-1034. https://doi.org/10.1021/nl049670j

[188] W.J. Galush, S.A. Shelby, M.J. Mulvihill, A. Tao, P.D. Yang, J.T. Groves, A nanocube plasmonic sensor for molecular binding on membrane surfaces, Nano Lett. 9 (2009) 2077-2082. https://doi.org/10.1021/nl900513k

Materials Research Forum LLC
doi: http://dx.doi.org/10.21741/9781644900130-3

[189] S.L. Zhu, F. Li, C.L. Du, Y.Q. Fu, A localized surface plasmon resonance nanosensor based on rhombic Ag nanoparticle array, Sens. Actuator B Chem. 134 (2008) 193-198. https://doi.org/10.1016/j.snb.2008.04.028

[190] W. Zhou, Y.Y. Ma, H.A. Yang, Y. Ding, X.G. Luo, A label-free biosensor based on silver nanoparticles array for clinical detection of serum p53 in head and neck squamous cell carcinoma, Int. J. Nanomed. 6 (2011) 381-386. https://doi.org/10.2147/IJN.S13249

[191] G.A. Sotiriou, T. Sannomiya, A. Teleki, F. Krumeich, J. Voros, S.E. Pratsinis, Non-toxic dry-coated nanosilver for plasmonic biosensors, Adv. Funct. Mater. 20 (2010) 4250-4257. https://doi.org/10.1002/adfm.201000985

[192] L. Chen, H. Xie, J. Li, Electrochemical glucose biosensor based on silver nanoparticles/multiwalled carbon nanotubes modified electrode, J. Solid state Electrochem. 16 (2012) 3323-3329. https://doi.org/10.1007/s10008-012-1773-9

[193] V. Kravets, Z. Almemar, K. Jiang, K. Culhane, R. Machado, G. Hagen, A. Kotko, I. Dmytruk, K. Spendier, A. Pinchuk, Imaging of biological cells using luminescent silver nanoparticles, Nanoscale Res. Lett. 11 (2016) 30(1-9).

[194] A.L.C.M. daSilva, M.G. Gutierres, A. Thesing, R.M. Lattuada, J. Ferreira, SPR biosensors based on gold and silver nanoparticle multilayer films, J. Br. Chem. Soc. 25 (2014) 928-934.

[195] G.A. Sotiriou, S.E. Pratsinis, Engineering nanosilver as an antibacterial, biosensor and bioimaging material, Curr. Opin. Chem. Eng. 1 (2011) 3-10. https://doi.org/10.1016/j.coche.2011.07.001

[196] B. Khalilzadeh, M. Hasanzadeh, S. Sanati, L. Saghatforoush, N. Shadjou, J.E.N. Dolatabadi, P. Sheikhzadeh, Preparation of a new electrochemical sensor based on cadmium oxide nanoparticles and application for determination of penicillamine, Int. J. Electrochem. Sci. 6 (2011) 4164-4175.

[197] J.E.N. Dolatabadi, M. de la Guardia, Applications of diatoms and silica nanotechnology in biosensing, drug and gene delivery and formation of complex metal nanostructures, Trends Anal. Chem. 30 (2011) 1538-1548. https://doi.org/10.1016/j.trac.2011.04.015

[198] J.E.N. Dolatabadi, O. Mashinchian, B. Ayoubi, A.A. Jamali, A. Mobed, D. Losic, Y. Omidi, M. de la Guardia, Optical and electrochemical DNA nanobiosensors, Trends Anal. Chem.30 (2011) 459-472. https://doi.org/10.1016/j.trac.2010.11.010

[199] A.A. Saei, P. Najafi-Marandi, A. Abhari, M. de la Guardia, J.E.N. Dolatabadi, Electrochemical biosensors for glucose based on metal nanoparticles, Trends Anal. Chem. 42 (2013) 216-227. https://doi.org/10.1016/j.trac.2012.09.011

[200] N.C. Bigall, T. Hartling, M. Klose, P. Simon, L.M. Eng, A. Eychmuller, Monodisperse platinum nanospheres with adjustable diameters from 10 to 100 nm: synthesis and distinct optical properties, Nano Lett.8(2008) 4588-4592. https://doi.org/10.1021/nl802901t

[201] E. Ramirez, L. Erades, K. Philippot, P. Lecante, B. Chaudret, Shape control of platinum nanoparticles, Adv. Func. Mater. 17 (2007) 2219-2228. https://doi.org/10.1002/adfm.200600633

[202] N.V. Long, N.D. Chien, T. Hayakawa, H. Hirata, G. Lakshminarayana, M. Nogami, The synthesis and characterization of platinum nanoparticles: a method of controlling the size and morphology, Nanotechnology 21 (2009) 035605 (1-16).

[203] D. Pedone, M. Moglianetti, E. de Luca, G. Bardi, P.P. Pompa, Platinum nanoparticles in medicine, Chem. Soc. Rev. 46 (2017) 4951-4975. https://doi.org/10.1039/C7CS00152E

[204] J.A. Creighton, D.G. Eadon, Ultraviolet–visible absorption spectra of the colloidal metallic elements, J. Chem. Soc. Faraday Trans. 87 (1991) 3881-3891. https://doi.org/10.1039/FT9918703881

[205] P. Chylekt, Light scattering by small particles in an absorbing medium, J. Opt. Soc. Am. 67 (1977) 561-563. https://doi.org/10.1364/JOSA.67.000561

[206] W.C. Mundy, J.A. Roux, A.M. Smith, Mie scattering by spheres in an absorbing medium, J. Opt. Soc. Am. 64 (1974) 1593-1597. https://doi.org/10.1364/JOSA.64.001593

[207] R.E. Benfield, A.P. Maydwell, J.M. van Ruitenbeek, D.A. van Leeuwen, Electronic spectra of metal cluster molecules, Z. Phzsik D 26 (1993) 4-7. https://doi.org/10.1007/BF01425600

[208] T. Yonezawa, Y. Gotoh, N. Toshima, Protecting structure model for nanoscopic platinum clusters protected by non-ionic surfactants: 13C nuclear magnetic resonance investigation, Reactive Polymer 23 (1994) 43-51. https://doi.org/10.1016/0923-1137(94)90001-9

[209] N. Toshima, K. Hirakawa, Polymer-protected Pt/Ru bimetallic cluster catalysts for visible-light-induced hydrogen generation from water and electron transfer dynamics, Appl. Surf. Sci. 121-122 (1997) 534-537. https://doi.org/10.1016/S0169-4332(97)00361-9

[210] A.L. Stepanov, A.N. Golubev, S.I. Nikitin, Y.N. Osin, A review on the fabrication and properties of platinum nanoparticles, Rev. Adv. Mater. Sci. 38 (2014) 160-175.

[211] Y. Zou, C. Xiang, L.X. Sun, F. Xu, Glucose biosensor based on electrodeposition of platinum nanoparticles onto carbon nanotubes and immobilizing enzyme with chitosan–SiO_2 sol–gel, Biosens. Bioelectron. 23 (2008) 1010-1016. https://doi.org/10.1016/j.bios.2007.10.009

[212] M. Yang, Y. Yang, Y. Liu, G. Shen, R. Yu, Platinum nanoparticles–doped sol–gel/carbon nanotubes composite electrochemical sensors and biosensors, Biosens. Bioelectron. 21 (2006) 1125-1131. https://doi.org/10.1016/j.bios.2005.04.009

[213] S. Hrapovic, Y. Liu, K.B. Male, J.H.T. Luong, Electrochemical biosensing platforms using platinum nanoparticles and carbon nanotubes, Anal. Chem. 76 (2004) 1083-1088. https://doi.org/10.1021/ac035143t

[214] G.M. Leteba, C.I. Lang, Synthesis of bimetallic platinum nanoparticles for biosensors, Sensors 13 (2013) 10358-10369. https://doi.org/10.3390/s130810358

[215] S.I. Tanaka, J. Miyazaki, D.K. Tiwari, T. Jin, Y. Inouye, Fluorescent platinum nanoclusters: synthesis, purification, characterization, and application to bioimaging, Angew. Chem. Int. Ed. 50(2011) 431-435. https://doi.org/10.1002/anie.201004907

[216] D. Chen, C. Zhao, J. Ye, Q. Li, X. Liu, M. Su, H. Jiang, C. Amatore, M. Selke, X. Wang, In situ biosynthesis of fluorescent platinum nanoclusters: toward self-bioimaging-guided cancer theranostics, ACS Appl. Mater. Interfaces 7 (2015) 18163-18169. https://doi.org/10.1021/acsami.5b05805

[217] J.W. Hu, J.F. Li, B. Ren, D.Y. Wu, S.G. Sun, Z.Q. Tian, Palladium-coated gold nanoparticles with a controlled shell thickness used as surface-enhanced Raman scattering substrate, J. Phys. Chem. C 111 (2007) 1105-1112. https://doi.org/10.1021/jp0652906

[218] J.M. McLellan, Y. Xiong, M. Hu, Y. Xia, Surface-enhanced Raman scattering of 4-mercaptopyridine on thin films of nanoscale Pd cubes, boxes, and cages, Chem. Phys. Lett. 417 (2006) 230-234. https://doi.org/10.1016/j.cplett.2005.10.028

[219] Y. Li, G. Lu, X. Wu, G. Shi, Electrochemical fabrication of two-dimensional palladium nanostructures as substrates for surface enhanced Raman scattering, J. Phys. Chem. B 110 (2006) 24585-24592. https://doi.org/10.1021/jp0638787

[220] H. Chen, G. Wei, A. Ispas, S.G. Hickey, A. Eychmuller, Synthesis of palladium nanoparticles and their applications for surface-enhanced Raman scattering and

electrocatalysis, J. Phys. Chem. C 114 (2010) 21976-21981.
https://doi.org/10.1021/jp106623y

[221] J. Cookson, The preparation of palladium nanoparticles, Platinum Metals Rev. 56(2012) 83-98. https://doi.org/10.1595/147106712X632415

[222] V.L. Nguyen, D.C. Nguyen, H. Hirata, M. Ohtaki, T. Hayakawa, M. Nogami, Chemical synthesis and characterization of palladium nanoparticles, Adv. Nat. Sci.: Nanosci. Nanotechnol. 1 (2010) 035012(1-5).

[223] Z. Li, X. Wang, G. Wen, S. Shuang, C. Dong, M.C. Paau, M.M.F. Choi, Application of hydrophobic palladium nanoparticles for the development of electrochemical glucose biosensor, Biosens. Bioelectron. 26 (2011) 4619-4623. https://doi.org/10.1016/j.bios.2011.04.057

[224] N. Cheng, H. Wang, X. Li, X. Yang, L. Zhu, Amperometric glucose biosensor based on integration of glucose oxidase with palladium nanoparticles/reduced graphene oxide nanocomposite, Am. J. Anal. Chem. 3 (2012) 312-319. https://doi.org/10.4236/ajac.2012.34043

[225] Z. Chang, H. Fan, K. Zhao, M. Chen, P. He, Y. Fang, Electrochemical DNA biosensors based on palladium nanoparticles combined with carbon nanotubes, Electroanalysis20(2008) 131-136.

[226] H. Heli, M. Hajjizadeh, A. Jabbari, A.A. Moosavi-Movahedi, Copper nanoparticles-modified carbon paste transducer as a biosensor for determination of acetylcholine, Biosens. Bioelectron. 24 (2009) 2328-2333. https://doi.org/10.1016/j.bios.2008.10.036

[227] J. Shen, L. Dudik, C.C. Liu, An iridium nanoparticles dispersed carbon based thick film electrochemical biosensor and its application for a single use, disposable glucose biosensor, Sensors Actuators B 125 (2007) 106-113. https://doi.org/10.1016/j.snb.2007.01.043

[228] C.P. da Silva, A.C. Franzoi, S.C. Fernandes, J. Dupont, I.C. Vieira, Development of biosensor for phenolic compounds containing PPO in β-cyclodextrin modified support and iridium nanoparticles, Enz. Microb. Technol. 52 (2013) 296-301. https://doi.org/10.1016/j.enzmictec.2012.12.001

[229] E. Magner, Trends in electrochemical biosensors, Analyst 123 (1998) 1967-1970. https://doi.org/10.1039/a803314e

[230] A.A. Saei, P.N. Marandi, A. Abhari, M. de la Guardia, J.E.N. Dolatabadi, Electrochemical biosensors for glucose based on metal nanoparticles, Trends Anal. Chem. 42 (2013) 216-227. https://doi.org/10.1016/j.trac.2012.09.011

[231] J.E.N. Dolatabadi, O. Mashinchian, B. Ayoubi, A.A. Jamali, A. Mobed, D. Losic, Y. Omidi, M de la Guardia, Optical and electrochemical DNA nanobiosensors, Trends Anal. Chem. 30 (2011) 459-472. https://doi.org/10.1016/j.trac.2010.11.010

[232] K. Greish, Enhanced permeability and retention of macromolecular drugs in solid tumors: a royal gate for targeted anticancer nanomedicines, J. Drug Target. 15 (2007) 457-464. https://doi.org/10.1080/10611860701539584

[233] M.E. Davis, Z. Chen, D.M. Shin, Nanoparticle therapeutics: an emerging treatment modality for cancer, Nat. Rev. Drug Discov. 7 (2008) 771-782. https://doi.org/10.1038/nrd2614

[234] R. Weissleder, Molecular imaging in cancer, Science 312 (2006) 1168-1171. https://doi.org/10.1126/science.1125949

[235] S. Lee, X.Y. Chen, Dual-modality probes for in vivo molecular imaging, Mol. Imaging 8 (2009) 87-100. https://doi.org/10.2310/7290.2009.00013

[236] S. Gioux, H.S. Choi, J.V. Frangioni, Image-guided surgery using invisible near-infrared light: fundamentals of clinical translation, Mol. Imaging 9 (2010) 237-255. https://doi.org/10.2310/7290.2010.00034

[237] J. Merian, J. Gravier, F. Navarro, I. Texier, Fluorescent nanoprobes dedicated to in vivo imaging: from preclinical validations to clinical translation. Molecules 17 (2012) 5564-5591. https://doi.org/10.3390/molecules17055564

[238] Z.Altintas, I.E.Tothills, Molecular biosensors: promising new tools for early detection of cancer, Nanobiosens. Disease Diagnos. 4 (2015) 1-10.

[239] S. Kunzelmann, C. Solscheid, M.R. Webb, Fluorescent biosensors: design and application to motor proteins, in: C.P. Toseland, N. Fili (Eds.), Fluorescent methods for molecular motors, Springer, Berlin, Germany, 2014, pp.25-47. https://doi.org/10.1007/978-3-0348-0856-9_2

[240] O. Tagit, N. Hildebrandt, Fluorescence sensing of circulating diagnostic biomarkers using molecular probes and nanoparticles, ACS Sens. 2 (2017) 31-45. https://doi.org/10.1021/acssensors.6b00625

[241] V. Helms, Fluorescence resonance energy transfer, in:V. Helms (Ed.), Principles of computational cell biology, Weinheim: Wiley-VCH, 2008, p. 202.

[242] D.C. Harris, Applications of spectrophotometry, in: Quantitative chemical analysis, eighthed, W.H. Freeman and Co., New York, 2010,pp. 419-444.

[243] C. Joo, H. Balci, Y. Ishitsuka, C. Buranachai, T. Ha, Advancesinsingle-molecule fluorescence methods for molecular biology, Ann. Rev. Biochem. 77 (2008) 51-76. https://doi.org/10.1146/annurev.biochem.77.070606.101543

[244] Y. Li, C.Y. Zhang, Analysis of microRNA-induced silencing complex-involved microRNA-target recognition by single-molecule fluorescence resonance energy transfer, Anal. Chem. 84 (2012) 5097-5102. https://doi.org/10.1021/ac300839d

[245] A. Periasamy, Fluorescence resonance energy transfer microscopy: a mini review, J. Biomed. Opt. 6 (2001) 287-291. https://doi.org/10.1117/1.1383063

[246] A. Shahzad, M. Knapp, I. Lang, G. Kohler, The use of fluorescence correlation spectroscopy (FCS) as an alternative biomarker detection technique: a preliminary study, J. Cell. Mol. Med.15 (2011) 2706-2711. https://doi.org/10.1111/j.1582-4934.2011.01272.x

[247] A.D. Kurdekar, L.A.A. Chunduri, S.M. Chelli, M.K. Haleyurgirisetty, E.P. Bulagonda, J. Zheng, I.K. Hewlett, V. Kamisetti, Fluorescent silver nanoparticle based highly sensitive immunoassay for early detection of HIV infection, RSC Adv. 7 (2017) 19863-19877. https://doi.org/10.1039/C6RA28737A

[248] M.M. Billingsley, R.S. Riley, E.S. Day, Antibody-nanoparticle conjugates to enhance the sensitivity of ELISA-based detection methods, PLoS ONE 12 (2017) e0177592 (1-15).

[249] P. Ciaurriz, F. Fernandez, E. Tellechea, J.F. Moran, A.C. Asensio, Comparison of four functionalization methods of gold nanoparticles for enhancing the enzyme-linked immunosorbent assay (ELISA), Beilstein J. Nanotechnol. 8 (2017) 244-253. https://doi.org/10.3762/bjnano.8.27

[250] F. Zhou, L. Yuan, H. Wang, D. Li, H. Chen, Gold nanoparticle layer: a promising platform for ultra-sensitive cancer detection, Langmuir, 27 (2011) 2155-2158. https://doi.org/10.1021/la1049937

[251] F. Zhou, M. Wang, L. Yuan, Z. Cheng, Z. Wu, H. Chen, Sensitive sandwich ELISA based on a gold nanoparticle layer for cancer detection, Analyst. 137(2012) 1779-1784. https://doi.org/10.1039/c2an16257a

[252] X. Xu, H. Li, D. Hasan, R.S. Ruoff, A.X. Wang, D.L. Fan, Near-field enhanced plasmonic-magnetic bifunctional nanotubes for single cell bioanalysis, Adv. Funct. Mater. 23 (2013) 4332-4338. https://doi.org/10.1002/adfm.201203822

[253] S. Nie, S.R. Emory, Probing single molecules and single nanoparticles by surface-enhanced Raman scattering, Science 275 (1997) 1102-1106. https://doi.org/10.1126/science.275.5303.1102

[254] R. Le, C. Eric,M. Meyer, P.G. Etchegoin, Proof of single-molecule sensitivity in surface enhanced Raman scattering (SERS) by means of a two-analyte technique, J. Phys. Chem. B 110 (2006) 1944-1948. https://doi.org/10.1021/jp054732v

Biosensors: Materials and Applications   Materials Research Forum LLC
Materials Research Foundations **47** (2019) 77-130   doi: http://dx.doi.org/10.21741/9781644900130-3

[255] S. Laing, L.E. Jamieson, K. Faulds, D. Graham, Surface-enhanced Raman spectroscopy for in vivo biosensing, Nat. Rev. Chem. 1 (2017) 0060.

[256] E. Wijaya, C. Lenaerts, S. Maricot, J. Hastanin, S. Habraken, J.P. Vilcot, R. Boukherroub, S. Szunerits, Surface plasmon resonance-based biosensors: from the development of different SPR structures to novel surface functionalization strategies, Curr. Opin. Sol. Stat. Mat. Sci. 15 (2011) 208-224. https://doi.org/10.1016/j.cossms.2011.05.001

[257] U. Prabhakar, H. Maeda, R.K. Jain, E.M. Sevick-Muraca, W. Zamboni, O.C. Farokhzad, S.T. Barry, A. Gabizon, P. Grodzinski, D.C. Blakey, Challenges and key considerations of the enhanced permeability and retention effect for nanomedicine drug delivery in oncology, Cancer res. 73 (2013) 2412-2417. https://doi.org/10.1158/0008-5472.CAN-12-4561

[258] A.F. Frellsen, A.E. Hansen, R.I. Jølck, P.J. Kempen, G.W. Severin, P.H. Rasmussen, A.Kjær, A.T.I. Jensen, T.L. Andresen, Mouse positron emission tomography study of the biodistribution of gold nanoparticles with different surface coatings using embedded copper-64, ACS Nano 10 (2016) 9887-9898. https://doi.org/10.1021/acsnano.6b03144

[259] S.B. Lee, G. Yoon, S.W. Lee, S.Y. Jeong, B.C. Ahn, D.K. Lim, J. Lee, Y.H. Jeon, Combined positron emission tomography and cerenkov luminescence imaging of sentinel lymph nodes using PEGylated radionuclide-embedded gold nanoparticles, Small 12 (2016) 4894-4901. https://doi.org/10.1002/smll.201601721

[260] M.M. Mahan, A.L. Doiron, Gold nanoparticles as X-ray, CT and multimodal imaging contrast agents: formulation, targeting and methodology, J. Nanomater. (2018) 5837276 (1-15). https://doi.org/10.1155/2018/5837276

[261] L. Yang, D.J. Watts, Particle surface characteristics may play an important role in phytotoxicity of alumina nanoparticles, Toxicol. Lett. 158(2005) 122-132. https://doi.org/10.1016/j.toxlet.2005.03.003

[262] K. Donaldson, D. Brown, A. Clouter, R. Duffin, W. MacNee, L. Renwick, L. Tran, V. Stone, The pulmonary toxicology of ultrafine particles, J.Aerosol Med.15(2002) 213-220. https://doi.org/10.1089/089426802320282338

[263] H. Bahadar, F. Maqbool, K. Niaz, M. Abdollahi, Toxicity of nanoparticles and an overview of current experimental models, Iran. Biomed. J. 20(2016) 1-11.

[264] E.E. Connor, J. Mwamuka, A. Gole, C.J. Murphy, M.D. Wyatt, Gold nanoparticles are taken up by human cells but do not cause acute cytotoxicity, Small 1(2005) 325-327. https://doi.org/10.1002/smll.200400093

[265] E. Boisselier, D. Astruc, Gold nanoparticles in nanomedicine: preparations, imaging, diagnostics, therapies and toxicity, Chem. Soc. Rev. 38(2009) 1759-1782. https://doi.org/10.1039/b806051g

[266] C.M. Goodman, C.D. McCusker, T. Yilmaz, V.M. Rotello, Toxicity of gold nanoparticles functionalized with cationic and anionic side chains, Bioconjug. Chem. 15(2004) 897-900. https://doi.org/10.1021/bc049951i

[267] H.K. Patra, S. Banerjee, U. Chaudhuri, P. Lahiri, A.K. Dasgupta, Cell selective response to gold nanoparticles, Nanomedicine 3(2007) 111-119. https://doi.org/10.1016/j.nano.2007.03.005

[268] J. Conde, M. Larguinho, A. Cordeiro, L.R. Raposo, P.M. Costa, S. Santos, M.S. Diniz, A.R. Fernandes, P.V. Baptista, Gold-nano beacons for gene therapy: evaluation of genotoxicity, cell toxicity and proteome profiling analysis, Nanotoxicology 8(2014) 521-532. https://doi.org/10.3109/17435390.2013.802821

[269] Y. Pan, A. Leifert, D. Ruau, S. Neuss, J. Bornemann, G. Schmid, W. Brandau, U. Simon, W. Jahnen-Dechent, Gold nanoparticles of diameter 1.4 nm trigger necrosis by oxidative stress and mitochondrial damage, Small 5(2009) 2067-2076. https://doi.org/10.1002/smll.200900466

[270] K.T. Kim, T. Zaikova, J.E. Hutchison, R.L. Tanguay, Gold nanoparticles disrupt zebrafish eye development and pigmentation, Toxicol. Sci. 133(2013) 275-288. https://doi.org/10.1093/toxsci/kft081

[271] A. Gerber, M. Bundschuh, D Klingelhofer, D.A. Groneberg, Gold nanoparticles: recent aspects for human toxicology, J. Occup. Med. Toxicol. 8(2013) 32(1-6). https://doi.org/10.1186/1745-6673-8-32

[272] Y.P. Jia, B.-Y. Ma, X.W. Wei, Z.-Y. Qian, The in vitro and in vivo toxicity of gold nanoparticles, Chinese Chem. Lett. 28 (2017) 691-702. https://doi.org/10.1016/j.cclet.2017.01.021

[273] X. Chen, H.J. Schluesener, Nanosilver: a nanoproduct in medical application, Toxicol. Lett. 176(2008) 1-12. https://doi.org/10.1016/j.toxlet.2007.10.004

[274] S.M. Hussain, K.L. Hess, J.M. Gearhart, K.T. Geiss, J.J. Schlager, In vitro toxicity of nanoparticles in BRL 3A rat liver cells, Toxicol. In Vitro 19(2005) 975-983. https://doi.org/10.1016/j.tiv.2005.06.034

[275] R. Foldbjerg, D.A. Dang, H. Autrup, Cytotoxicity and genotoxicity of silver nanoparticles in the human lung cancer cell line, A549, Arch. Toxicol. 85(2011) 743-750. https://doi.org/10.1007/s00204-010-0545-5

[276] A. Haase, J. Tentschert, H. Jungnickel, P. Graf, A. Mantion, F. Draude, J. Plendl, M.E. Goetz, S. Galla, A. Masic, Toxicity of silver nanoparticles in human macrophages: uptake, intracellular distribution and cellular responses, J. Phys. Conf. Ser. 304 (2011) 012030 (1-15). https://doi.org/10.1088/1742-6596/304/1/012030

[277] J. Tang, L. Xiong, S. Wang, J. Wang, L. Liu, J. Li, F. Yuan, T. Xi, Distribution, translocation and accumulation of silver nanoparticles in rats, J. Nanosci. Nanotechnol. 9 (2009) 4924-4932. https://doi.org/10.1166/jnn.2009.1269

[278] X. Feng, A. Chen, Y. Zhang, J. Wang, L. Shao, L. Wei, Central nervous system toxicity of metallic nanoparticles, Int. J. Nanomed. 10 (2015) 4321-4340.

[279] F.J. Rang, J. Boonstra, Causes and consequences of age-related changes in DNA methylation: arole for ROS? Biology (Basel) 3 (2014) 403-425. https://doi.org/10.3390/biology3020403

[280] K. Brieger, S. Schiavone, F.J. Miller Jr., K.H. Krause, Reactive oxygen species: from health to disease, Swiss Med. Wkly. 142 (2012) w13659 (1–14).

[281] A. Nel, T. Xia, L. Madler, N. Li, Toxic potential of materials at the nanolevel, Science 311 (2006) 622-627. https://doi.org/10.1126/science.1114397

[282] C. Hanley, A. Thurber, C. Hanna, A. Punnoose, J. Zhang, D.G. Wingett, The influences of cell type and ZnO nanoparticle size on immune cell cytotoxicity and cytokine induction, Nanoscale Res. Lett. 4 (2009) 1409-1420. https://doi.org/10.1007/s11671-009-9413-8

[283] T.C. Long, N. Saleh, R.D. Tilton, G.V. Lowry, B. Veronesi, Titanium dioxide (P25) produces reactive oxygen species in immortalized brain microglia (BV2): implications for nanoparticle neurotoxicity, Environ. Sci. Technol. 40 (2006) 4346-4352. https://doi.org/10.1021/es060589n

[284] V. Freyre-Fonseca, N.L. Delgado-Buenrostro, E.B. Gutiérrez-Cirlos, C.M. Calderón-Torres, T. Cabellos-Avelar, Y. Sanchez-Pérez, E. Pinzón, I. Torres, E. Molina-Jijon, C. Zazueta, J. Pedraza-Chaverri, C.M. Garcia-Cuellar, Y.I. Chirino, Titanium dioxide nanoparticles impair lung mitochondrial function, Toxicol. Lett. 202 (2011) 111-119. https://doi.org/10.1016/j.toxlet.2011.01.025

[285] E. Huerta-Garcia, J.A. Perez-Arizti, S.G. Marquez-Ramirez, N.L. Delgado-Buenrostro, Y.I. Chirino, G.G. Iglesias, R. Lopez-Marure, Titanium dioxide nanoparticles induce strong oxidative stress and mitochondrial damage in glial cells, Free Radic. Biol. Med. 73(2014) 84-94. https://doi.org/10.1016/j.freeradbiomed.2014.04.026

[286] N.B. Golovina, L.M. Kustov, Toxicity of metal nanoparticles with a focus on silver, Mendeleev Commun. 23 (2013) 59-65. https://doi.org/10.1016/j.mencom.2013.03.001

[287] A.M. Schrand, M.F. Rahman, S.M. Hussain, J.J. Schlager, D.A. Smith, A.F. Syed, Metal-based nanoparticles and their toxicity assessment, Wiley Interdiscip. Rev. Nanomed. Nanobiotechnol. 2 (2010) 544-568. https://doi.org/10.1002/wnan.103

[288] R. Singla, A. Guliani, A. Kumari, S.K. Yadav, Metallic nanoparticles, toxicity issues and applications in medicine,in: S. Yadav (Ed.) Nanoscale materials in targeted drug delivery, theragnosis and tissue regeneration, Springer, Singapore, 2016, pp. 41-80. https://doi.org/10.1007/978-981-10-0818-4_3

Biosensors: Materials and Applications
Materials Research Foundations **47** (2019) 131-156

Materials Research Forum LLC
doi: http://dx.doi.org/10.21741/9781644900130-4

Chapter 4

Layered Double Hydroxide Based Biosensors

Nadeem Baig·*, Azeem Rana

Chemistry Department, King Fahd University of Petroleum and Minerals, Dhahran 31261, Saudi Arabia

*nadeembaig@kfupm.edu.sa; nadeembaig38@gmail.com

Abstract

Layered double hydroxide is a unique layered material with numerous fascinating characteristics. The tunable composition, interlayer spaces, uniform distribution of cations, exchangeable anions and swelling characteristics make them extraordinary material. The desired LDH can be prepared through many routes with low-cost precursors. For biosensor fabrication, the LDH is also getting great attention due to its active sites, biocompatible environment, facile loading of the catalyst and enzyme stability. Moreover, the exchangeable anions also provide room for the entrance of other charged biomolecules Layered double hydroxide improved the biosensor sensitivity by pre-concentrating the analyte into the layers of the LDH. This book chapter focused on the properties of LDH, fabrication of LDH based biosensors and their electroanalytical applications for sensing of various analytes.

Keywords

Chemically Modified Electrodes, Clay, Biosensor Fabrication, Electrochemical Sensor, Biomolecule

Contents

1.   Introduction

Electrochemical methods are considered attractive and competitive with other methods due to their simplicity, cost-effectiveness, easy handling and simple instrumentation [1,2]. Many modern materials day by day are being introduced to improve the sensitivity and the selectivity of electrochemical sensors. These materials are not limited to carbon nanotubes [3], carbon nanofibers [4], graphene/reduced graphene oxide [5,6], conductive polymers [7], MOF [8], metals and metal oxide nanoparticles [9].

Layered double hydroxide is among the most studied advance material and it is extensively explored as a catalyst, fire retardant additive, cement additive, CO_2 adsorbent, drug delivery host, ion exchange, in the field of energy and material chemistry [10]. Layered double hydroxide is a unique material which has brucite like layers in which some of the divalent cations are replaced by the trivalent cations to produce a layer of positive charge [11]. These layers of positive charge are balanced by the negatively charged anions. These negatively charged anions are present in between the positively charged layers of the layered double hydroxide. The general formula of the layered

Biosensors: Materials and Applications
Materials Research Foundations **47** (2019) 131-156

Materials Research Forum LLC
doi: http://dx.doi.org/10.21741/9781644900130-4

double hydroxide can be demonstrated as $[M^{2+}_{1-x}M^{3+}_x (OH)_2]^{x+} (A^{n-})_{x/n} \cdot mH_2O$, where M^{+2} is divalent and M^{+3} is the trivalent metal cations, the A^{n-} is the interlayers anions [12]. The value X may varied between 0.22 to 0.33 [13]. The possible M^{2+} cations can be Co^{2+}, Cu^{2+}, Mg^{2+}, Zn^{2+}, Mn^{2+}, Ga^{2+}, Ni^{2+}, and M^{3+} can be Fe^{3+}, Al^{3+}, Cr^{3+}, Mn^{+3} [14]. The charge density of the layered double hydroxide is influenced by the ratio of the M^{2+}/M^{3+} cations. Despite M^{2+} and M^{3+} in certain cases, M^{+1} and M^{+4} are also present such as Li^+ and Ti^{4+} [15]. The possible interlayer anions (A^{n-}) can be NO_3^-, Cl^-, OH^-, SO_4^{-2}, CO_3^{2-}, and other organic anions [14]. Figure 1 displaying the presence of anions and water molecules in between the layers of LDH. Layered double hydroxide is a unique material due to their exchangeable anions in between the layers and LDH has a great attraction in the intercalation chemistry. The desired physical and chemical properties can be created in the layered double hydroxide by tuning the cations in the layers and the anions in the interlayers of the layered double hydroxide. This behavior of LDH facilitates to develop a nanoarchitectures and host-guest assemblies with certain characteristics [16]. Layered double hydroxide is the anionic clay. The delamination of the layered double hydroxide is difficult compared to the cationic clay. This behavior of the LDH is due to the high charge density and high anion presence which result into the strong electrostatic interaction between the sheets. However, the catalytic activity of the material also depends upon the synthetic route. The layered double hydroxide material due to its high importance was synthesized by numerous routes [17] including hydrothermal, sol-gel method [18], urea hydrolysis [19], exchange methods [20], and co-precipitation method [21]. The most commonly used methods are ion exchange and the co-precipitation methods due to their simplicity. The charged biomolecules can also enter the LDH through the ion exchange method. The hydrothermal method is generally adopted where the co-precipitation or ion exchange method is not feasible especially when the organic species of low affinity required to intercalate into LDH [13].

Several studies in the literature are evident that layered double hydroxide is a fascinating material for electrode modification along with other advanced materials [22–24]. During electroanalysis, the pre-concentration step is playing a crucial role in trace level quantification of the analyte. The inherent characteristic of the layered double hydroxide allows the modified surface to confine the analyte near to the electrode surface in a small volume. This behavior of the layered double hydroxide has a substantial effect on the sensitivity and the selectivity of the electrodes. It enables the modified electrodes for trace level quantification [25]. The layered double hydroxide effectively facilitates the fabrication of the biosensor by providing the facile immobilization of the biomolecule recognition element into the electrode modified surface.

Biosensors: Materials and Applications   Materials Research Forum LLC
Materials Research Foundations **47** (2019) 131-156   doi: http://dx.doi.org/10.21741/9781644900130-4

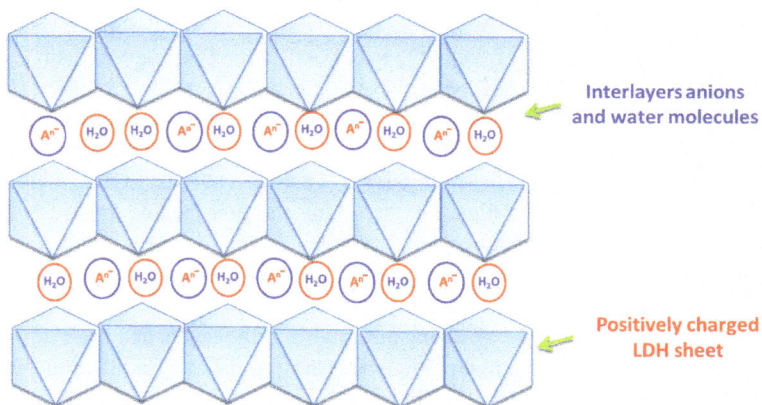

Figure 1 Layered double hydroxide structure containing anions and water molecules.

2.   Prominent and unique features of layered double hydroxide modified electrodes

Layered double hydroxide has some fascinating features which make them attractive contenders for electrode modification. The precursors of layered double hydroxide are cheap and it can be prepared through many synthetic routes [26]. Positively charged layers of the LDH provided a better interaction with the polar and the negatively charged biomolecules due to electrostatic force of attraction [25]. The LDH provides a better opportunity to load the catalyst. The layered double hydroxide has relatively weak interlayer bonding which imparts the expansion characteristic to the LDH [27]. Moreover, it provides a biocompatible environment for the trapped biomolecules due to its high water content. These factors provided the feasible environment for the loaded enzyme to retain its activity for a longer time. The porous and the layered nature of the LDH assist the fabricated electrochemical sensor to accumulate the analyte near to the electrode surface. This behavior of the LDH contributes significantly to improve the sensitivity and selectivity of the fabricated sensors [13].

Biosensors: Materials and Applications   Materials Research Forum LLC
Materials Research Foundations **47** (2019) 131-156   doi: http://dx.doi.org/10.21741/9781644900130-4

3.   Layered double hydroxide based biosensors

Development of biosensors for the assay of target analytes have been extensively studied for real-life applications [28]. The electrochemical biosensors based on enzymatic transduction is one of the major categories because of cost-effectiveness, simplicity, and compactness [29]. A biosensor consists of two main components.

a) Bio-recognition unit

b) Transducer

A biosensor is constructed for more selective and sensitive analysis. A biosensor responds to the only specific analyte and less affected by the potential interferences. The bio-recognition is any biomolecule which may be a nucleic acid, protein or carbohydrate which can recognize the targeted analyte from the complex matrix. The targeted analyte attached to the bio-recognition unit. At certain potential, the electrochemical reaction takes place. The transducer converts this electrochemical reaction into the electrical signals. The bio-recognition unit of the biosensor imparts the high selectivity to the biosensor.

One of the major challenges for the fabrication of the biosensors is the effective immobilization of the biomolecule within a suitable matrix in order to prevent its leaching, thus ensuring maintenance of the biological activity, stability under storage and accessibility of the molecular target to the active site [30]. There are various inorganic structures available for immobilization of biomolecules, however, two dimensional layered solid such as layered double hydroxides (LDHs) are one of the best choices for this purpose. They have unique layer and interlayer structure with tunable textural morphology [31].

For the preparation of active bioinorganic LDH based biosensors, various soft methodologies for synthesis can be applied which did not affect the bioactivity of the intercalated molecules. Such as adsorption, delamination, covalent grafting, cross-linking, anion exchange reaction, direct co-precipitation, and electrodeposition [31,32]. The preparation of LDH from precursors is usually done by co-precipitation in basic solution, ion-exchange method, surface synthesis, metal precursor hydrolysis, template-based synthesis, separate nucleation, or by thermal decomposition of a retardant base [33–35]. Great care should be taken to get the LDH of specific composition, particle size and surface area [36]. Because of the fact that little variation in the LDH may cause prominent effect on its characteristics and its interaction with a biomolecule to form a bio-hybrid assembly [32,37]. The possible forces that operate in the LDH hybrid compounds are van der Waals forces, hydrogen bonding or ionic bonds. These forces are

Biosensors: Materials and Applications   Materials Research Forum LLC
Materials Research Foundations 47 (2019) 131-156   doi: http://dx.doi.org/10.21741/9781644900130-4

responsible to bound the embedded organic and inorganic compounds into LDH for the development of biosensors [38,39].

As aforementioned, the biosensors are the small tailored devices which can selectively recognize the analyte due to their biological part [40,41]. In the biosensor, the analyte reacts with the bio-recognition component either by inhibiting the enzyme activity or distorting its conformation [42]. Enzyme coupled biosensor display a catalytic activity by selectively recognizing the analyte which can cause an electron transfer [43]. The intercalation behavior of the layered double hydroxide provided the opportunity to trap different biomolecules into the layers of the LDH for the construction of LDH based biosensors. Furthermore, the swelling, hydrophilic, and porous behavior of the LDH provides a stable and biocompatible environment for the immobilization of the enzyme. The negatively charged biomolecules is immobilized easily due to the anion exchange behavior of the material. All these characteristics of the material make them an ideal choice for the construction of the biosensors [13].

4.   Fabrication of LDH based biosensors

The LDH based bio-hybrid modified electrodes are prepared by thin film coating on the electrode surface for the electroanalytical application [44,45]. There are various methods described in the literature for the preparation of electrochemical LDH based biosensors such as solvent casting, layer by layer assembly and electrodeposition. These methods are discussed briefly in the following section (Fig. 2).

4.1   Solvent casting

For the fabrication of electrochemical sensors, the most effective and frequently applied method is the solvent casting method. For the preparation of LDH based biosensor, the colloidal suspension of the LDH with biomolecule is generally prepared first. Afterward, the drop of the colloidal suspension containing LDH and biomolecule is casted on the surface of the electrode and allowed to dry. The drying step is generally completed at low temperature for approximately 12 hours to avoid the severe drying of the modifying materials [46–49]. The enzyme stability can be improved by cross-linking with glutaraldehyde [50–52]. In case of LDH/composite films, the LDH can be prepared by simple mixing with different components. Nanocomposite polymeric membrane, graphene, nafion, ionic liquid or gold [53] nanoparticles are employed to improve the electrical conductivity and electrocatalytic activity of the sensor [54].

Materials Research Forum LLC
doi: http://dx.doi.org/10.21741/9781644900130-4

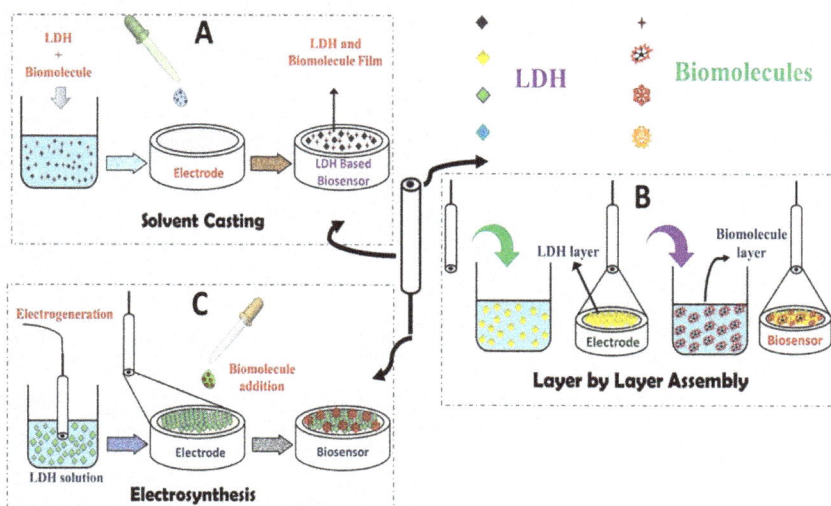

Figure 2 Fabrication methods of LDH based biosensors: (A) solvent casting method, (B) Layer by layer assembly and (C) electrosynthesis.

4.2   Layer by layer assembly

The layer by layer self-assembly is an efficient process in order to get heterogeneous LDH thin film on the electrode surface by alternate stacking of exfoliated LDH and polyanions [55]. To get an LbL assembly of the of LDH and enzyme is done by alternate immersion of the electrode into different solutions or dispersion of LDH and enzyme [56,57].

4.3   Electrogeneration (electrosynthesis)

The simplest strategy to get an LDH thin film on the electrode surface is by electrodeposition. The electrodeposition can be achieved by applying certain or sweeping potentials. In this process, electrochemical reduction of nitrate produces hydroxyl groups on the surface of the electrode that in turn favors the attachment of LDH layer on the electrode surface [50]. In the electrogeneration process, small ill-crystallized LDH particles homogenously deposit on the electrode surface that acts as a connector for the

Biosensors: Materials and Applications   Materials Research Forum LLC
Materials Research Foundations **47** (2019) 131-156   doi: http://dx.doi.org/10.21741/9781644900130-4

immobilization of protein. In some cases, the sacrificial polymeric template is used to generate the active site for the enzyme attachment. The inorganic framework obtained by the template-assisted method have higher porosity and resultant biosensor displays greater sensitivity towards target analyte [58].

4.4   Carbon paste electrode

Carbon paste electrode is extensively used in electroanalytical chemistry [53] due to its simplicity, facile fabrication and variety of composition can be used. Graphite is commonly used in the paste electrode due to its low background current and wide potential window. LDH based carbon-based biosensors were also explored for sensing of various analytes. The LDH based and carbon-based biosensor can be developed by mixing the LDH bio-hybrid material with graphite powder and paraffin oil as a binder in a mortar (Fig. 3). Then the prepared carbon paste material is packed into the cavity of electrode holder and applied for the biosensing application [59–61].

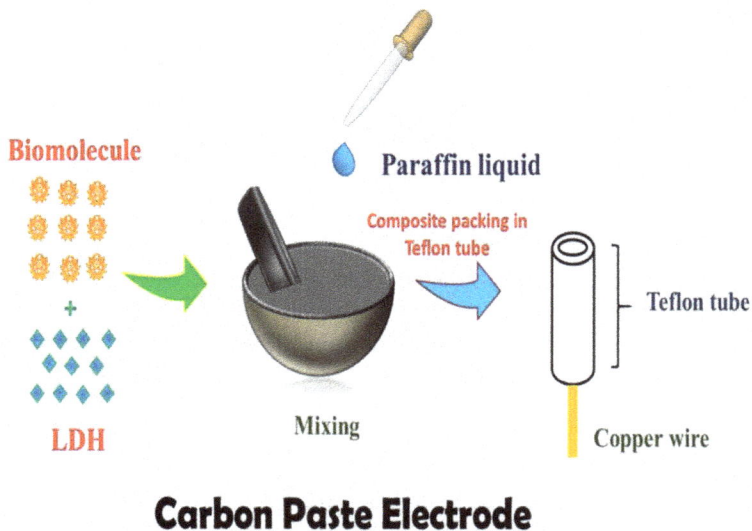

Figure 3 Schematic representation of fabrication of LDH based carbon paste biosensors.

Biosensors: Materials and Applications   Materials Research Forum LLC
Materials Research Foundations **47** (2019) 131-156   doi: http://dx.doi.org/10.21741/9781644900130-4

5. Electroanalytical applications of LDH based biosensors

Biosensors are vastly applied to various fields including biomedical, environmental and food applications. Biosensors must be highly specific for target analytes, reclaimable and its response should not be affected by the physical factors such as temperature and pH [62]. A sensor mainly consists of two components i.e. receptor and transducers. The receptor interacts with physical or chemical stimulant and transmits the gathered data in the pattern of the electrical signal to the transducer. The transducer boosts the signal to get useful analytical information which can be presented in electronic form [63,64]. In case of the biosensor, the transducer is attached to the biological material that acts as receptors such as enzymes, peptides, oligonucleotides, antibodies, etc. [65]. The LDH interact with the biological material and form a biohybrid biosensor. These LDH based biosensors response to the specific analyte due to their extraordinary selectivity [66].

5.1 Glucose oxidase based biosensor

Enzyme/LDH biosensors are the biohybrids that are obtained by the immobilization of glucose oxidase in the host structure of the LDH and target various sort of analytes by using amperometric methods. Colombari and coworkers utilized ferrocene derivatives (ferrocene-carboxylate (Fc-COOH) and ferrocene sulfonic acid (Fc-SO$_3$H)) intercalated into Mg/Al LDH for immobilizing the glucose oxidase (GOx) by drop casting method. The biosensor based on Fc-COOH displayed higher sensitivity towards glucose oxidation under optimized conditions. GOx/Fc-COOH-LDH modified glassy carbon electrode (GCE) displayed satisfactory results under flow injection analysis (FIA) conditions that made this biosensor suitable for real-life application [67]. In another work, the same research group modified the surface of the platinum electrode with Ni, Al LDH by electrochemical reduction at a constant potential. The deposited layer of LDH was modified by applying oxidation potential in order to decrease the interferences from ascorbic acid (AA) or uric acid (UA) during the detection of glucose in real samples. Drop casting method was employed to further modify the surface of the electrode with GOx enzyme. The resultant Pt-LDHox-GOx displayed the higher affinity towards glucose detection and no response was observed for the interfering species present in real life samples of fruit juices [50].

Mousty et. al. [52] for the first time reported the use of nanohybrid LDH material consist of ferrocene anions (Fc-PSO$_3$ and Fc-(PSO$_3$)$_2$) as a host to immobilize the GOx on the surface of the electrode. The intercalated anions act as mediators and help to transport the electrons between FAD center in the enzyme and the electrode surface. The novel biosensor displayed good sensitivity under anaerobic conditions as compared to other LDH based glucose biosensors. The reaction of glucose with FAD produce the reduced

FADH$_2$, that in turn oxidize FMCA$^+$ to regenerate FMCA as a result FAD is generated again to complete the catalytic cycle [68], as shown in figure 4.

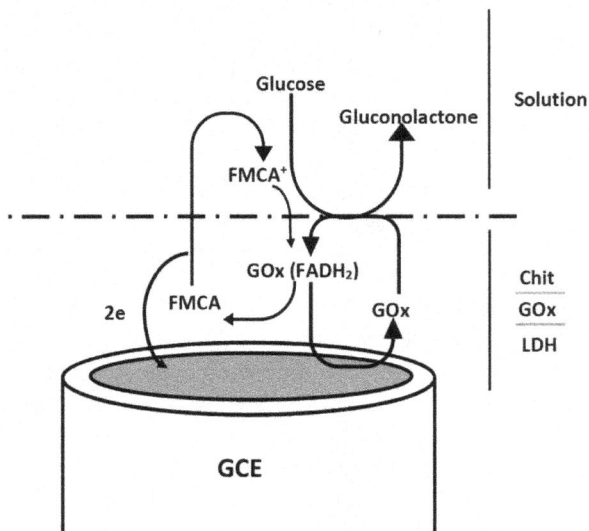

Figure 4 Schematic representation of the proposed glucose biosensor to explain its construction and the electrocatalytic mechanism on the surface of the fabricated electrode.

5.2   Tyrosinase based LDH biosensors

Polyphenol oxidases (PPOs) are widely distributed enzyme among plants that contain four copper atoms per molecule. PPOs carry the binding sites for two aromatic compounds and oxygen that binds the monophenol molecules and carry out their o-hydroxylation to produce o-diphenols. It also has the ability to further oxidize the o-diphenols to produce o-quinones. Tyrosine is an amino acid that contains a phenolic ring that may be oxidized to o-quinone by the action of PPOs that's why PPOs may also be referred as tyrosinase. That is why throughout this chapter, the term tyrosinase will refer to the enzyme activity that oxidizes monophenols to o-diphenols. D. Shan et al. [44]

developed an electrochemical biosensor by immobilizing tyrosinase into LDH. The activity and sensitivity of the LDH based biosensor were compared with laponite based biosensors. The greater permeability and lower diffusion resistance of the PPO/LDH biosensor cause amplification of detection signal, as a result, the higher sensitivity and lower detection limit of PPO/LDH was obtained as compared to PPO/laponite towards the oxidation of catechol.

Transketolase (TK) is an intracellular enzyme that is very crucial in the metabolism of all living cells. The conventional methodology for monitoring the activity of TK enzyme utilizes D-xylulose-5-phosphate as donor substrate and D-ribose-5-phosphate as acceptor substrate. However, D-xylulose-5-phosphate is very difficult to synthesize and no longer commercially available. TK has the ability to produce tyrosine derivatives in the electrolyte medium [69] that can be easily detected by the PPO/LDH electrochemical biosensor as an indicator of TK activity. The selectivity of the biosensor towards tyrosine and its derivatives was enhanced by covering the electrode surface by perm-selective membranes such as nafion [70].

A. Soussou and coworkers [71] developed a biosensor by immobilization of tyrosinase immobilized on CoAl LDH and utilized for detection of polyphenols extracted from green tea. The same research group produced a similar biosensor for the polyphenols present in green tea extract, where tyrosinase was hosted by the hydrocalumite (Ca_2AlCl-LDH) support. The outcomes revealed that the modified surface show linear response within the concentration range of 2.4 x 10^{-5} – 2.4 µM and display higher sensitivity (1.5µA/ng·mL^{-1}) as compared to that of former biosensor developed by the same group [51]. The mixture of CoAl LDH nanomaterial was utilized to modify the surface of the goldscreen-printed electrode (Au-SPE). Afterward, tyrosinase was immobilized on the modified surface of Au-SPE by drop casting method. The tyrosinase enzyme-modified electrode oxidizes the polyphenols into o-quinone, which electrochemically reduced to o-diphenol. The current produced by the reduction peak enable the biosensor to detect the presence of polyphenols in the electrolyte solution. The author's suggested that biosensor have the ability to be applied to real pharmaceutical, environmental and food samples [48].

5.3   Heme-based LDH biosensors

5.3.1 Hemoglobin (Hb) based LDH biosensors

Hemoglobin is a metallo protein that is a crucial component of the many living organism blood and function as an oxygen carrier. Due to its eminent structure and viability, Hb is widely studied in electrochemistry to develop a biosensor. The graphene (GR) due to its

Biosensors: Materials and Applications   Materials Research Forum LLC
Materials Research Foundations **47** (2019) 131-156   doi: http://dx.doi.org/10.21741/9781644900130-4

two-dimensional structure with higher flexibility, electrical conducted and surface area proved to be the best candidate that provides specific microenvironment to immobilize the protein on the surface of LDH based electrodes. Sun at el. [72] modified the surface of carbon ionic liquid electrode (CILE) by employing GR-LDH nanocomposite and immobilizes Hb. Chitosan (Chit) film was coated on the surface of a biosensor in order to stop the separation of bionanohybrid from the surface. The electrochemical investigation of CTS/Hb-GR-LDH/CILE proved that GR and LDH facilitate the direct electron transfer between Hb and LDH. CTS/Hb-GR-LDH/CILE biosensor showed higher catalytic behavior towards the reduction of trichloroacetic acid (TCA). The higher electrocatalytic activity of the biosensor was ascribed as the ability of Hb to decrease the electroreduction activation energy of TCA. The developed electrode displayed a small decrease in its response for 10.0 mM TCA for 30 days that demonstrated the higher stability and good biocompatibility of the developed biosensor. Recently, the nanohybrid of GR-LDH nanocomposite decorated with graphitic C_3N_4 nanoparticle was used as a host to immobilize Hb on the CILE to get CTS/Hb-C_3N_4-GR-LDH/CILE biosensor. The synergetic effect of components of transducer resulted in greater heterogeneous electron transfer rate for reduction of TCA. The CTS/Hb-C_3N_4-GR-LDH/CILE biosensor shows lower detection limit for reduction of TCA i.e. 50 µM as compared to CTS/Hb-GR-LDH/CILE detection limit that was 534 µM. Developed biosensor CTS/Hb-C_3N_4-GR-LDH/CILE also showed a greater resistance towards the higher concentration of inorganic or organic interfering agent while detecting TCA in real water samples [54].

5.3.2 Myoglobin (Mb) based LDH biosensors

Myoglobin is a heme-containing protein that is abundantly found in the muscles of the diving mammals. A higher quantity of the myoglobin allows the mammals to hold their breath while diving for a longer time [73]. It is also found in the human body specifically at the time of muscle injury [74,75]. Mb in the form of nanocomposite with various materials can be utilized to develop a biosensor for the sensitive detection of important analytes such as H_2O_2, trichloroacetic acid, nitrite, etc. [76–78]. Mb can be immobilized on the surface of LDH to get a biosensor due to a specific interaction between Mb and LDH. An ionic liquid 1-carboxyl-methyl-3-methylimidazolium tetrafluoroborate (CMMIMBF$_4$) was used to functionalize the Mg$_2$Al-LDH to immobilize Mb on the surface of LDH modified CILE. The ionic liquids play a major role in the acceleration of electron kinetics due to higher ionic conductivity. The Mb-CMMIMBF$_4$-LDH/CILE electrode show acceptable inter day and intra day reproducibility while measuring the electro reduction of TCA in real life samples [79].

5.3.3 Cytochrome c (Cyt c) based LDH biosensors

Cytochrome C is a heme protein that is loosely bounded by the inner wall of mitochondria. It is very crucial in Adenosine triphosphate ATP synthesis and control the programmed cell death during apoptosis [80,81]. It is an important part of the electron transfer route that exchange charge between Cyt c oxidase and Cyt c reductase to undergo redox reactions [82]. Cyt c can be potentially employed to get third generation protein based electrochemical biosensors, in order to detect different sort of molecules such as glucose, nitrite, nitric oxide, etc. [83–85]. To exploit electron transfer property of the Cyt c Yin and coworkers developed Cyt c-LDH-Nafion/Au biosensor to target trace amount of nitrite in food samples. The study of the electrochemical behavior of Cyt c-LDH-Nafion/Au revealed that adsorption controlled process is occurring on the electrode surface. It was observed that Nafion-LDH composite not only acts as a suitable platform to bind the Cyt c, it also acts as a mediator to accelerate the electron transfer between Cyt c and Au electrode. The mechanism of nitrite oxidation of developed biosensor involve oxidation of Fe^{+2}-Cyt c to Fe^{+3}-Cyt c, afterward further oxidation of Fe^{+3}-Cyt c occurs at very high potential to produce Fe^{+4}-Cyt c. The highly reactive Fe^{+4}-Cyt c species oxidize nitrite to nitrate and reduce back to Fe^{+2}-Cyt c via Fe^{+3}-Cyt c [86].

5.3.4 Horseradish peroxidase (HRP) based LDH biosensors

Horseradish peroxidase (HRP) is an important heme-containing enzyme and its isoenzyme C is extensively studied for its catalytic action. It is a highly stable and low-cost enzyme which is widely applied to biomedical and environmental applications. Due to its characteristic features now HRP find its way in biosensor applications [87,88]. Chen et al. [89] immobilized the HRP on Ni-Al LDH to study the electrocatalytic reduction of H_2O_2 and trichloroacetic acid (TCA). In the developed biosensor, HPR shows good compatibility with the LDH and displayed satisfactory results while detecting the H_2O_2 in real pharmaceutical samples and TCA in tap water sample. Cholesterol is a chief component of steroid hormones, bio precursors, nerve cells and brain cells and its higher concentration inside the human body can cause various health issues. In order to eliminate the interferences of coexisting electroactive substances and to detect the cholesterol at trace level, a bi-enzymatic biosensor was developed. Cholesterol oxidase (ChOx) and HRP were immobilized on the chitosan (Chit) LDH support and the resultant biosensor ChOx/HRP/LDHs-Chit /GCE was obtained. The components of formulated biosensor work in a systematic way i.e. ChOx react with cholesterol in the presence of oxygen and produce H_2O_2. The HRP reduces the H_2O_2 and gives the detectable electrical signal [90].

In another work, Kong et al. [56] immobilized the two heme-containing enzyme (hemoglobin (HB) and HRP) on the surface of treated indium–tin oxide-coated glass(ITO) substrate by using a layer by layer process. Morphological studies demonstrated that the $(LDH/HB/LDH/HRP)_n$ ultra-thin film (UTF) displayed the ability of long-range stacking order due to a superlattice nanostructure, that can hold and stabilize the protein structures in the LDH monolayer arrangement. The $(LDH/HB/LDH/HRP)_2$ UTF demonstrated higher sensitivity and low detection limit towards oxidation of catechol as compared to the single heme-protein based biosensor. The electrochemical reactions occurring at the surface of the biosensor can be described as follow: where Catechol* is denoted for the catechol free radical.

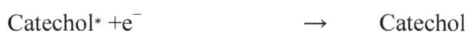

$$Heme–Fe(III) + C + e^- \quad \longrightarrow \quad Heme–Fe(II) + Catechol*$$

$$Heme–Fe(II) \quad \longrightarrow \quad Heme–Fe(III) + e^-$$

$$Catechol* + e^- \quad \longrightarrow \quad Catechol$$

Faster electron transfer rate of a biosensor badly affect its sensitivity. HRP immobilized on the surface of Co-Al LDH/GCE surface was unable to resolve the peaks of Fe_{III}/Fe_{II} redox couple. It was attributed to Co present in LDH that act as electron relay matrix and facilitates the electron hopping and direct electron transfer between HRP and GCE. In order to tackle this issue sodium dodecylbenzenesulfonate (SDBS) that block the cation active redox sites and HRP-SDBS-LDH-GCE was successfully used for detection of 2-chlorophenol in drinking water samples [91]. In order to increase the sensitivity of the HRP based biosensor, Wang and coworkers [92] used CoFe-LDH and carbon dots (C-Dots) mixture as a host to immobilize the HRP. The synergistic effect of CoFe-LDH, C-Dots, and HRP was used for sensing of H_2O_2. The HRP/C-Dots/LDHs/GCE biosensor is highly selective in the presence of possible interfering agents such as uric acid (UA), ascorbic acid(AA) and epinephrine (EP).

5.4 Acetylcholinesterase (AChE) based LDH biosensors

Acetylcholinesterase is an important enzyme that catalyzes the dismantling of neurotransmitters such as acetylcholine and other choline esters. AChE is widely used to develop biosensors that are applied for the sensing of pesticides and monitoring of food quality. The lamellar structure, large surface area and high anion exchange capability of the LDH act as host to immobilize the AChE. The brief structural study revealed that a biohybrid obtained due to the reaction between LDH and AChE which result in the covalent bond formation between them displayed good bioanalytical applications [93]. Gong et al. [94] developed an AChE based biosensor immobilized on LDH in order to detect organophosphate pesticides (OPs). Methyl parathion (MP) was employed as a model for inhibition biosensor study and optimization of analytical conditions for flow injection analysis (FIA). The amperometric detection by using FIA displayed that AChE/LDH/GCE biosensor can be automated and provided real-time outcomes. Another research group used Cu-Mg-Al calcined layered double hydroxide (CLDH) to immobilize the AChE. CLDH can carry a large amount of AChE due to the higher effective surface area, however low electrical conductivity is the major drawback of CLDH. To overcome this limitation firstly a nanocomposite of graphene-gold nanoparticles (Gr-AuNPs) was deposited on the surface of GCE. Afterward, a combination of CLDH-AChE was immobilized on the Gr-AuNP/GCE to get a highly sensitive biosensor CLDH-AChE/Gr-AuNP/GCE for the detection of chlorpyrifos (class of OPs) with a very low limit of detection [95].

6.   Miscellaneous

Apart from the aforementioned biomolecules, there are also other biomolecules that were used with LDH to develop biosensors for various electroanalytical applications. For example, the biomolecules such as DNA carrying hydroxyl, amine or thiol groups have the ability to chemically attach with the metal oxides present in LDHs. To get a low-cost and label-free DNA impedimetric biosensor, Baccar et al. [96] utilized LDH to immobilize DNA probe consist of 20 nucleotide bases. The chemically synthesized DNA probe of 100 bases binds specifically with the probe attached to the LDH surface and produces a relevant electrical signal. The developed DNA based LDH biosensor present a strong base to detect the single mutation occurring in DNA strand.

An electrochemical immune biosensor was established by dispersing graphene and Ni-Al LDH nanocomposite in Chit and dropping on the surface of GCE to get LDH/GR/GCE. Afterward, the hollow gold nanoparticles (HAuNPs) and monoclonal chlorpyrifos antibody were immobilized on the electrode surface to get anti-chlorpyrifos/HAuNPs/LDH/GR/GCE. Finally, the electrode was incubated with 0.5% Bovine serum albumin (BSA) to increase the specificity of the electrode by attachment of BSA with non-specific binding sites and active groups. The BSA/anti-chlorpyrifos/HAuNPs/LDH/Gr/GCE electrode displays highly specific response towards chlorpyrifos due to immunoreactions with the antibody. The BSA/anti-chlorpyrifos/HAuNPs/LDH/GR/GCE biosensor demonstrate lower detection limit i.e. 0.052 ng/ml as compared to CLDH-AChE/Gr-AuNP/GCE biosensor [95]. The developed immunosensorcan be regenerated by immersing in glycine-HCl buffer (pH 2.8). The regeneration process allows the reuse of biosensor with a minor decrease in response. The decrease in response of biosensor is attributed to degradation of a biosensor in the regeneration process and a decrease in a number of the active site for the immune reaction [57]. Recently, a thiamine (vitamin B1) biosensor is established based on entrapping of transketolase (TK) enzyme-substrate intermediate. The TK enzyme originated from E.coli bacteria was immobilized on the surface of LDH/GCE by drop casting method. Thiamine pyrophosphate (ThDP) is a cofactor that has the ability to activate TK enzyme. The ThDP binds the TK in TK/LDH/GCE biosensor and produces amperometric signals with a very high sensitivity of 3831 mAM^{-1}cm^{-2} [49]. The figure of merits of various LDH based biosensors was compared in table 1.

Conclusion and future perspective:

From the previously reported work, it is evident that layered double hydroxide based electrochemical biosensor is getting great attention. The layered double hydroxide unique features may help to develop some valuable biosensors for various electroanalytical applications. The characteristics of the layered double hydroxide can be controlled through various synthetic routes. Moreover, the variety of cations and the anions assist to attain diversity in LDH material with certain characteristics. Biosensor efficiency and activity depend upon three main factors [97]:

- o Targeted analyte and its product diffusion through the bio-membrane
- o Enzyme activity within the immobilization matrix
- o Electrochemical transduction step efficiency

Layered double hydroxide in this regard is an ideal material for the immobilization of the enzyme. The layered and porous architecture of LDH provided the stable immobilization of enzyme and better stability. However, the electrochemical activity of the LDH is not good. It can be improved by making its hybrid composite with other materials including graphene, CNTs and metal nanoparticles. Layered double hydroxide has good future in the field of biosensors. However, more efforts and detailed studies are required to bring them to commercialization.

Acknowledgments

The authors gratefully acknowledge the support provided by the Chemistry Department at King Fahd University of Petroleum and Minerals (KFUPM).

References

[1]   A.N. Kawde, N. Baig, M. Sajid, Graphite pencil electrodes as electrochemical sensors for environmental analysis: a review of features, developments, and applications, RSC Adv. 6 (2016) 91325–91340. https://doi.org/10.1039/C6RA17466C

[2]   S. Kanchi, M.I. Sabela, P.S. Mdluli, Inamuddin, K. Bisetty, Smartphone based bioanalytical and diagnosis applications: A review., Biosens. Bioelectron. 102 (2018) 136–149. https://doi.org/10.1016/j.bios.2017.11.021

[3]   J. Wang, Carbon-nanotube based electrochemical biosensors: A review, Electroanalysis. 17 (2005) 7–14. https://doi.org/10.1002/elan.200403113

[4]   Y. Yue, G. Hu, M. Zheng, Y. Guo, J. Cao, S. Shao, A mesoporous carbon nanofiber-modified pyrolytic graphite electrode used for the simultaneous determination of dopamine, uric acid, and ascorbic acid, Carbon N. Y. 50 (2012) 107–114. https://doi.org/10.1016/j.carbon.2011.08.013

[5]   N. Baig, A.-N. Kawde, A cost-effective disposable graphene-modified electrode decorated with alternating layers of Au NPs for the simultaneous detection of dopamine and uric acid in human urine, RSC Adv. 6 (2016) 80756–80765. https://doi.org/10.1039/C6RA10055D

[6]   N. Baig, A.-N. Kawde, A novel, fast and cost effective graphene-modified graphite pencil electrode for trace quantification of l -tyrosine, Anal. Methods. 7 (2015) 9535–9541. https://doi.org/10.1039/C5AY01753J

[7]   M. Gerard, A. Chaubey, B.D. Malhotra, Application of conducting polymers to biosensors, Biosens. Bioelectron. 17 (2002) 345–359. https://doi.org/10.1016/S0956-5663(01)00312-8

[8]   Y. Wang, Y. Wu, J. Xie, X. Hu, Metal–organic framework modified carbon paste electrode for lead sensor, Sensors Actuators B Chem. 177 (2013) 1161–1166. https://doi.org/10.1016/j.snb.2012.12.048

[9]   A.-N. Kawde, M. Aziz, N. Baig, Y. Temerk, A facile fabrication of platinum nanoparticle-modified graphite pencil electrode for highly sensitive detection of hydrogen peroxide, J. Electroanal. Chem. 740 (2015) 68–74. https://doi.org/10.1016/j.jelechem.2015.01.005

[10]   Y. Cao, G. Li, X. Li, Graphene/layered double hydroxide nanocomposite: Properties, synthesis, and applications, Chem. Eng. J. 292 (2016) 207–223. https://doi.org/10.1016/j.cej.2016.01.114

Biosensors: Materials and Applications   Materials Research Forum LLC
Materials Research Foundations **47** (2019) 131-156   doi: http://dx.doi.org/10.21741/9781644900130-4

[11] Y. Kuang, L. Zhao, S. Zhang, F. Zhang, M. Dong, S. Xu, Morphologies, preparations and applications of layered double hydroxide micro-/nanostructures, Materials (Basel). 3 (2010) 5220–5235. https://doi.org/10.3390/ma3125220

[12] G. Mishra, B. Dash, S. Pandey, Layered double hydroxides: A brief review from fundamentals to application as evolving biomaterials, Appl. Clay Sci. 153 (2018) 172–186. https://doi.org/10.1016/j.clay.2017.12.021

[13] N. Baig, M. Sajid, Applications of layered double hydroxides based electrochemical sensors for determination of environmental pollutants: A review, Trends Environ. Anal. Chem. 16 (2017) 1–15. https://doi.org/10.1016/j.teac.2017.10.003

[14] N. Mao, C.H. Zhou, D.S. Tong, W.H. Yu, C.X. Cynthia Lin, Exfoliation of layered double hydroxide solids into functional nanosheets, Appl. Clay Sci. 144 (2017) 60–78. https://doi.org/10.1016/j.clay.2017.04.021

[15] Q. Wang, D. Ohare, Recent advances in the synthesis and application of layered double hydroxide (LDH) nanosheets, Chem. Rev. 112 (2012) 4124–4155. https://doi.org/10.1021/cr200434v

[16] K. Yan, G. Wu, W. Jin, Recent advances in the synthesis of layered, double-hydroxide-based materials and their applications in hydrogen and oxygen evolution, Energy Technol. 4 (2016) 354–368. https://doi.org/10.1002/ente.201500343

[17] M. Sarfraz, I. Shakir, Recent advances in layered double hydroxides as electrode materials for high-performance electrochemical energy storage devices, J. Energy Storage. 13 (2017) 103–122. https://doi.org/10.1016/j.est.2017.06.011

[18] A.R. González, Y.J.O. Asencios, E.M. Assaf, J.M. Assaf, Dry reforming of methane on Ni–Mg–Al nano-spheroid oxide catalysts prepared by the sol–gel method from hydrotalcite-like precursors, Appl. Surf. Sci. 280 (2013) 876–887. https://doi.org/10.1016/j.apsusc.2013.05.082

[19] X. Wu, Y. Du, X. An, X. Xie, Fabrication of NiFe layered double hydroxides using urea hydrolysis—Control of interlayer anion and investigation on their catalytic performance, Catal. Commun. 50 (2014) 44–48. https://doi.org/10.1016/j.catcom.2014.02.024

[20] F. Leroux, J.-P. Besse, Polymer interleaved layered double hydroxide: A new emerging class of nanocomposites, Chem. Mater. 13 (2001) 3507–3515. https://doi.org/10.1021/cm0110268

[21] M.R. Othman, N.M. Rasid, W.J.N. Fernando, Effects of thermal treatment on the micro-structures of co-precipitated and sol–gel synthesized (Mg–Al) hydrotalcites,

Microporous Mesoporous Mater. 93 (2006) 23–28.
https://doi.org/10.1016/j.micromeso.2006.02.007

[22]   B. Habibi, F.F. Azhar, J. Fakkar, Z. Rezvani, Ni–Al/layered double hydroxide/Ag
       nanoparticle composite modified carbon-paste electrode as a renewable electrode
       and novel electrochemical sensor for hydrogen peroxide, Anal. Methods. 9 (2017)
       1956–1964. https://doi.org/10.1039/C6AY03421G

[23]   J. Zhou, M. Min, Y. Liu, J. Tang, W. Tang, Layered assembly of NiMn-layered
       double hydroxide on graphene oxide for enhanced non-enzymatic sugars and
       hydrogen peroxide detection, Sensors Actuators B Chem. 260 (2018) 408–417.
       https://doi.org/10.1016/j.snb.2018.01.072

[24]   Y. Ma, Y. Wang, D. Xie, Y. Gu, H. Zhang, G. Wang, Y. Zhang, H. Zhao, P.K.
       Wong, NiFe-layered double hydroxide nanosheet arrays supported on carbon cloth
       for highly sensitive detection of nitrite, ACS Appl. Mater. Interfaces. 10 (2018)
       6541–6551. https://doi.org/10.1021/acsami.7b16536

[25]   D. Tonelli, E. Scavetta, M. Giorgetti, Layered-double-hydroxide-modified
       electrodes: electroanalytical applications, Anal. Bioanal. Chem. 405 (2013) 603–
       614. https://doi.org/10.1007/s00216-012-6586-2

[26]   J.J. Bravo-Suárez, E.A. Páez-Mozo, S.T. Oyama, Review of the synthesis of
       layered double hydroxides: a thermodynamic approach, Quim. Nova. 27 (2004)
       601–614. https://doi.org/10.1590/S0100-40422004000400015

[27]   G. Fan, F. Li, D.G. Evans, X. Duan, Catalytic applications of layered double
       hydroxides: recent advances and perspectives, Chem. Soc. Rev. 43 (2014) 7040–
       7066. https://doi.org/10.1039/C4CS00160E

[28]   J. Castillo, S. Gáspár, S. Leth, M. Niculescu, A. Mortari, I. Bontidean, V.
       Soukharev, S.A. Dorneanu, A.D. Ryabov, E. Csöregi, Biosensors for life quality -
       Design, development and applications, Sensors Actuators, B Chem. 102 (2004)
       179–194. https://doi.org/10.1016/j.snb.2004.04.084

[29]   C.R. Ispas, G. Crivat, S. Andreescu, Review: recent developments in enzyme-
       based biosensors for biomedical analysis, Anal. Lett. 45 (2012) 168–186.
       https://doi.org/10.1080/00032719.2011.633188

[30]   I. Bazin, S.A. Tria, A. Hayat, J.-L. Marty, New biorecognition molecules in
       biosensors for the detection of toxins, Biosens. Bioelectron. 87 (2017) 285–298.
       https://doi.org/10.1016/j.bios.2016.06.083

[31]   Q. Wang, D. O'Hare, Recent advances in the synthesis and application of layered
       double hydroxide (LDH) nanosheets, Chem. Rev. 112 (2012) 4124–4155.
       https://doi.org/10.1021/cr200434v

Biosensors: Materials and Applications   Materials Research Forum LLC
Materials Research Foundations **47** (2019) 131-156 doi: http://dx.doi.org/10.21741/9781644900130-4

[32] C. Mousty, V. Prévot, Hybrid and biohybrid layered double hydroxides for electrochemical analysis, Anal. Bioanal. Chem. 405 (2013) 3513–3523. https://doi.org/10.1007/s00216-013-6797-1

[33] Y. Zhao, F. Li, R. Zhang, D.G. Evans, X. Duan, Preparation of layered double-hydroxide nanomaterials with a uniform crystallite size using a new method involving separate nucleation and aging steps, Chem. Mater. 14 (2002) 4286–4291. https://doi.org/10.1021/cm020370h

[34] S.P. Newman, W. Jones, Synthesis, characterization and applications of layered double hydroxides containing organic guests, New J. Chem. 22 (1998) 105–115. https://doi.org/10.1039/a708319j

[35] J. He, M. Wei, B. Li, Y. Kang, D.G. Evans, X. Duan, Preparation of layered double hydroxides, Layer. Double Hydroxides. 119 (2006) 89–119. https://doi.org/10.1007/430_006

[36] D.G. Evans, X. Duan, Preparation of layered double hydroxides and their applications as additives in polymers, as precursors to magnetic materials and in biology and medicine, Chem. Commun. (2006) 485–496. https://doi.org/10.1039/B510313B

[37] D. Tonelli, E. Scavetta, M. Giorgetti, Layered-double-hydroxide-modified electrodes: Electroanalytical applications, Anal. Bioanal. Chem. 405 (2013) 603–614. https://doi.org/10.1007/s00216-012-6586-2

[38] F. Leroux, C. Taviot-Guého, Fine tuning between organic and inorganic host structure: new trends in layered double hydroxide hybrid assemblies, J. Mater. Chem. 15 (2005) 3628. https://doi.org/10.1039/b505014f

[39] M.-A. Thyveetil, P. V. Coveney, H.C. Greenwell, J.L. Suter, Computer simulation study of the structural stability and materials properties of dna-intercalated layered double hydroxides, J. Am. Chem. Soc. 130 (2008) 4742–4756. https://doi.org/10.1021/ja077679s

[40] F. Karim, A.N.M. Fakhruddin, Recent advances in the development of biosensor for phenol: a review, Rev. Environ. Sci. Bio/Technology. 11 (2012) 261–274. https://doi.org/10.1007/s11157-012-9268-9

[41] Y. Zhang, J. Shen, H. Li, L. Wang, D. Cao, X. Feng, Y. Liu, Y. Ma, L. Wang, Recent progress on graphene-based electrochemical biosensors, Chem. Rec. 16 (2016) 273–294. https://doi.org/10.1002/tcr.201500236

[42] H. Kaur, R. Kumar, J.N. Babu, S. Mittal, Advances in arsenic biosensor development – A comprehensive review, Biosens. Bioelectron. 63 (2015) 533–545. https://doi.org/10.1016/j.bios.2014.08.003

[43] S. Gupta, C.N. Murthy, C.R. Prabha, Recent advances in carbon nanotube based electrochemical biosensors, Int. J. Biol. Macromol. 108 (2018) 687–703. https://doi.org/10.1016/j.ijbiomac.2017.12.038

[44] D. Shan, S. Cosnier, C. Mousty, Layered double hydroxides : An attractive material for electrochemical biosensor design, Anal. Chem. 75 (2003) 3872–3879

[45] B. Saifullah, M.Z.B. Hussein, Inorganic nanolayers: structure, preparation, and biomedical applications., Int. J. Nanomedicine. 10 (2015) 5609–33. https://doi.org/10.2147/IJN.S72330

[46] X. Guo, F. Zhang, D.G. Evans, X. Duan, Layered double hydroxide films: synthesis, properties and applications, Chem. Commun. 46 (2010) 5197. https://doi.org/10.1039/c0cc00313a

[47] L. Cui, H. Yin, J. Dong, H. Fan, T. Liu, P. Ju, S. Ai, A mimic peroxidase biosensor based on calcined layered double hydroxide for detection of H_2O_2, Biosens. Bioelectron. 26 (2011) 3278–3283. https://doi.org/10.1016/J.BIOS.2010.12.043

[48] A. Soussou, I. Gammoudi, F. Morote, A. Kalboussi, T. Cohen-Bouhacina, C. Grauby-Heywang, Z.M. Baccar, Efficient immobilization of tyrosinase enzyme on layered double hydroxide hybrid nanomaterials for electrochemical detection of polyphenols, IEEE Sens. J. 17 (2017) 4340–4348. https://doi.org/10.1109/JSEN.2017.2709342

[49] M. Halma, B. Doumèche, L. Hecquet, V. Prévot, C. Mousty, F. Charmantray, Thiamine biosensor based on oxidative trapping of enzyme-substrate intermediate, Biosens. Bioelectron. 87 (2017) 850–857. https://doi.org/10.1016/j.bios.2016.09.049

[50] E. Scavetta, L. Guadagnini, A. Mignani, D. Tonelli, Anti-interferent properties of oxidized nickel based on layered double hydroxide in glucose amperometric biosensors, Electroanalysis. 20 (2008) 2199–2204. https://doi.org/10.1002/elan.200804310

[51] A. Soussou, I. Gammoudi, A. Kalboussi, C. Grauby-Heywang, T. Cohen-Bouhacina, Z.M. Baccar, Hydrocalumite thin films for polyphenol biosensor elaboration, IEEE Trans. Nanobioscience. 16 (2017) 650–655. https://doi.org/10.1109/TNB.2017.2736781

[52] C. Mousty, C. Forano, S. Fleutot, J.C. Dupin, Electrochemical study of anionic ferrocene derivatives intercalated in layered double hydroxides: Application to glucose amperometric biosensors, Electroanalysis. 21 (2009) 399–408. https://doi.org/10.1002/elan.200804407

Biosensors: Materials and Applications   Materials Research Forum LLC
Materials Research Foundations **47** (2019) 131-156   doi: http://dx.doi.org/10.21741/9781644900130-4

[53] A.-N. Kawde, M.A. Aziz, M. El-Zohri, N. Baig, N. Odewunmi, cathodized gold nanoparticle-modified graphite pencil electrode for non-enzymatic sensitive voltammetric detection of glucose, Electroanalysis. 29 (2017) 1214–1221. https://doi.org/10.1002/elan.201600709

[54] T. Zhan, Z. Tan, X. Wang, W. Hou, Hemoglobin immobilized in g-C3N4 nanoparticle decorated 3D graphene-LDH network: Direct electrochemistry and electrocatalysis to trichloroacetic acid, Sensors Actuators B Chem. 255 (2018) 149–158. https://doi.org/10.1016/J.SNB.2017.08.048

[55] J. Han, X. Xu, X. Rao, M. Wei, D.G. Evans, X. Duan, Layer-by-layer assembly of layered double hydroxide/cobalt phthalocyanine ultrathin film and its application for sensors, J. Mater. Chem. 21 (2011) 2126–2130. https://doi.org/10.1039/C0JM02430A

[56] X. Kong, X. Rao, J. Han, M. Wei, X. Duan, Layer-by-layer assembly of bi-protein/layered double hydroxide ultrathin film and its electrocatalytic behavior for catechol, Biosens. Bioelectron. 26 (2010) 549–554. https://doi.org/10.1016/J.BIOS.2010.07.045

[57] L. Qiao, Y. Guo, X. Sun, Y. Jiao, X. Wang, Electrochemical immunosensor with NiAl-layered double hydroxide/graphene nanocomposites and hollow gold nanospheres double-assisted signal amplification, Bioprocess Biosyst. Eng. 38 (2015) 1455–1468. https://doi.org/10.1007/s00449-015-1388-5

[58] V. Prevot, C. Forano, A. Khenifi, B. Ballarin, E. Scavetta, C. Mousty, A templated electrosynthesis of macroporous NiAl layered double hydroxides thin films, Chem. Commun. 47 (2011) 1761–1763. https://doi.org/10.1039/C0CC04255B

[59] M. Darder, M. López-Blanco, P. Aranda, F. Leroux, E. Ruiz-Hitzky, Bio-nanocomposites based on layered double hydroxides, Chem. Mater. 17 (2005) 1969–1977. https://doi.org/10.1021/cm0483240

[60] J. Labuda, M. Hudáková, Hexacyanoferrate-anion exchanger-modified carbon paste electrodes, Electroanalysis. 9 (1997) 239–242. https://doi.org/10.1002/elan.1140090310

[61] E. Han, D. Shan, H. Xue, S. Cosnier, Hybrid material based on chitosan and layered double hydroxides: Characterization and application to the design of amperometric phenol biosensor, Biomacromolecules. 8 (2007) 971–975. https://doi.org/10.1021/bm060897d

[62] P. Mehrotra, Biosensors and their applications - A review., J. Oral Biol. Craniofacial Res. 6 (2016) 153–9. https://doi.org/10.1016/j.jobcr.2015.12.002

[63] J. Ali, J. Najeeb, M. Asim Ali, M. Farhan Aslam, A. Raza, Biosensors: Their
      fundamentals, designs, types and most recent impactful applications: A Review, J.
      Biosens. Bioelectron. 08 (2017) 1–9. https://doi.org/10.4172/2155-6210.1000235

[64] A.A. Parwaz Khan, A. Khan, A.M. Asiri, Graphene and graphene oxide polymer
      composite for biosensors applications, in: Electr. Conduct. Polym. Polym.
      Compos., Wiley-VCH Verlag GmbH & Co. KGaA, Weinheim, Germany, 2018:
      pp. 93–112. https://doi.org/10.1002/9783527807918.ch5

[65] B. Bhushan, Biosensors: surface structures and materials., Philos. Trans. A. Math.
      Phys. Eng. Sci. 370 (2012) 2267–8. https://doi.org/10.1098/rsta.2012.0011

[66] C. Mousty, O. Kaftan, V. Prevot, C. Forano, Sensors and Actuators B : Chemical
      Alkaline phosphatase biosensors based on layered double hydroxides matrices :
      Role of LDH composition, Sensors Actuators B Chem. 133 (2008) 442–448.
      https://doi.org/10.1016/j.snb.2008.03.001

[67] M. Colombari, B. Ballarin, I. Carpani, L. Guadagnini, A. Mignani, E. Scavetta, D.
      Tonelli, Glucose biosensors based on electrodes modified with ferrocene
      derivatives intercalated into Mg/Al layered double hydroxides, Electroanalysis. 19
      (2007) 2321–2327. https://doi.org/10.1002/elan.200703985

[68] Y. Xu, X. Liu, Y. Ding, L. Luo, Y. Wang, Y. Zhang, Y. Xu, Preparation and
      electrochemical investigation of a nano-structured material Ni^{2+} / MgFe layered
      double hydroxide as a glucose biosensor, Appl. Clay Sci. 52 (2011) 322–327.
      https://doi.org/10.1016/j.clay.2011.03.011

[69] F. Charmantray, V. Hélaine, A. Làsikovà, B. Legeret, L. Hecquet,
      Chemoenzymatic synthesis of l-tyrosine derivative for a transketolase assay,
      Tetrahedron Lett. 49 (2008) 3229–3233.
      https://doi.org/10.1016/J.TETLET.2008.03.099

[70] M.S.-P. Lopez, F. Charmantray, V. Helaine, L. Hecquet, C. Mousty,
      Electrochemical detection of transketolase activity using a tyrosinase biosensor,
      Biosens. Bioelectron. 26 (2010) 139–143.
      https://doi.org/10.1016/j.bios.2010.05.023

[71] A. Soussou, I. Gammoudi, F. Moroté, M. Mathelié-Guinlet, A. Kalboussi, Z.M.
      Baccar, T. Cohen-Bouhacina, C. Grauby-Heywang, Amperometric polyphenol
      biosensor based on tyrosinase immobilization on coal layered double hydroxide
      thins films, Procedia Eng. 168 (2016) 1131–1134.
      https://doi.org/10.1016/J.PROENG.2016.11.371

[72] W. Sun, Y. Guo, Y. Lu, A. Hu, F. Shi, T. Li, Z. Sun, Electrochemical biosensor
      based on graphene, Mg2Al layered double hydroxide and hemoglobin composite,

Electrochim. Acta. 91 (2013) 130–136.
https://doi.org/10.1016/J.ELECTACTA.2012.12.088

[73] Masanori Sono, Mark P. Roach, and Eric D. Coulter, J.H. Dawson, Heme-Containing Oxygenases, (1996). https://doi.org/10.1021/CR9500500

[74] J. Ishii, J. Wang, H. Naruse, S. Taga, M. Kinoshita, H. Kurokawa, M. Iwase, T. Kondo, M. Nomura, Y. Nagamura, Y. Watanabe, H. Hishida, T. Tanaka, K. Kawamura, Serum concentrations of myoglobin vs human heart-type cytoplasmic fatty acid-binding protein in early detection of acute myocardial infarction, Clin. Chem. 43 (1997)

[75] P. Brancaccio, G. Lippi, N. Maffulli, Biochemical markers of muscular damage, Clin. Chem. Lab. Med. 48 (2010) 757–67.
https://doi.org/10.1515/CCLM.2010.179

[76] J. Yoon, T. Lee, B. Bapurao G., J. Jo, B.-K. Oh, J.-W. Choi, Electrochemical H_2O_2 biosensor composed of myoglobin on MoS_2 nanoparticle-graphene oxide hybrid structure, Biosens. Bioelectron. 93 (2017) 14–20.
https://doi.org/10.1016/J.BIOS.2016.11.064

[77] C. Ruan, T. Li, Q. Niu, M. Lu, J. Lou, W. Gao, W. Sun, Electrochemical myoglobin biosensor based on graphene–ionic liquid–chitosan bionanocomposites: Direct electrochemistry and electrocatalysis, Electrochim. Acta. 64 (2012) 183–189. https://doi.org/10.1016/J.ELECTACTA.2012.01.005

[78] A.T.E. Vilian, V. Veeramani, S.-M. Chen, R. Madhu, C.H. Kwak, Y.S. Huh, Y.-K. Han, Immobilization of myoglobin on Au nanoparticle-decorated carbon nanotube/polytyramine composite as a mediator-free H2O2 and nitrite biosensor, Sci. Rep. 5 (2016) 18390. https://doi.org/10.1038/srep18390

[79] J. Lou, Y. Lu, T. Zhan, Y. Guo, W. Sun, C. Ruan, Application of an ionic liquid-functionalized Mg_2Al layered double hydroxide for the electrochemical myoglobin biosensor, Ionics (Kiel). 20 (2014) 1471–1479. https://doi.org/10.1007/s11581-014-1088-1

[80] M. Hüttemann, P. Pecina, M. Rainbolt, T.H. Sanderson, V.E. Kagan, L. Samavati, J.W. Doan, I. Lee, The multiple functions of cytochrome c and their regulation in life and death decisions of the mammalian cell: From respiration to apoptosis., Mitochondrion. 11 (2011) 369–81. https://doi.org/10.1016/j.mito.2011.01.010

[81] Y.-L.P. Ow, D.R. Green, Z. Hao, T.W. Mak, Cytochrome c: functions beyond respiration, Nat. Rev. Mol. Cell Biol. 9 (2008) 532–542.
https://doi.org/10.1038/nrm2434

[82] S. Zaidi, M.I. Hassan, A. Islam, F. Ahmad, The role of key residues in structure, function, and stability of cytochrome-c, Cell. Mol. Life Sci. 71 (2014) 229–255. https://doi.org/10.1007/s00018-013-1341-1

[83] E. Pashai, G. Najafpour Darzi, M. Jahanshahi, F. Yazdian, M. Rahimnejad, An electrochemical nitric oxide biosensor based on immobilized cytochrome c on a chitosan-gold nanocomposite modified gold electrode, Int. J. Biol. Macromol. 108 (2018) 250–258. https://doi.org/10.1016/J.IJBIOMAC.2017.11.157

[84] Y. Haldorai, S.-K. Hwang, A.-I. Gopalan, Y.S. Huh, Y.-K. Han, W. Voit, G. Sai-Anand, K.-P. Lee, Direct electrochemistry of cytochrome c immobilized on titanium nitride/multi-walled carbon nanotube composite for amperometric nitrite biosensor, Biosens. Bioelectron. 79 (2016) 543–552. https://doi.org/10.1016/J.BIOS.2015.12.054

[85] M. Eguílaz, C.J. Venegas, A. Gutiérrez, G.A. Rivas, S. Bollo, Carbon nanotubes non-covalently functionalized with cytochrome c: A new bioanalytical platform for building bienzymatic biosensors, Microchem. J. 128 (2016) 161–165. https://doi.org/10.1016/J.MICROC.2016.04.018

[86] H. Yin, Y. Zhou, T. Liu, L. Cui, S. Ai, Y. Qiu, L. Zhu, Amperometric nitrite biosensor based on a gold electrode modified with cytochrome c on Nafion and Cu-Mg-Al layered double hydroxides, Microchim. Acta. 171 (2010) 385–392. https://doi.org/10.1007/s00604-010-0444-8

[87] N.C. Veitch, Horseradish peroxidase: A modern view of a classic enzyme, Phytochemistry. 65 (2004) 249–259. https://doi.org/10.1016/j.phytochem.2003.10.022

[88] C. Regalado, B.E. García-Almendárez, M.A. Duarte-Vázquez, Biotechnological applications of peroxidases, Phytochem. Rev. 3 (2004) 243–256. https://doi.org/10.1023/B:PHYT.0000047797.81958.69

[89] X. Chen, C. Fu, Y. Wang, W. Yang, D.G. Evans, Direct electrochemistry and electrocatalysis based on a film of horseradish peroxidase intercalated into Ni–Al layered double hydroxide nanosheets, Biosens. Bioelectron. 24 (2008) 356–361. https://doi.org/10.1016/J.BIOS.2008.04.007

[90] S.-N. Ding, D. Shan, T. Zhang, Y.-Z. Dou, Performance-enhanced cholesterol biosensor based on biocomposite system: Layered double hydroxides-chitosan, J. Electroanal. Chem. 659 (2011) 1–5. https://doi.org/10.1016/J.JELECHEM.2011.04.003

[91] L. Fernández, I. Ledezma, C. Borrás, L.A. Martínez, H. Carrero, Horseradish peroxidase modified electrode based on a film of Co–Al layered double hydroxide

modified with sodium dodecylbenzenesulfonate for determination of 2-chlorophenol, Sensors Actuators B Chem. 182 (2013) 625–632. https://doi.org/10.1016/J.SNB.2013.02.109

[92] Y. Wang, Z. Wang, Y. Rui, M. Li, Horseradish peroxidase immobilization on carbon nanodots/CoFe layered double hydroxides: Direct electrochemistry and hydrogen peroxide sensing., Biosens. Bioelectron. 64 (2015) 57–62. https://doi.org/10.1016/j.bios.2014.08.054

[93] S. Hidouri, Z.M. Baccar, H. Abdelmelek, T. Noguer, J.-L. Marty, M. Campàs, Structural and functional characterisation of a biohybrid material based on acetylcholinesterase and layered double hydroxides, Talanta. 85 (2011) 1882–1887. https://doi.org/10.1016/J.TALANTA.2011.07.026

[94] J. Gong, Z. Guan, D. Song, Biosensor based on acetylcholinesterase immobilized onto layered double hydroxides for flow injection/amperometric detection of organophosphate pesticides, Biosens. Bioelectron. 39 (2013) 320–323. https://doi.org/10.1016/j.bios.2012.07.026

[95] C. Zhai, Y. Guo, X. Sun, Y. Zheng, X. Wang, An acetylcholinesterase biosensor based on graphene–gold nanocomposite and calcined layered double hydroxide, Enzyme Microb. Technol. 58–59 (2014) 8–13. https://doi.org/10.1016/J.ENZMICTEC.2014.02.004

[96] Z.M. Baccar, D. Caballero, R. Eritja, A. Errachid, Development of an impedimetric DNA-biosensor based on layered double hydroxide for the detection of long ssDNA sequences, Electrochim. Acta. 74 (2012) 123–129. https://doi.org/10.1016/J.ELECTACTA.2012.04.031

[97] D. Shan, S. Cosnier, C. Mousty, Layered double hydroxides: An attractive material for electrochemical biosensor design, Anal. Chem. 75 (2003) 3872–3879. https://doi.org/10.1021/ac030030v

Biosensors: Materials and Applications
Materials Research Foundations **47** (2019) 157-210

Materials Research Forum LLC
doi: http://dx.doi.org/10.21741/9781644900130-5

Chapter 5

Electrochemical Nanobiosensors for Cancer Diagnosis

Anu Bharti[1], Shilpa Rana[1], Nirmal Prabhakar*

Department of Biochemistry, Panjab University, Chandigarh, India-160014

* nirmalprabhakar@gmail.com

[1]Both authors equally contributed

Abstract

Modern lifestyle has invited many devastating diseases like cancer to mankind. Cancer has become a huge threat to the population and its diagnosis being expensive and time-consuming gives an indication of the need to develop cost-effective and reliable detection methods. This chapter focuses on various electrochemical nano biosensors established for the detection of commonly occurring cancers (lung, breast, prostate and colorectal). It has been divided into four major parts that emphasize the detailed biosensor fabrication strategies utilized for the detection of biomarkers related to particular cancer type. Recently developed electrochemical techniques based on different bio-recognition elements (antibody, enzyme, nucleic acid, aptamer, phage and lectin) and nanomaterials have been explored with their advantages and limitations.

Keywords

Electrochemical Biosensors, Lung Cancer, Breast Cancer, Prostate Cancer, Colorectal Cancer, Biomarkers

Contents

1.   Introduction

Cancer is considered as the major cause of mortality among the population with 14.1 million cancer cases and 8.2 million deaths worldwide in the year 2012 [1]. Overall, men contribute to 7.4 million and women contribute to 6.7 million of all cancer cases. Various types of cancer have been diagnosed which affects more than 60 human organs. Lung cancer is found to be the most common cancer followed by breast, prostate and colorectal cancer according to their prevalence. According to the American Cancer Society, the estimated number of new cancer patients and deaths in the US for the year 2018 are given in Table 1 [2].

Biosensors: Materials and Applications   Materials Research Forum LLC
Materials Research Foundations **47** (2019) 157-210   doi: http://dx.doi.org/10.21741/9781644900130-5

Table 1. Estimated Number of different Cancer Cases and Deaths by Sex, US, 2018.

Cancer type	Estimated no. of new cases			Estimated no. of deaths		
	Both sexes	**Male**	**Female**	**Both sexes**	**Male**	**Female**
Overall	1,735,350	856,370	878,980	609,640	323,630	286,010
Lung	234,030	121,680	112,350	154,050	83,550	70,500
Breast	268,670	2,550	266,120	41,400	480	40,920
Prostate	164,690	164,690	-	29,430	29,430	
Colon	97,220	49,690	47,530	50,630	27,390	23,240
Rectum	43,030	25,920	17,110	-	-	-

Cancer is a form of abnormal cells that show unusual growth beyond specific limits and gets metastasized to the surrounding organs of the body. Cancer generally arises from pre-cancerous lesion to a malignant tumour due to numerous factors including a person's genetic makeup and exposure to various carcinogens present in their food and environment. As the person gets older, the risk of cancer development increases due to the poor performance of the cellular repair mechanism. Moreover, the incidence of cancer varies depending upon gender, genetics, ethnic and geographical patterns [3]. The cancer burden could be reduced to some extent by improving the lifestyle and evading the risk factors. Early detection of cancer may also be useful for improving the treatment methods and reducing the mortality rate.

Conventional methods for cancer diagnosis rely on various morphological and imaging techniques. Morphological studies may involve cytology and histopathology of affected cells and tissues, respectively whereas imaging techniques like mammography, endoscopy, MRI, X-ray and ultrasound are used to distinguish between healthy and cancerous tissues. These diagnostic methods are associated with low sensitivity, specificity, high treatment cost and are not able to fulfil the patient's needs. Furthermore, these methods are based on the phenotype of the tumour whereas cancer involves a complex array of many alterations at a genetic level resulting in tumour formation [4]. The molecules which are differentially expressed during cancer development are considered as potential diagnostic biomarkers and these have the capability of improving the early detection as well as treatment methods [5,6]. Immunohistochemistry (IHC), enzyme-linked immunosorbent assay (ELISA), radioimmunoassay (RIA), northern blotting and real time-PCR (polymerase chain reaction) are various biomarker-based methods which are being used routinely. Although these techniques are efficient in terms of specificity as well as sensitivity but are time-consuming, expensive and require special laboratory skills in biomolecule extraction and purification [7]. Hence, the focus of this

Biosensors: Materials and Applications   Materials Research Forum LLC
Materials Research Foundations **47** (2019) 157-210   doi: http://dx.doi.org/10.21741/9781644900130-5

chapter is toward exploring diagnostic tools which are highly sensitive, less invasive, inexpensive and point of care (POC) diagnosis to improve the screening complexity. For this, biosensors are suitable devices with desired sensitivity, selectivity, reusability, portability and simplicity with the ease of direct assessment of physiological fluids [8–11].

The biosensor is an analytical device that detects the target analyte by binding to its bio-recognition element giving rise to some measurable signals which are detected by transducer element. Transducers convert the detected molecular signals into electrical or digital signals which can be quantified and analyzed (Figure 1). Biosensors can be categorized into optical, mass sensitive, electrochemical and calorimetric on the basis of the transducer system. Among all the biosensors, electrochemical biosensors are considered a promising alternative due to their easy usability, quick response, cost-effectiveness and low background [12,13]. The other properties of the electrochemical method include high sensitivity, the possibility of miniaturization and cost-effectiveness that make them worth for early-stage cancer diagnosis and therapy monitoring.

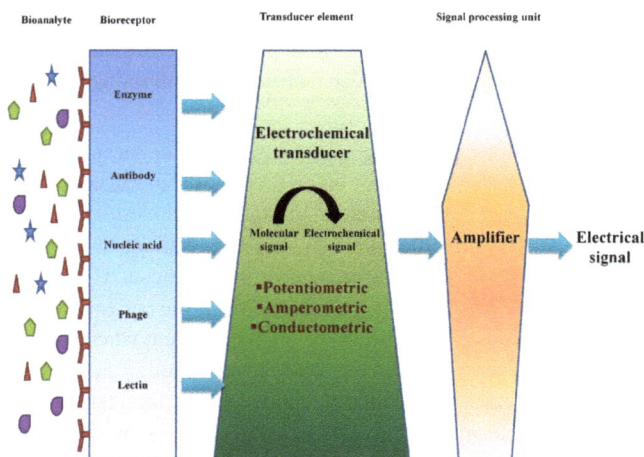

Figure 1: Schematic illustration of an electrochemical biosensor.

Herein, various electrochemical methods with the aim of diagnosing commonly occurring cancers such as lung cancer, breast cancer, colorectal cancer and prostate cancer using different biomarkers have been discussed (Table 2).

Biosensors: Materials and Applications
Materials Research Foundations **47** (2019) 157-210

Materials Research Forum LLC
doi: http://dx.doi.org/10.21741/9781644900130-5

Table 2. Different biomarkers involved in the specific detection of cancer.

Biomarker	Nature	Cancer
Vascular endothelial growth factor (VEGF)	Signal protein	Lung cancer, prostate cancer
Sex determining region Y-box 2 (SOX2)	Transcriptional factor	Lung cancer, prostate cancer
Epithelial cell adhesion molecule (EpCAM)	Transmembrane glycoprotein	Prostate cancer
Neuron specific enolase (NSE)	Enzyme	Lung cancer
17β-estradiol	Estrogen steroid hormone	Lung cancer
Endothelial growth factor receptor (EGFR)	Transmembrane protein	Lung cancer
Nicotinamide adenine dinucleotide (NADH)	Coenzyme	Lung cancer
Hypoxia inducible factor 1α (HIF1α)	Transcriptional factor	Lung cancer
Vimentin	Type III intermediate filament protein	Lung cancer
Melanoma associated antigen (MAGE) A2 & A11	Antigens	Lung cancer
Sarcosine	Amino acid derivative	Prostate cancer
Prostate specific antigen (PSA)	Protein	Prostate cancer
Engrailed 2 (EN2)	Transcriptional factor	Prostate cancer
Prostate specific membrane antigen (PSMA)	Transmembrane protein	Prostate cancer
Fetuin	Glycoprotein	Prostate cancer
Leucine-rich alpha-2-glycoprotein 1 (LRG1)	Glycoprotein	Colorectal cancer
p53	Tumor suppressor protein	Colorectal cancer
Chemokine ligand 5 (CXCL5)	Chemokine protein	Colorectal cancer
Carcinoembryonic antigen *(CEA)*	Glycoprotein	Colorectal cancer
Cancer antigen 15-3 (CA15-3)	Protein	Breast cancer
Human epidermal growth factor receptor 2 (HER 2)	Growth promoting protein	Breast cancer
Mucin-1 (MUC-1)	Glycoprotein	Breast cancer

Metalloproteinase-2 (MMP-2)	Enzyme	Colorectal cancer
Genetic Biomarker		
miRNA-21	Short non coding RNA	Lung cancer, breast cancer
miRNA-221	Short non coding RNA	Lung cancer
miRNA-141, 375, 182-5p	Short non coding RNA	Prostate cancer
miRNA-205, 155	Short non coding RNA	Breast cancer
let 7 miRNA	Short non coding RNA	Lung cancer
Breast cancer-1 (BRCA-1)	Tumor suppressor gene	Breast cancer
MCF-7*	Carcinoma cell line	Breast cancer
SKBR3*	Carcinoma cell line	Breast cancer
Gluthione S-transferase P 1(GSTP 1)	Gene	Prostate cancer
BRAF $^{\#}$V600E	Proto-oncogene	Colorectal cancer
K-ras	Downstream component of the EGFR signaling network	Colorectal cancer
HCT116 & HT 29*	Colon carcinoma cell line	Colorectal cancer
CT26*	Colon carcinoma cell line	Colorectal cancer

*-Breast cancer cell line, $^{\#}$- v-raf murine sarcoma viral oncogene homolog B1

A variety of biomolecules such as enzymes, antibodies and nucleic acids can be used as recognition elements against the target analyte for biosensor development [9]. Initially, some enzymes such as alkaline phosphatase and horseradish peroxidase (HRP) enzymes were frequently used for biosensing strategies but due to a deficiency in providing sufficient signal amplification to achieve ultrasensitive detection of biomarker, these have discouraged the researchers to use the enzymatic methods. Moreover, operational stability and specificity is also an important concern while working with enzymes. Therefore, in order to improve the biosensing methods, antibody-based techniques were preferred over enzymatic methods due to their better specificity, sensitivity and homogenous nature. They can be used directly or indirectly for biomarker detection but are also associated with some limitations. Antibodies are also sensitive to the environment and regeneration as well as reproducibility is also a major issue while working with antibody as bioreceptors. In contrast, nucleic acids provide highly specific and sensitive detection along with a faster, simpler, and cheaper analytical method [14]. Hence, to understand the link between cancer and genetics, nucleic acid hybridization method is a promising alternative to the above-mentioned assays.

Biosensors: Materials and Applications   Materials Research Forum LLC
Materials Research Foundations **47** (2019) 157-210   doi: http://dx.doi.org/10.21741/9781644900130-5

2.   Lung Cancer

Among all the cancers, lung cancer is the most prevalent and life-threatening disease. The small lung cancer, non-small lung cancer and lung carcinoid tumour are the three main categories of lung cancer. The non-small lung cancer and small lung cancer constitutes approximately 85% and 10-15% respectively amongst all lung cancers. Although for detecting lung cancer, there are several standardized conventional methods such as fluorescence bronchoscopy, virtual bronchoscopy and electromagnetic navigation bronchoscopy, these are expensive invasive techniques and need appropriate implementation. There still remains a need for more efficient, reliable and cost-effective detection technique. These limitations have been resolved with the development of electrochemical biosensors for lung cancer detection.

2.1   Antibody-based biosensor

To meet the point of care (POC) requirements and challenge the commonly used detection technique known as ELISA, the development of immunosensors have revolutionized the diagnostic field. It is based on the principle that the interaction of biomarker specific antibodies to biomolecule on electrode surface alters the electron transfer rate of the electrode during electrochemical biosensing. Immunosensors basically incorporate the ELISA like strategy, for example, Choudhary et al. [15] prepared an immunosensor using graphite electrode fabricated with carbon nanotube (CNT)- chitosan (CHI) films to immobilize the antigens (MAGE A-2 or MAGE A-11). In the presence of the analyte, these antigens which correlate with the metastatic stage of lung cancer bind to the anti-MAGE antibodies that are expressed as a defence against the MAGE antigens during cancer existence (Figure 2). The current biosensor had a limit of detection (LOD) as 5fg/mL–50ng/mL for MAGE A-2 and MAGE A-11 respectively. However, the performance of this biosensor was not assessed in real samples. In another study, Hussain et al. [16] utilized the ELISA-like strategy by immobilizing anti-HIF1α antibody on functionalized polymer [2,2:5,2-terthiophene-3-(p-benzoic acid)] (pTTBA) grafted on a layer of AuNPs (gold nanoparticles) and the signal label consists of hydrazine and secondary antibody attached to gold nanoparticles. This nanobiosensor amperometrically senses the HIF1α expression in lung cancer cells with a detection limit as 5.35±0.02pM/mL. Nevertheless, HIF1α is not a reliable and specific biomarker for lung cancer detection as that of neuron-specific enolase (NSE) whose normal serum value ranges from 5-12ng/mL [17]. In addition, nanocomposites are also essential for transferring of electrons providing conductivity, hence, Wang et al. [18] described a detection strategy for monitoring the NSE levels by using excellent conductive hydrogel consisting of ligand (1,3,5-benzenetricarboxylic acid) and metal ion (Fe^{3+}) grafted with

Biosensors: Materials and Applications   Materials Research Forum LLC
Materials Research Foundations **47** (2019) 157-210   doi: http://dx.doi.org/10.21741/9781644900130-5

AuNPs and immobilized with anti-NSE antibody. The proposed immunosensor had 0.2pg/mL as LOD as measured by square wave voltammetry (SWV) detection technique. To decrease the complexity in preparing and detecting tagged labels for signals, label-free sensors are highly preferred. For example, Wei et al. [19] created a label-free sensor using graphene-Au nanocomposites for detecting NSE at LOD 0.05ng/mL. The anti-NSE antibody grafted immunosensor displayed 0.05ng/mL as the detection limit. Apart from sensing NSE, 17β-estradiol has also been found to be associated with lung carcinoma. Dai and Liu [20] developed a biosensor for sensing 17β-estradiol (a biomarker of lung cancer) in the simulated urine sample using anti-estrogen antibody immobilized on Au electrode. Aydin and Sezginturk [21] fabricated an anti-SOX2 (Sex-determining region Y-box 2) antibody grafted CTES modified Indium tin oxide (ITO) electrode for sensing the SOX2 as a marker for lung cancer detection with LOD 7fg/mL. Fan et al. [22] designed a commercial microfluidic paper-based analytical device (μPAD) keeping in mind the POC requirements for detecting lung cancer by sensing the NSE levels. The low cost and disposability of μPAD allow its integration with POCT system. The electrochemical sensor was fabricated with amino-linked graphene (NH_2-G), thionine and AuNPs to immobilize the anti-NSE antibody and resulted in the detection limit of 10pg/mL. However, the immunosensors incorporating antibodies is an expensive way and rarely have been commercialized in spite of being largely used in research laboratories.

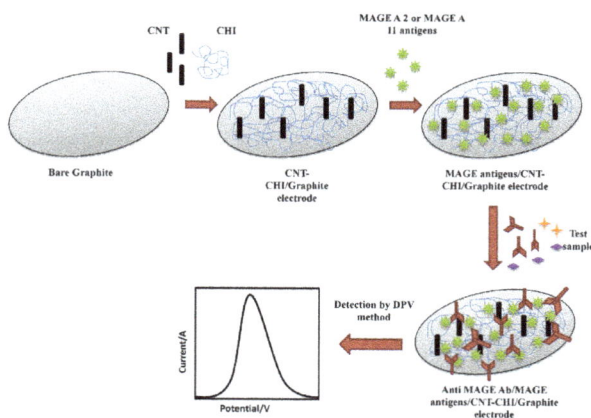

Figure 2: Schematic diagram illustrating methodology for biosensor development for detection of MAGE in lung cancer diagnosis.

2.2   Nucleic acid-based biosensor

The nucleic acid-based bioassay involves preferential binding of immobilized oligonucleotides probe with its complementary target sequence by hybridization, thereby, producing target dependent response. The nucleic acid-based biosensors have the ability to detect DNA or RNA nucleotides even if they differ by one nucleotide and hence can be used for diagnosis with enhanced sensitivity. In a study by Tian et al. [23] a label-free electrochemical biosensor for resolving lung cancer-related let-7 miRNAs (let-7a, let-7f, and let-7g) which have only one-nucleotide variations were developed. The label-free strategy involved hybridization of the hairpin probe with the target miRNA initiating the CNT based rolling-circle amplification. This produces super large DNA fragments resulting in an amplified signal and enhanced sensitivity. The detection limit of the sensor was 1.2fM. Another novel design by Liu et al. [24] utilized a ferrocene (F_c) labelled DNA tetrahedron as a recognition element for the lung cancer-specific miRNA-21 grafted onto the AuNPs modified Au electrode and achieved the detection limit to 10 pM as detected by the differential pulse voltammetry (DPV) technique. Xu et al. [25] reported a DNA biosensor to detect EGFR exon 19 deletions. The electrode surface fabricated with capture probe allowed the detection of target DNA hybridized with reporter probe. Though this sensor focuses on a clinical sample by using an enzyme named lambda exonuclease for generating ssDNA from dsDNA enabling its sensitive biosensing, the strategy involved is quite complex. Su et al. [26] reported oligonucleotides based sensor fabricated with capture DNA on hierarchical flower-like Au nanostructures grafted on ITO electrodes for the detection of the miRNA-21 biomarker. The $Ru(NH3)_6^{3+}$ signal molecule allowed the detection to be as low as 1fM. Voccia et al. [27] reported a genosensor fabricated with DNA capture probe to hybridize the target miRNA-221. The sensing of RNA extracted from different cancer lines resolved the discrimination between cancerous and non-cancerous cells with detection limit 0.7pM. Bo et al. [28] reported a biosensor to detect miRNA-21 for detection of lung cancer. The strategy involves grafting the capture probes on the Au electrode which on miRNA detection hybridizes with it and initiates the cleavage by duplex specific nuclease. Once activated, the capture probe-DNA hybrid co-localized with AuNPs forms, thus, allowing the detection using chronocoulometry (CC) method (Figure 3). The reduced CC response was observed with increasing miRNA concentration. The observed detection limit was 6.8aM. Besides this, Liu et al. [29] developed a biosensor based on hybrid DNA hydrogel for detecting miRNA-21. They used silanized ITO electrode to fabricate hybrid hydrogel comprising of linear polyacrylamide linked with DNA and F_c-tagged reporter DNA probe and found detection limit as 5nM. Overall, nucleic acid biosensors have immensely contributed in paving the path for diagnostic approaches but the need of pretreatment step renders its preference.

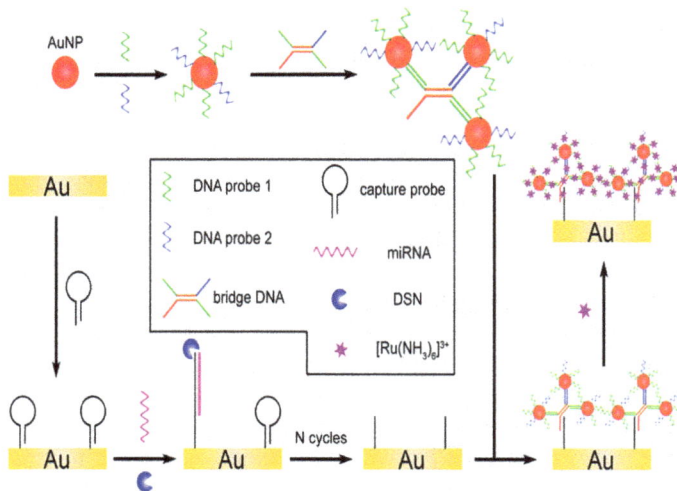

Figure 3: Illustration of the electrochemical approach for triple amplified detection of miRNA-21.

2.3   Biomimetic based biosensor

Aptamer-based biosensors came as a breakthrough for antibody-based sensors as aptamers are chemically synthesized after their *in vitro* selection to mimic the antibody-making it cost effective. Amouzadeh, Shamsipur and Farzin [30] designed an aptasensor by fabricating the screen-printed electrode with mesoporous carbon-Au nanocomposite and aptamer sequence for detecting VEGF 165 lung cancer marker. The detection limit obtained for human serum sample was 1.0pg/mL using electrochemical impedance spectroscopy (EIS) technique. Another study by Zamay et al. [31] utilized a DNA aptamer for molecular biosensing vimentin expression in peripheral blood of 100 samples of healthy and lung cancer patients.

2.4   Other

Ahmed et al. [32] used an approach for detecting the phosphorylation of the EGFR which is marked as an indicator of lung cancer. They studied different lung carcinoma cell lines to observe the phosphorylation of target EGFR protein using Au electrode. The principle involves decreased affinity of proteins towards the Au electrode on phosphorylation-

Biosensors: Materials and Applications   Materials Research Forum LLC
Materials Research Foundations **47** (2019) 157-210   doi: http://dx.doi.org/10.21741/9781644900130-5

induced conformational changes which are directly detected electrochemically with a detection limit of 10ng/μL. However, such an approach does not provide a high throughput analysis. Another study by Pu et al. [33] assessed EGFR related oncogenic mutations using an electrochemical biosensor. The electric field-induced release and measurement (EFIRM) method being efficient in capturing short sequences in biofluids was used to detect exon 19 deletion in saliva and plasma samples of non-small lung cancer patients. The sensor was fabricated in a sandwich-like configuration using HRP labelled anti-fluorescein antibody against detector probe labelled with fluorescein isothiocyanate. The target-capture probe hybrid on Au electrode was then detected by its conjugation to labelled detector probe. Akhtar et al. [34] detected altered intracellular concentration ratio of NADH and NAD^+ (an indicator of lung cancer) in tumorigenic lung epithelial (A549) cell samples with nanobiosensor. They fabricated the glassy carbon electrode with activated graphene oxide and then grafted the EDTA linked PEI to the sensor probe and analyzed NADH concentration with LOD 20.0±1.1nM

Hence, diagnosis of lung cancer using electrochemical biosensors targeting various biomarkers and using different approaches has enlightened the path for the development of novel sensor. In addition, more strong and excellent strategies can be made an application to prepare a miniaturized commercial device for diagnostic purposes. Table 3 listed the electrochemical biosensors utilized in detection of lung cancer.

Table 3: List of electrochemical biosensors utilized in detection of lung cancer.

Biosensing platform	Biomarker	Detection method	Linear range	LOD	Ref.
RuHex/AuNPs-bridge DNA/capture probe/Au electrode	miRNA-21	CA	10^{-17}-10^{-11} M	6.8aM	[27]
DNA hybrid hydrogel/ITO/PET electrode	miRNA-21	DPV	10nM-50μM	5nM	[28]
Anti-SOX2 Ab/CTES/ITO	SOX2	CV & EIS	25-2000fg/mL	7fg/mL	[20]
Anti-NSE Ab/NH$_2$-G/Thi/AuNPs	NSE	DPV	1-500ng/mL	10pg/mL	[21]
Anti-estrogen Ab/SAM/Au electrode	17β-estradiol	DPV	2.25-2250pg/mL	-	[19]
AP-anti-IgG/rGO/AuNPs	NSE	DPV	0.1-2000ng/mL	0.05ng/mL	[18]
Anti-NSE Ab/AuNPs/hydrogel/GCE	NSE	SWV	1pg/mL-200ng/mL	0.26pg/mL	[17]
Au electrode	EGFR	DPV	-	10ng/μL	[31]

Btn DNA probe/streptavidin/Au electrode	miRNA-221	EIS	1–100pM	0.7pM	[26]
HRP-Ab/detector probe-fluorescein isothiocyanate/capture probe-pyrrole/Au electrode	EGFR mutations	EFIRM	-	-	[32]
RuHex/miRNA/DNA probe/Au nanostructure/ITO	miRNA-21	CA	1fM-10pM	1fM	[25]
PEI-EDTA/AGO/GCE	NADH	CV & QCM	0.05-500µM	20.0±1.1nM	[33]
sec-Ab$_2$-Hyd-AuNPs/ anti HIF1α Ab$_1$/ pTTBA/ AuNPs/ GCE	HIF1α	CA	25-350pM/mL	5.35±0.02 pM/mL	[15]
DNA aptamer/Au electrode	Vimentin	EIS	-	-	[34]
Reporter probe-target/capture probe/Au electrode	EGFR mutations	Amperometric	-	-	[24]
anti-VEGF aptamer/ OMC–Au$_{nano}$/SPE	VEGF(165)	EIS	10-300pg/mL	1.0pg/mL	[29]
F$_c$- DNA probe/AuNPs/Au	miRNA-21	DPV	100pM-1µM	10pM	[23]
Hairpin probe/CNTs/GCE	let 7miRNAs	DPV	0-1000fM	1.2fM	[22]
Antigens/CNT-CHI/Graphite	Anti-MAGE A2 and anti-MAGE A11	DPV	5fg/mL–50ng/mL	5fg/mL & 50ng/mL	[14]

AuNPs: Gold nanoparticles; ITO: Indium-tin oxide; PET: Polyethylene terephthalate; CTES: Carboxyethylsilanetriol sodium salt; NH$_2$-G: Amino functional graphene; Thi: Thionine; SAM: Self assembled monolayer; AP: Alkaline phosphatase; rGO: Reduced graphene oxide; GCE: Glassy carbon electrode; Btn: Biotinylated; HRP: Horseradish peroxide; RuHex: Hexaammineruthenium(III) chloride; AGO: Activated graphene oxide; PEI: Polyethylenimine; EDTA: Ethylenediaminetetraacetic acid; pTTBA: 2,2:5,2-terthiophene-3-(p-benzoic acid); Hyd: Hydrazine sulfate; SPE: Screen printed electrode; OMC–Au$_{nano}$: Ordered mesoporous carbon–gold nanocomposite; F$_c$: Ferrocene; CNTs: Carbon nanotubes; CHI: Chitosan; CA-Chronoamperometry; SWV- Square wave voltammetry; CC- Chronocoulometry; QCM: Quartz crystal microbalance; EFIRM: Electric field-induced release and measurement

3. Breast Cancer

Breast cancer is highly prevalent and is the leading cause of cancer-related deaths among women. It was estimated that nearly 1.67 million new cases of breast cancer were diagnosed and 522,000 cases of deaths occurred worldwide in 2012 [35]. The incidence rates are highest in North America, Australia, New Zealand, Northern and Western

Biosensors: Materials and Applications   Materials Research Forum LLC
Materials Research Foundations **47** (2019) 157-210   doi: http://dx.doi.org/10.21741/9781644900130-5

Europe whereas Africa and Asia are having the least number of breast cancer cases [36]. Increase in the number of cancer cases is mainly due to limited sources of early detection as well as treatment methods. So, breast cancer detection at an early stage would be beneficial for the development of new therapeutics and to decrease the mortality rate. There are certain molecules which are highly over-expressed in breast cancer patients including glycoproteins, DNA, miRNA, etc. [37]. For the specific detection of these molecules in breast cancer, various nanobiosensor have been discussed in this section and are enlisted in Table 4.

3.1   Enzyme-based biosensor

The basic principle of enzymatic biosensors is the measurement of the signal generated after the formation of substrate-enzyme complex. The change in the electrochemical signal is directly proportional to the amount of substrate/analyte present in the sample. Limited studies have been found for the detection of breast cancer using enzymes as bioreceptor molecules. An amperometric biosensor to check the level of extracellular H_2O_2 liberated from the breast cancer cells using sequence-specific peptides has been constructed by Zhao et al. [38]. To improve the orientation of enzymes immobilized on the electrode surface, use of peptide ligands could be useful for easy accessibility of their substrate and to improve their catalytic activity. H_2O_2 is a by-product of many oxidative metabolic pathways found to be highly stable reactive oxygen species. During bioelectrode fabrication, HRP enzyme was immobilized on to the peptide modified electrode surface in a suitable orientation which improved the electron transfer reactivity of the enzyme and could be useful for the sensitive detection of H_2O_2 with LOD up to 3.0 x 10^{-8}M. Hence, the designed biosensor could be potentially applied for physiological, pathological as well as environmental applications in future and many different peptide ligands can be explored for other detection studies.

3.2   Antibody-based biosensor

Numerous immunosensing methods have been reported for the electrochemical detection of breast cancer using breast cancer type 1 and 2 susceptibility proteins (BRCA-1, BRCA-2) [39], mucin-1 (MUC-1) [40], cancer antigen 15-3 (CA15-3) [41] and human epidermal growth factor receptor (HER-2) [42–44] etc. BRCA-1 and BRCA-2 are tumour suppressor genes that help in the repair of damaged DNA. Development of breast cancer is associated with specific mutations in BRCA-1 and BRCA-2 gene sequence. A dual amplification sandwich immunosensor (DASI) for BRCA-1 detection was developed using N-doped graphene along with hydroxypropyl chitosan and CO_3O_4 mesoporous nanosheets [45]. DASI is an ultrasensitive technique which can capture more antigens on the electrode surface and showed a detection limit of 0.33pg/mL with high accuracy.

Tabrizi et al. [46] formulated a sandwich type immunosensor using rGO/SPE as a fabrication platform and rGO-TPA/FeHCFnano Anti-HCT as a label for the sensitive detection of SKBR-2 breast cancer cells as depicted in Figure 4. SKBR-2 cells possess HER-2 antigen on their plasma membrane which stimulates cell proliferation and malignant growth and highly specific for breast cancer detection. The biosensor exhibited a good limit of detection up to 21cells/mL but is time consuming as it require 80 min. of incubation. In a study by Mouffouk et al. [47] polymeric nanoparticles based electrochemical immunosensor has been constructed for MUC-1 detection. MUC-1 is a high molecular weight protein present in most of the human epithelial surfaces and its upregulation in cancerous tissues leads to the non-specific expression over the cell surfaces. The polymeric micelles enclosed F_c acting as a tracer molecule get released during antigen-antibody (Ag-Ab) binding, resulting in the signal enhancement and able to detect up to 10cells/mL in less than 1min. This method is advantageous due to the presence of polymeric nanoparticles which are highly stable and non-toxic. Moreover, it is cost-effective, require very less amount of sample and can detect low abundant samples, as a single binding event of Ag-Ab can release thousands of tracer molecules into the solution which enhances the sensitivity of the developed method. A similar method was reported for the detection of HER-2 biomarker of breast cancer using disposable inkjet printed electrochemical biosensing platform [48]. The immunoassay was constructed by step by step injection of HER-2 protein, biotinylated antibody and HRP label to produce a complete working 8-electrode array along with an integrated counter and a reference electrode. The sensor exhibited a detection limit of 12pg/mL within 15min. The inkjet printed electrochemical methods could be further used for multiple sample analysis and clinical diagnosis.

3.3 Nucleic acid-based biosensor

A number of researchers employed nucleic acid as bioreceptor molecules for breast cancer detection [49–52]. Such a method with a direct and selective diagnosis of target molecules in the presence of many other non-specific proteins and nucleic acid sequences is highly required. Dopamine (DA), tannic acid (TA) and polyethylene glycol (PEG) are associated with highly hydrophilic and antifouling properties which make them useful to construct a reliable detection method. So, Chen, Liu and Chen [53] fabricated a PEG/TA/pDA platform by layer by layer surface deposition for the detection of the BRCA-1 gene. AuNPs along with thiol-modified probe DNA was then electrochemically deposited on the prepared PEG/TA/pDA platform for hybridization detection of BRCA-1 upto 0.05fM. The PEG and AuNPs act as enhancer molecule to prove its potential for clinical serum sample analysis. Other methods using PEG as an antifouling agent were also explored by Wang et al. [54] for BRCA-1 detection with a limit of detection upto

Biosensors: Materials and Applications   Materials Research Forum LLC
Materials Research Foundations **47** (2019) 157-210   doi: http://dx.doi.org/10.21741/9781644900130-5

1.72fM. Cui et al. [55] has also exploited a DNA biosensor for the detection of BRCA-1 using zwitter ionic peptide self-assembled monolayer (SAM) as an antifouling substrate.

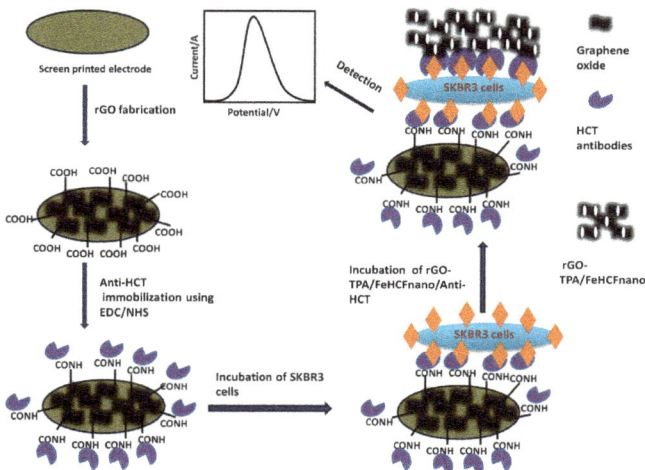

Figure 4: Schematic diagram of the sandwich type electrochemical immunosensor developed for the detection of SKBR-2 breast cancer cells.

miRNAs shows great specificity for their complementary probe sequence and are easily accessible in biological fluids. Some researchers applied cDNA probe as a recognition element for miRNA detection specifically targeting breast cancer. Kilic et al. [56] developed an electrochemical biosensing platform for breast cancer detection. In this study, oxidation signals of alpha-naphthol (α-NAP) were measured for the detection of miRNA-21 by DPV method. miRNA-21 is found in the circulatory fluids and highly over-expressed in breast cancer patients. The capture probe was covalently immobilized on pencil graphite electrode (PGE) using EDC-NHS as a coupling agent and hybridized with its 3' biotinylated complementary target. After binding extravidin labelled alkaline phosphatase to the biotinylated target, electro inactive alpha naphthyl phosphate (the substrate) gets converted to electroactive α-NAP (the product). The obtained oxidation signals of α-NAP were used to check miRNA-21 level as low as 6pM from total RNA. Due to higher sensitivity, accuracy and less time devoting procedure, it is considered as an important alternative to the previously available miRNA screening methods like northern blotting. This method shows a better response in terms of specificity and reproducibility as compared to guanine-based detection methods. The developed method

Biosensors: Materials and Applications   Materials Research Forum LLC
Materials Research Foundations **47** (2019) 157-210   doi: http://dx.doi.org/10.21741/9781644900130-5

utilized biotin labeled target molecule which would make the procedure complicated for real sample analysis. Furthermore, Hong et al. [57] reported an electrochemical DNA concatamers based biosensing method for direct detection of circulating miRNA-21 in human serum. In this report, DNA concatamers acts as carriers for signal enhancement that was obtained from the self assembly of two auxillary probes AP1 and AP2. In the presence of its complementary target miRNA hairpin looped capture probe gets unfolded and binds to DNA concatamers through complementary base pairing resulting in the signal amplification. The developed biosensor can detect upto 100aM of target miRNA-21 with high accuracy and low complexity. Another biosensing strategy based on anti-miRNA-155/Au-SPE was developed for attomolar level detection of miRNA-155 in breast cancer patients with minimum complexity and great selectivity against CA15-3 biomarker [58]. An electrochemical magnetosensor for the simultaneous detection of breast cancer markers, miRNA-21 and miRNA-205 has been described in an experiment by Torrente-Rodreguez et al. [59]. It has been found that up regulation of miRNA-21 alone may not be able to discriminate between various receptors, in that case, miRNA-205 is only down regulated in triple negative breast cancer (TNBC) patients. So, this dual magneto sensor has proved to be highly sensitive and specific which could detect upto 0.6nM of target miRNA within 2h without any amplification. Furthermore, label free detection of biomolecules is highly advantageous than label based due to minimum cost and easy fabrication. Many other studies have been done for the label free detection of breast cancer biomarkers using nucleic acid bioreceptors [60,61]. Tian et al. [62] for the first time used polypyrrole coated AuNPs super lattice as a support material and toludine blue as a redox indicator for label free electrochemical detection of miRNA-21. The biosensing platform detected 78aM of miRNA-21 with a linear range from 100aM to 1nM in 60min. The supperlattice enhanced the surface area and electrical conductivity of proposed biosensor along with improvement in the amount of probe immobilized for the interaction with target molecule. It is a highly sensitive, selective and reproducible method which could be applied for direct assessment of miRNA-21 in real samples during clinical analysis. Another biosensor based on toludine blue as imprinted polymer receptor has been also employed for breast cancer biomarker CA15-3 detection by Ribeiro et al. [63].

Advancement in the biosensing methods for breast cancer detection has been observed in recent years. A disposable microfluidic electrochemical device (μFED) was constructed for the detection of breast cancer using estrogen receptor alpha (ERα) biomarker [64]. Increased level of ERα in breast epithelium is indicative of cancer development in about 50-80% of breast cases. ERα acts as a nuclear hormone receptor in association with other transcriptional factors, activates gene expression and affects cell proliferation as well as

Biosensors: Materials and Applications   Materials Research Forum LLC
Materials Research Foundations **47** (2019) 157-210   doi: http://dx.doi.org/10.21741/9781644900130-5

differentiation in the target tissues. A single device composed of a reference electrode, counter electrode and eight carbon-based working electrodes modified with ERα specific DNA sequence i.e. estrogen response elements (DNA-ERE), was employed in this study. The anti-ERα-antibody and HRP modified paramagnetic particles (MP-Ab-HRP) were conjugated with ERα to form ERα-MP-Ab-HRP bioconjugate which is further incubated over the surface of DNA-ERE modified (Figure 5). A mixed solution of H_2O_2 and hydroquinone was also injected to the microfluidic device and used for amperometric detection of ERα with applied voltage -0.2V vs Ag/AgCl and detection limit of 10fg/mL was achieved with good sensitivity, reproducibility and recovery percentage in a range from 94.7–108%. The μFED is cost effective due to use of less expensive materials in manufacturing and is available at a cost less than US$ 0.20 per device and can be easily prepared in less than 2h.

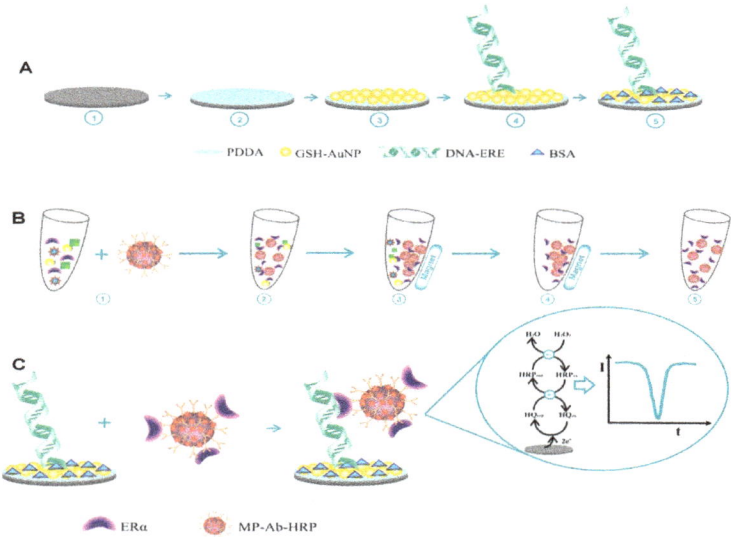

Figure 5: Diagrammatical representation for the electrochemical detection of ERα (A) Carbon-based screen-printed electrode modification with ERα specific DNA sequence (DNA-ERE); (B) Capture of ERα target molecule; (C) Electrochemical detection of the signals generated by sandwich type platform containing H_2O_2 and hydroquinone.

3.4   Biomimetic biosensor

Aptamer is preferred alternative over other bioreceptors due to their binding efficiency, very small size, synthetic nature and free of immunogenicity. Moreover, aptamer shows high stability to high temperature and can be easily modified. Sheng et al. [65] reported a rolling circle amplification method for the detection of MCF-7 cells using enzyme-catalyzed polymerization reaction. A hairpin-structured SYL3C aptamer specific for cancer cells were used as a probe for sensing of MCF-7 cells with LOD 12cells/mL. RCF method used for breast cancer diagnosis could improve the signal amplification by triggering the enzyme catalytic polymerization resulting in the release of cancer cells and the binding of multiple streptavidin-labeled HRP. Due to the complexity of development and involvement of enzymatic procedures, the biosensing method is not very effective. Afterwards, an electrochemical aptasensor with the utilization of free running DNA walker has been constructed for MCF-7 cells diagnosis [66]. DNA walker machines works by strand displacement cascade, cleaving the target oligonucleotide strands into shorter segments which produces some measurable signals. These signals were measured by CC under optimal conditions with a linear concentration range from 0-500cells/mL with LOD of 47cells/mL. With the use of DNAzymes, this method holds great potential due to its label free, cost effective and environmental friendly designing. Moreover, use of nanomaterials could enhance its sensitivity as well as integration with electrode chip could make them a portable and affordable device for cancer detection.

A number of sandwich type aptasensor for MCF-7 detection has been reported earlier which are associated with modification cost required for functional group labelling [67–70]. These functional groups somehow could also affect the affinity of aptamer molecules for their target analytes. So, a sandwich electrochemical aptasensor was designed to sense MCF-7 cells with polyadenylated   (polyA) MUC-1 specific aptamer modified gold electrode and polyA aptamer/AuNPs/GO hybrid which significantly enhanced the signal response [71]. It shows LOD 8cells/mL with a detection range from $10–10^5$cells/mL and found to be a highly valuable point of care diagnostic device for cancer. Polyadenine shows very high affinity with gold and makes the biosensing studies cost effective without the use of various functional groups like thiol, amine and sulfuryl etc.

Recently, Tian et al. [72] developed an electrochemical method for detection of MCF-7 cells based on reduced graphene oxide -gold nanoparticles composite as a fabrication platform and CuO nanozyme as a catalyst. The proposed biosensor applied MUC-1 specific aptamer for determination of MCF-7 cells with LOD 27cells/mL. CuO nanozyme is associated with a catalytic reduction of H_2O_2 as that of peroxidase enzyme and can improve the detection limit of prepared cytosensor.

Table 4: List of electrochemical biosensors developed for the determination of breast cancer biomarkers.

Biosensing platform	Biomarker	Detection method	Linear range	LOD	Ref.
Ex-Ap/Btn-miRNA-21/Anti-miRNA-21/EDC-NHS/PGE	miRNA-21	DPV	0.17-6.7mg/mL	0.67mg/mL	[55]
AntiBRCA-1/BMIM.BF4/MCN-TB/GCE	BRCA-1	CA	0.01-15ng/mL	3.97pg/mL	[38]
MCH/CP+ miRNA-21+AP1+AP2/SPGE	miRNA-21	DPV	100aM-100pM	100aM	[56]
NGS-Ab/GCE	CA15-3	DPV	0.1-20U/mL	0.012U/mL	[40]
HRP labeled Apt/MCF-7/Thio-Apt/Au electrode	MCF-7	CV	$100\text{-}1 \times 10^7$ cells	100cells	[69]
S1-Thi-PtFe/MCF-7/S1/GO/Au/GCE	MCF-7	DPV	$100\text{-}5 \times 10^7$ cells/mL	38cells/mL	[66]
HRP/Peptide modified Au electrode	MCF-7	CA	$1.0 \times 10^{-7}\text{-}1.0 \times 10^{-4}$ M	3.0×10^{-8} M	[37]
Anti-HER2/Au/SPCE	HER-2	LSV	15-100ng/mL	4.4ng/mL	[42]
Anti-HER2/Fe$_3$O$_4$NP/ Au electrode	HER-2	DPV	0.01-10ng/mL 10-100ng/mL	0.995pg/mL	[41]
GCE/HPCS@N-GS/Ab-Ag@Co$_3$O$_4$/BSA/BRCA-1/Ab-Ag@Co$_3$O$_4$	BRCA-1	CV	0.001–35ng/mL	0.33pg/mL	[44]
MCH/ssDNA/rGO/AuNPs/GCE	BRCA-1	EIS	$3.0 \times 10^{-20}\text{-}1 \times 10^{-12}$M $1.0 \times 10^{-12}\text{-}1 \times 10^{-7}$M	1×10^{-20}M	[48]
Apt-Thi-AuNPs/SiO$_2$@MWCNT/MUC-1/Apt-AuNPs-PoPD/Au electrode	MUC-1	DPV	1-100nM	1pM	[68]
PANI/RCA reaction/MCF-7/Hairpin aptamer P1/Au electrode	MCF-7	SWV	$20\text{-}5 \times 10^6$ cells/mL	12cells/mL	[64]
Strep-HRP/Btn ds-RNA/p19/Chitin-MBs/SPdCE	miRNA-21 miRNA-205	CA	2.0-10.0nM	0.6nM	[58]
S1 Probe/PEG/AuNPs/GCE	BRCA-1	EIS	50fM-1nM	1.72fM	[53]
MBCPE/Fe$_3$O$_4$@Ag/ssDNA	BRCA-1	EIS	$1.0 \times 10^{-16}\text{-}1.0$	$3.0 \times$	[49]

			$\times 10^{-8}$M	10^{-17}M	
DNA walker/D-RNA/Au electrode	MCF-7	CC	0-500cells/mL	47cells/mL	[65]
Thiolated-anti-miRNA-155/Au-SPE	miRNA-155	EIS,SWV	10aM-1nM	5.7aM	[57]
ssDNA probe/MWCNT-COOH/GCE	miRNA-21	DPV	0.1-500pM	84.3fM	[59]
C1/MCH/Peptide/Au electrode	BRCA-1	EIS	1fM-10pM	0.3fM	[54]
ssDNA probe/Au/PEG/TA/pDA/GCE	BRCA-1	EIS	0.1fM-10pM	0.05fM	[52]
polyA MUC-1 Apt/AuNPs/GO/MCF-7/Apt MUC1/ Au electrode	MCF-7	DPV	$10-10^5$ cells/mL	8cells/mL	[70]
HS-ssDNA/Au electrode	BRCA-1	EIS	1-200nM/L	0.15nM/L	[60]
CD24-c/AuNPs-GO/GCE ERBB2-c/AuNPs-GO/GCE	HER-2	CA	0.37-10nM	0.23nM 0.16nM	[50]
rGO-TPA/FeHCF$_{nano}$/ Anti-HCT SKBR3/Anti-HCT/rGO/SPE	SKBR3	DPV	5×10^2 -3×10^4 cells/mL	21cells/mL	[45]
ssDNA/CHI-NVTO/ITO	BRCA-1	EIS	1.0×10^{-15}M $\sim 1.09 \times 10^{-6}$M	1.09×10^{-16}M	[51]
H_2O_2-HQ/Btn Ab/HER-2/Ab/Inkget printed Au electrode (WEA)	HER-2	Amperometric	0.002-12.5ng/mL	12pg/mL	[47]
Anti-MUC1/GO-COOH-SPCE	MUC-1	DPV	0.1- 2U/mL	0.04U/mL	[39]
Protein imprinted poly TB film/Au SPE	CA15-3	DPV	0.10-100U/mL	0.10U/mL	[62]
Ag/LB/HER2/BSA/PB/GCE	HER-2	DPV	5×10^{-4}-50ng/mL	2×10^{-5} ng/mL	[43]
H_2O_2/CuO nanozyme/MCF-7/MUC-1 Apt/rGO-AuNPs/GCE	MCF-7	DPV	$50-7 \times 10^{-3}$cells/mL	27cells/mL	[71]
TB/miRNA-21/ssRNA/AuNS/GCE	miRNA-21	DPV	100aM-1nM	78aM	[61]
H_2O_2-HQ/ERα-MP-Ab-HRP/DNA-ERE/ µFED	MCF-7	Amperometric	16.6-513.3fg/mL	10fg/mL	[63]

Ex-Ap: Extravidin-alkaline phosphatase; BMIM.BF4: Butyl-3-methylimidazolium bromide; MCN-TB: Mesoporous carbon nanosphere-Toluidine blue; MCH/CP: 6-mercapto-1-hexanol/Capture probe DNA; AP: Auxiliary probe; SPGE: Screen printed gold electrode; NGS: Nitrogen doped graphene sheets; Thio-Apt: Thiolated aptamer; S1: MUC-1 Aptamer; SPCE:

Biosensors: Materials and Applications   Materials Research Forum LLC
Materials Research Foundations **47** (2019) 157-210   doi: http://dx.doi.org/10.21741/9781644900130-5

Screen printed carbon electrode; LSV: Linear sweep voltammogram; Fe_3O_4NP: Iron oxide nanoparticles; BSA: Bovine serum albumin; HPCS: Hydroxypropyl Chitosan; SiO_2@MWCNT: Silica@multiwalled carbon nanotubes; PoPD: Poly(o-phenylenediamine); PANI: Polyaniline; RCA: Rolling circle amplification; P1: Capturing probe; Strep: Streptavidin; Chitin-MBs: Chitin modified magnetic beads; SPdCE: Dual screen printed carbon electrodes; S1 Probe: Immobilized capture probe; PEG: Polyethylene glycol; MBCPE: Magnetic bar carbon paste electrode; D-RNA: A chimeric DNA/RNA oligonucleotide with a cleavage point; C1: Capture probe; TA: Tannic acid; pDA: Polydopamine; polyA: Polyadenylated; HS-ssDNA: Probe DNA; CD-24c: Complementary probe; TPA: Tetra sodium1,3,6,8-pyrenetetrasulfonic acid; $FeHCF_{nano}$: Iron hexacyanoferrate nanocomposite; Anti-HCT: Primary herceptin antibody; SKBR-3-Sloan-Kettering breast cancer; NVTO: (Nb,V) codoped TiO_2; H_2O_2-HQ: Hydrogen peroxide-hydroquinone; WEA: Working electrode array; LB: Label bioconjugate; PB: Platform bioconjugate; AuNS: Gold nanoparticles superlattices; ERα: Estrogen receptor alpha; MP: Modified paramagnetic particles; ERE: Estrogen response element; μFED: Microfluidic electrochemical device

4. Prostate Cancer

Prostate carcinoma is the most widespread threat amongst men affecting approximately every 1 in 7 men worldwide. Since therapeutic strategies can only be implemented after the identification of the disease and due to usual absence of signs and symptoms screening helps to find cancers at early stages. Several serological biomarkers exist for early detection and monitoring of prostate cancer including proteins (PSA, PSMA, VEGF, EN2, SOX2) miRNAs (miRNA-375, 141, 182-5p), glycoprotein (fetuin) and amino acid derivative (sarcosine). Several biorecognition elements (antibodies, aptamers, lectins, enzymes, DNA probes etc.) and nanocomposites have been utilized to enhance the performance of the biosensors (Table 5).

4.1 Enzyme-based biosensor

The catalytic activities of enzymes and their high specificity for the target have encouraged the researchers to develop enzyme-based biosensors. Rebelo et al. [73] prepared an enzyme-based electrochemical biosensor against sarcosine biomarker present in elevated levels in prostate cancer using sarcosine oxidase immobilized SPE with LOD of 16nM in a urine sample. However, with an improved LOD (0.1pM) and storage stability (180 days), another biosensor with an immobilized sarcosine oxidase enzyme on cMWCNT/CHI/copper nanoparticles (CuNPs)/Au electrode was developed [74]. Despite all, these biosensors require optimal conditions for the enzyme's activity and are usually expensive due to isolation procedures for the enzyme.

4.2   Antibody-based biosensor

Numerous established methods such as ELISA, RIA and fluorescent immunoassay are very sensitive methods for diagnosis but are laborious, expensive and require sophisticated instrumentation and trained staff. To overcome such limitations, Immunoassay based biosensors have proved to be the potential real-time diagnostic devices offering improvement of patients compliance and clinical outcomes and cut off in clinical costs. A variety of enzyme-labeled immunosensors has been developed for sensitive detection of PSA. Salimi et al. [75] reported an immunosensor for detecting PSA in serum and prostate carcinoma cells of human patients. They fabricated the GCE electrode with MWCNTs– ionic liquids (ILs)–thionine nanocomposite to immobilize the HRP-labeled anti-PSA antibody and the electrochemical response against the target was monitored using DPV technique. The detection limit obtained was 20pg/mL with a signal to noise ratio as 3. With an improved detection limit, another study by Tran et al. [76] reported an immunosensor based on ELISA-like detection of prostate cancer-specific miRNA-141 biomarker. The strategy involved hybridization of the oligonucleotides capture probe and complementary miRNA sequence on the reduced graphene oxide (rGO) and CNT based electrode. The resulting RNA-DNA hybrid was then conjugated to the antibody for ELISA-like configuration using HRP-conjugated secondary antibody. This novel method showed LOD down to 10fM. However, the activity of HRP enzyme gets affected by temperature, pH, humidity and other factors, therefore, metals are excellent replacements for enzymes. For example, Zhang et al. [77] used the platinum nanoparticles encapsulated in infinite coordination polymer (ICP) grafted on polyamidoamine (PAMAM) layer with anti-PSA antibody to form a sandwich-like-immunosensor as shown in Figure 6 for voltammetric detection of PSA in real serum samples and achieved a detection limit of 0.3pg/mL. Since the irreversible conjugation of biomolecules on the electrode hinders the regeneration of the electrode that necessitates the need of efficient reusable biosensor. On account of this Hong et al. [78] reported a reusable immunosensor based on poly(N-isopropylacrylamide) (PNIPAAm) polymerized on Au surface having immobilized HRP-labeled antibody against PSA. The amperometric detection resulted in the detection limit of as low as 0.9pg/mL. Other immunosensors based on sandwich-like-immunoassay with enzymatic labels as glucose oxidase and HRP had a detection limit of 0.0012ng/mL and 0.5136ng/mL respectively [79,80]. In the last five years, the highly ultrasensitive immunosensor with minimum LOD (i.e. 0.24fg/mL) has been designed by Pal and Khan [81] fabricating the platinum electrode with AuNPs supported graphene oxide nanocomposites to immobilize the monoclonal anti-PSA antibody. Recently, the antibody based immunosensor was developed by Khan et al. [82] using human saliva as an analyte source for detecting PSA.

Biosensors: Materials and Applications   Materials Research Forum LLC
Materials Research Foundations **47** (2019) 157-210   doi: http://dx.doi.org/10.21741/9781644900130-5

They constructed a paper based graphene-polymer (PS_{67}-b-PAA_{27})-Au biosensor and immobilized the anti-PSA antibody onto the electrode via dithiobis (succinimidyl propionate). The amount of PSA binding to the biosensor caused changes in the electrical resistance of the biosensor thus allowing the detection of PSA in human saliva with a faster response time of 3-5 minutes with LOD of 40fg/mL. Blel et al. [83] used the functionalization strategy to improve the sensitivity against the target PSA with the construction of two sensors: one functionalized with 3- glycidoxypropyltrimethoxysilane and second with iron oxide nanoparticles functionalized with 3-aminopropyltriethoxysilane. Both immunosensors found to have LOD in the order of 10fg/mL. Now-a-days, label-free immunosensors [84–86] have taken over ELISA based immunosensors as they can directly measure the antigen-antibody interactions without the need of examining enzymatic activity. For example, label-free electrochemical immunosensor by Aydin and Sezginturk [21], fabricated using carboxyethylsilanetriol (CTES) modified ITO electrode immobilized with an anti-SOX2 antibody, allowed LOD to be 7fg/mL in the linear range 25fg/mL-2pg/mL. In order to be more accurate about the detection of prostate cancer, Sharafeldin et al. [87], constructed an ultrasensitive sensor by examining two prostate specific biomarkers PSA and PSMA simultaneously allowing high selectivity. This dual marker detection was due to the iron oxide nanoparticles loaded onto graphene nanosheets applied on screen carbon printed electrode functioning as magnetic analyte isolation as well as label free detection. The LOD for PSA and PSMA were 15fg/mL and 4.8fg/mL respectively. Another study demonstrated by Pan et al. [88], lead to the simultaneous detection of two prostate biomarkers, VEGF and PSA, using graphene oxide-ssDNA biosensor immobilized with poly-$_L$-lactide (PLLA) nanoparticles modified with dual-antibodies against VEGF and PSA. Such immunosensors increases the reliability by proving accurate and rapid results with a detection limit as 50pg/mL and 1ng/mL for VEGF and PSA, respectively.

Electrochemiluminescence (ECL) based electrochemical immunosensor offer potential benefits such as the elimination of scattered light in samples and no background from photoexcitation thus improving the sensitivity of the sensor. Several ECL-based biosensors have been reported, for example, Wu et al. [89] developed ECL approach based on the bipolar electrode with its surface doped with thionine silica nanoparticles and $Ru(bpy)_3^{2+}$ as ECL emitter to detect PSA (LOD- 0.49ng/mL). In comparison to thionine silica nanoparticles, near-infrared emitting nanoparticles were advantageous due to minimal biological auto fluorescence and tissue absorption in 650-900 nm wavelength range. Hence, Gao et al. [90] synthesized near-infrared emitting $NaYF_4$:Yb, Tm/Mn upconverting nanoparticles and Au nanorods as donors and acceptors respectively and observed LOD as 3.16pg/mL. Another study by Ma et al. [91] reported an ECL based sensor comprising of AuNPs doped Pb(II)-β-cyclodextrin dropped on GCE immobilized with anti-PSA antibody. The LOD obtained was 0.34pg/mL.

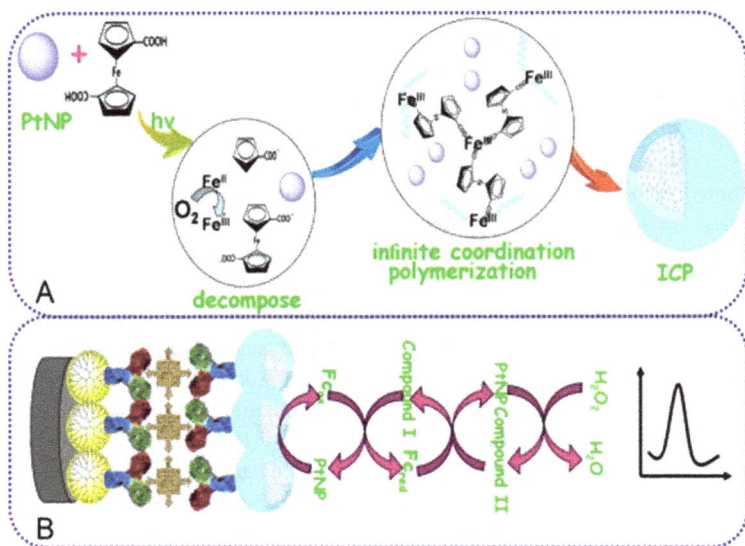

Figure 6: (A) Mechanism for the formation of 'all-in-one' redox-active PtNP@ICP nanocatalyst; and (B) schematic illustration of the electrochemical immunosensor by coupling with the 'all-in-one' redox-active PtNP@ICP nanocatalyst.

Nowadays, artificial antibodies or the so-called molecularly imprinted polymers (MIPs) are being used for developing biosensors due to their long shelf life, cost-effectiveness and high sensitivity/selectivity. One such electrochemical immunosensor was synthesized by Rebelo et al. [92] where the MIPs were doped on graphene layers to detect PSA in artificial serum as well as in several prostate cell lines to assess the expression of several genes expressed by healthy and prostate carcinoma cells with obtained LOD as 2ng/mL. However, the construction of these MIP-based biosensors still needs to be focused to head towards the miniaturization for achieving marketing value.

Although there are wide varieties of electrochemical immunosensors, only a few have made it to commercial success. This could be due to expensive production and storage problems associated with the appropriate biological receptors.

4.3   Lectin-based biosensor

Alterations in the glycosylation pattern of the glycoproteins are an indication of the tumour progression and metastatic spread. Thus, the ability of lectin-based immunosensors to determine the changes in a minimal amount of sample with high sensitivity in a quick manner are gaining attention in the detection of prostate cancer. Silva et al. [93] used Cratylia mollis lectin having an affinity for fetuin biomarker to prepare a label-free nanosensor assembled with CNTs and poly-l-lysine exhibiting 0.017µg/mL detection limit (Figure 7). Further Pihikova and colleagues immensely contributed to the exploration of lectin-based immunosensors using an antibody against PSA to detect the specific target PSA up to aM followed by ultrasensitive *in situ* glycan analysis by lectins on the same electrode surface, which otherwise would not have been possible with immobilized lectins [94–96]. In their recent work, two different types of lectins, *Sambucus nigra* agglutinin type I (SNA) and *Maackia amurensis* agglutinin II (MAA) were used for examining altered sialylation on PSA surface. This impedimetric biosensor had detection limit of 100ag/mL [23]. Though these biosensors effectively detected the prostate cancer but they offer low throughput analysis, have complex electrode fabrication, less reproducibility and further work is required to observe the substantial change in the sialylation patterns on PSA in prostate cancer patients. Moreover, they require more accurate measurements by eradicating frequent noises and misinterpretations.

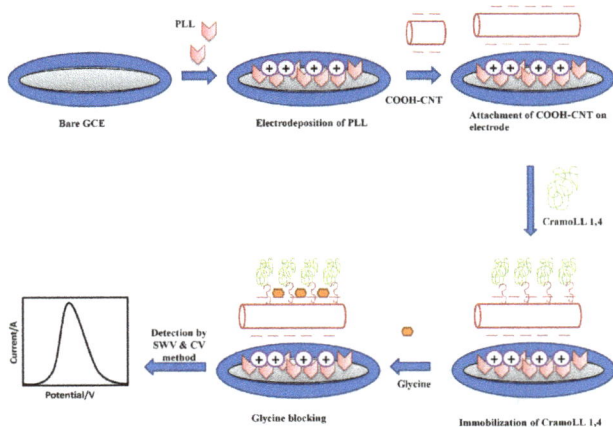

Figure 7:   Schematic illustration for the development of CramoLL 1,4 lectin nanoelectrode in the prostate cancer diagnosis.

4.4   Nucleic acid-based biosensor

Oligonucleotides have the higher binding efficiency with the target sequence and less immunogenicity due to a smaller size as compared to the antibodies. Ren et al. [97] developed a DNA assisted electrochemical biosensor based on proximity ligation assay. The Au interface modified with F_c-P was grafted with methylene blue (MB)-DNA1-antibody1. In the presence of target PSA, the hybridization of both DNA sequences gets triggered and MB-DNA1-antibody1 departs from the sensing surface leaving behind only F_c-P hairpin structure to produce a signal. Thus, the electrochemical sensor readouts an increase in the oxidation signal ratio of F_c to MB with LOD as 16pg/mL. To improve the assay time, flexibility and detection limit, Ren et al. [98] further designed another electrochemical DNA sensor based on proximity hybridization and achieved a detection limit of 4.3pg/mL. Although these were novel strategies to detect the biomarker but the detection limit was extremely narrow as compared to other electrochemical biosensors.

In the earlier work, Tran et al. [86] achieved a detection limit of 8fM with MWCNTs electropolymerized with two monomers, 3-(5-hydroxy-1,4-dioxo-1,4-dihydro naphthalene-2(3)-yl) propanoic acid (JUGA) and 5-hydroxy-1,4-naphthoquinone (JUG), by immobilizing oligonucleotides probe on the sensing interface. The presence of miRNA-141 biomarker allowed hybridization with oligonucleotides probe leading to increase in the current thus allowing detection of a prostate cancer biomarker in the sample. Lee et al. [99] used the electrochemical cell fabricated with electrodeposited AuNPs onto the Au sensing surface. The prostate cancer biomarker named EN2 is a transcriptional factor whose unusual expression marks the irregularity of the transcription process ultimately resulting in cancer. The EN2 was captured using the DNA probe with LOD of 5.62fM.   Saheb, Patterson and Josowicz [100] described the use of electrochemical detection of the hypermethylated GSTP-1 (prostate cancer biomarker) using ssDNA detector to recognize the methylation pattern of the complementary DNA strand from GSTP-1 promoter region. The method is based on hybridization of the complementary DNA sequences (i.e. non-methylated and methylated sequences) with the ssDNA probe immobilized on the polypyrrole-platinum-microelectrodes. The CV examined the difference in the area of the exposed complementary and non-complementary DNA target. Although nucleic acid-based biosensors are able to detect a complementary target with less response time. Still, necessity of pretreatment step makes it less preferable choice for point of care testing. Hence, Chang et al. [101] established a reusable biosensor for miRNA-182-5p prostate marker detection based on DNA cross configuration. The designed electrochemical biosensor converts the single miRNA to multiple ssDNA after cycling and strand displacement reaction amplification to enhance the detection sensitivity. This PAMAM-CNTs-platinum nanomaterials constructed electrochemical biosensor had 0.5fg/mL as the limit of detection.

4.5   Biomimetic biosensor

Furthermore, antibody immobilized biosensors also offer limitations such as expensive, inconvenience, stability and need for expertise manpower for production of antibodies. To overcome these, oligonucleotides based sensors are preferred. Aptamers being one of many choices have the potential to capture the target with strong affinity and specificity thus widening the detection range. The aptamer-based electrochemical biosensors include multiple nanomaterials and binding steps for the aptamer fabrication on the electrode to achieve the sensitive detection limit for PSA [102–106]. Yang et al. [107] used DNA assisted aptamer immobilization strategy to detect PSA using dsDNA intercalator with a Fc redox label and observed LOD as 1.5fM. Another study by Tzouvadaki et al. [108] reported a first worldwide DNA aptasensor based on memristive effect achieving detection limit down to 23aM. The aptasensor is comprised of silicon nanowires functionalized with DNA aptamers against PSA that is capable in lowering the detection limit making it an ultrasensitive detection technique. To improve the recognition properties of aptasensor, Jolly et al. [102] used molecularly imprinted polymers (electropolymerised polydopamine) with aptamers instead of aptamer alone. The detection of PSA using apta-MIP surface showed three-fold higher detection (1pg/mL) than previously reported PSA based aptasensor. Settu et al. [109] developed the screen-printed carbon graphene electrode to immobilize the amine modified DNA aptamer for the detection of EN2. Although EN2 can be potentially used for diagnosing breast, ovarian and bladder cancer, it has been reported that EN2 also releases in the urine of the prostate cancer patients. The present biosensor allowed the detection of EN2 in the urine with LOD as 33.8nM and in the linear range from 35 to 185nM. Crulhas et al. [110] prepared a silicon coated gold electrode immobilized with aptamer for PSA and VEGF and MB as redox marker to detect the cancer. The aptamer-protein proximity unfolds the DNA hairpin allowing the transfer of the electron and ultimately altering the current response.

4.6   Phage-based biosensor

Mohan et al. [111] used PEDOT films integrated with phage bearing two ligands having an affinity for prostate marker PSMA. One of the two ligands was genetically encoded and the other was synthesized chemically to wrap around the phage. The two ligands allowed better detection due to dual ligands having an affinity for the same biomarker. The observed detection limit in synthetic urine was 100pM. Despite using two ligands for

Materials Research Forum LLC

doi: http://dx.doi.org/10.21741/9781644900130-5

the same target, the LOD achieved is not optimal and hence could be improved for an enhanced response for PSMA or other prostate specific biomarkers.

4.7   Fabricated biochips

Towards the advancement of commercialization and the miniaturization of the detection biosensors, efforts are being progressively made for the development of fabricated biochips and microfluidic-based paper-based analytical devices. Chiriacò et al. [112] reported a biochip device comprising of two chambers: one for free PSA and second for total PSA determination using antibodies for the detection of PSA in the sample. The formation of Ag-Ab complex on EIS chip changed the capacitance and resistance in the electron flow that allows the percentage ratio detection of the amount of free and total PSA, which further elucidated the difference between malignant and benign prostate hyperplasia. The detection limit obtained was 1ng/mL. Li et al. [80] designed an electrochemical paper-based 3D analytical origami device. Firstly, the origami device comprising of the auxiliary pad and sample tab was prepared using wax printing on paper sheets and then fabricated with AuNPs electrodeposited with manganese oxide nanowires. Afterwards a monoclonal capture antibody against PSA was immobilized on the fabricated device which binds to PSA and then detected using carbon nanospheres/glucose oxidase-McAb2 (the signal anti-PSA antibody) by catalyzing the GOx (enzyme label) enzyme-based reaction of forming gluconic acid and H_2O_2 from glucose in presence of oxygen and subsequently MnO_2 oxidized 3,3',5,5'-tetramethylbenzidine (TMB; redox terminator) upon reduction of H_2O_2 to H_2O. This enzymatic electrochemical sensor based on redox cycling was applied to detect PSA with a detection limit of 0.0012ng/mL. In recent years, Uludag et al. [113] envisioned a biochip for the detection of PSA with LOD as 0.2ng/mL. The strategy is based on the detection of interaction of PSA to the antibody immobilized onto the electrode by HRP modified AuNPs. Mohamadi et al. [114] envisioned a velocity valley (vv) chip to assess the state of the prostate cancer. The strategy involves the capture of the prostate cancerous cells with EpCAM conjugated superparamagnetic nanoparticles followed by the electrochemical lysis of the entrapped cells. The lysate was further sensed for PSA mRNA using peptide nucleic acid probe specific for PSA. Hence, vv chip integrated with electrochemical sensing strategy to detect specific PSA mRNA sequences in patient samples proved valid, though, it still requires experimentation with more clinical samples to assess its further applications. Moscovici, Bhimji and Kelley [115] designed a glass chip fabricated with chrome and gold layer followed by functionalization with EpCAM antibody which rapidly detected the prostate cancer cells. Xu et al. [116] developed a super wettable microchip with gold nanodendrites deposited on Ti/Au thin film to detect miRNA-375, miRNA-141, and PSA. For miRNA sensing, the microchip comprised of redox-reporter-modified DNA probe which on target binding undergoes a conformational change thus altering the position of the Fc region of the reporter producing current change. Whereas for PSA detection in goat serum, the microchip was immobilized with PSA aptamer so that binding of PSA to the aptamer, there are changes in the position of the reporter relative to the sensing interface thus altering the current response. The limits of detection using miRNAs and PSA as biomarkers were 0.8nM and 1pM respectively.

Table 5: List of electrochemical biosensors utilized in detection of prostate cancer.

Biosensing platform	Biomarker	Detection method	Linear range	LOD	Ref.
Sarcosine oxidase/cMWCNT/CHI/CuNPs/Au electrode	Sarcosine	Amperometric	0.1-100µM	0.1pM	[74]
Anti-PSA Ab/GRP-PS67-b-PAA27-Au electrode	PSA	PCB	0.1pg/mL-100ng/mL	40fg/mL	[82]
nAu/Au/Ti-ITO	miRNA-141, 375 & PSA	DPV & CV	miRNA-10nM-10µM & PSA-10pM-10nM	miRNA-0.8nM & PSA-1pM	[116]
Aptamer/Au/Si	PSA & VEGF	SWV	PSA-0.08-100ng/mL & VEGF- 0.15-100ng/mL	PSA-0.08ng/mL & VEGF-0.15ng/mL	[110]
Fe_3O_4NP/3-aminopropyltriethoxysilane SAM	PSA	SWV	-	50pg/mL	[83]
Anti-PSA Ab/AuNPs/GO	PSA	CV & EIS	0.001fg/mL-0.02µg/mL	0.24fg/mL	[81]
Nano Au/PAMAM-CNTs-Pt/GCE	mi-RNA-182-5p	SWV & CV	-	0.5fM	[101]
DNA aptamer/graphene/SPCE	EN2	CV	35-185nM	38.5nM	[109]
Anti SOX-2/CTES/ITO	SOX2	CV & EIS	25fg/mL-2pg/mL	7fg/mL	[20]
Anti-PSA Ab_2/MUDA modified GNRs & Anti-PSA $Ab_{1/}$ PAA modified NaYF4:Yb,Tm/Mn UCNPs	PSA	ECL-RET	3.75-938pg/mL	3.16pg/mL	[90]

Ab$_2$-Fe$_3$O$_4$@GO particles/rGO/SPCA	PSA & PSMA	Amperomet ric	PSMA-9.8fg/mL-10pg/mL & PSA-61fg/mL-3.9pg/mL	PSA-15fg/mL & PSMA-4.8fg/mL	[87]
Aptamer/rGO-MWCNT/AuNPs	PSA	DPV, EIS	0.005-20ng/mL, 0.005-100ng/mL	1pg/mL	[105]
Anti-PSA Ab/HRP -AuNPs/SiO$_2$/Ti/Au electrode	PSA	CV	-	0.2ng/mL	[113]
Lectin/PSA/CF/anti-PSA/SAM5/Au electrode	PSA	EIS	100ag/mL-1µg/mL	100ag/mL	[95]
Anti-PSA DNA aptamers/Si NWs	PSA	Memristive aptasensor	33aM–330fM	23aM	[108]
Aptamer-nanospears Au/Au electrode	PSA	DPV & CV	0.125-200ng/mL	50pg/mL	[103]
Cramoll1,4/CNTs/PLL/G CE electrode	Fetuin	SWV & CV	0.5-25µg/mL	0.017µg/m L	[93]
GO/ssDNA/PLLA NPs/Au electrode	VEGF & PSA	DPV & CV	VEGF- 0.05-100ng/mL & PSA-1-100ng/mL	50pg/mL-VEGF & 1ng/mL-PSA	[88]
Anti-PSA Ab/Ag@Pb(II)-β-CD/GCE	PSA	ECL, DPV & CV	0.001-50ng/mL	0.34pg/m L	[91]
HRP-anti PSA/PSA/biotin anti-PSA/avidin/NH$_2$-PEG$_3$-biotin/SAM1/Au electrode	PSA	EIS	2-24ng/mL	0.51ng/m L	[79]
MIP/Graphite/epoxy resin/Cupper electrical wire	PSA	Potentiomet ric	2.0-89.0ng/mL	2ng/mL	[92]
Biotinylated Ab/Streptavidin/BSA-Biotin/PNIPAAm/Au electrode	PSA	Amperomet ric	10pg/mL-10ng/mL	0.9ng/mL	[78]
Aptamer-MIP/Au electrode	PSA	EIS	100pg/mL-100ng/mL	1pg/mL	[102]
PNA probe/NME	PSA expressing circulating tumor cells	DPV	-	-	[114]
ZNRs/rGO-PWE	PSA	CV	0.001–110ng/mL	0.35pg/m L	[85]
Aptamer/ssDNA/Au electrode	PSA	DPV	0.05-50ng/mL	0.5pg/mL	[107]

Anti-PSA Ab/CdS/ZNTs/AuPd/ITO	PSA	EIS	1.0pg/mL-50ng/mL	0.3pg/mL	[106]
Fc-labeled hairpin DNA/Au electrode	PSA		0.01-200ng/mL	4.3pg/mL	[98]
DNA probe/AuNPs/Au electrode	EN2	EIS	-	5.62fM	[99]
PSA capture probe/Au-ITO hybrid BPE	PSA	ECL	5×10^{-4}-100ng/mL	0.49ng/mL	[89]
Sarcosine oxidase/carbon/SPE	Sarcosine	EIS	10-100nM	16nM	[73]
Anti-PSA Ab/PAMAM/GCE	PSA	DPV	0.001-60ng/mL	0.3pg/mL	[77]
Anti-PSA/graphene/AuNPs/GCE	PSA	CV	0-10ng/mL	0.59ng/mL	[84]
ODN-141-P/o-MWCNT/rGO/GSPE	miRNA-141	SWV	10fM–1nM	10fM	[76]
McAb/BSA-MnO_2/AuNPs/paper fiber	PSA	DPV	0.005-100ng/mL	0.0012ng/mL	[80]
Aptamer/MPA/Au electrode	PSA	QCM & EIS	-	-	[117]
MB-DNA_1-Ab_1/F_c-P/Au electrode	PSA	ACV	0.05-100ng/mL	Fc signal-43 pg/mL & MB signal- 48 pg/mL	[97]
Anti-PSA doped polypyrrole/AuNWs	PSA	DPV	10fg/mL-10ng/mL	0.3fg/mL	[97]
GSTP-1 promoter probe DNA/pTPT/Polypyrrole/Pt electrode	Hypermethy lated Gluthione S-transferase P1 (GSTP-1)	CV	-	-	[100]
Anti-PSA(total)/BSA/protein A/SAM3/Au electrode	PSA	EIS	1-10ng/mL	1ng/mL	[112]
poly(JUG-co-JUGA)/o-MWCNT/ODN probe	miRNA-141	SWV	100pM -1fM	8fM	[86]
Phage wrapped secondary ligands/PEDOT/Au electrode	PSMA	EIS	-	100pM	[111]
Anti-EpCAM Ab/Au/Cr/Glass	EpCAM	DPV	-	--	[115]

HRP-PSA/anti-PSA/MWCNTs-ILs/Thi/GCE	PSA	DPV	0.2-40ng/mL	20pg/mL	[75]

cMWCNT: Carboxylated multi-walled carbon nanotube; CuNPs: Copper nanoparticles; GRP-PS67-b-PAA27: Graphene-polymer; nAu: Nanodendritic gold; Ti: Titanium; Si: silicone; PAMAM: Polyamidoamine; Pt: Platinum; MUDA: 11-mercaptoundecanoic acid; GNRs: Gold nanorods; PAA: Poly(acrylic acid) ; UCNPs: Upconverting nanoparticles; Fe_3O_4: Iron oxide; SPCA: Screen printed carbon assay; SiO_2: Silicon oxide; CF: Carbo-free; PLL- Poly-l-lysine; PLLA: Poly-l-lactic acid; Ag: Silver; Pb: lead; β-CD: beta-cyclo dextrin; MIP: Molecularly imprinted polymer; PNIPAAm: Poly(N-isopropylacrylamide); PNA: Peptide nucleic acid; NME: Nanostructured microelectrodes; ZNRs: Zinc nanorods; PWE: Paper working electrode; ZNTs: Zinc oxide nanotubes; AuPd: Gold-palladium; BPE: Biopolar electrode; ODN-P: Oligonucleotide probe; GSPE: Gold screen printed electrode; Mc: Monoclonal; MnO_2: Manganese oxide; MPA: 3-mercaptopropionic acid; JUG: 5-hydroxy-1,4-naphthoquinone; JUGA: 3-(5-hydroxy-1,4-dioxo-1,4-dihydronaphthalen-2(3)-yl) propanoic acid; Cr: Chromium; PEDOT: Poly(3,4-ethylenedioxythiophene); ILs: Ionic liquids.

5.   Colorectal Cancer

Colorectal cancer is the fourth most occurring cancer and the fourth main cause of cancer-related deaths all over the world. It accounts for more than 9% of all cancer-related deaths and is equally diagnosed among men and women [117]. Incidents of colorectal cancer are higher in developed countries with western culture. Weight, diet and physical exercises are highly associated with the possibility of colorectal cancer progression. Risk of developing colorectal cancer is 1 in 22 for men and 1 in 24 for women in their lifetime. Colorectal cancer generally starts with the development of polyps which is associated with the growth of the inner lining of the colon or rectum. From the past few years, the death rate due to colorectal cancer has been decreased due to screening and removal of colorectal polyps. Nowadays, colonoscopy with biopsy confirmation using cancer tissue or blood is the most effective diagnostic method for colorectal cancer. This technique is invasive with the risk of colon wall perforation, highly expensive and requires lengthy clinical processes. Electrochemical biosensing method by using biomarker molecules is a highly effective method for early-stage colorectal cancer detection.

Commonly used tumour biomarkers in colorectal cancer detection are glycoproteins (CEA, CA19-9, TIMP1), genetic (p53, KRAS, BRAF and *PIK3CA* genes mutation) and miRNAs etc [118–120]. Among all the available biomarkers, CEA (carcinoembryonic antigen) is highly associated with colorectal cancer. CEA is a glycoprotein which is produced in the bloodstream by the cells of gastrointestinal tract after birth at a very low level. Increased concentration of CEA is associated with many other carcinomas like

Biosensors: Materials and Applications   Materials Research Forum LLC
Materials Research Foundations **47** (2019) 157-210   doi: http://dx.doi.org/10.21741/9781644900130-5

breast, ovarian, lung but especially in colorectal cancer. Serum level of CEA in colorectal cancer is found to be always higher than healthy individuals with approximately 5ng/mL of concentration [121]. It has been found to be a useful biomarker to identify staging, the spread of the disease and reoccurrence after any treatment or surgical resection. A vast variety of biosensing methods has been studied in the past few years for CEA detection [121–126]. Apart from this, many other electrochemical biosensing methods have been specifically targeted for colorectal cancer detection only. Some diagnostic approaches for colorectal cancer are being discussed below based on the bioreceptor molecule and also indexed in Table 6.

5.1   Enzyme-based biosensor

Situ et al. [127] have demonstrated an enzyme-based electrochemical biosensor for the detection of BRAF mutation in colorectal cancer. BRAF mutation involves transversion from valine to glutamic acid in codon 600 (V600E). It affects the RAS-RAF-MAPK signalling pathway associated with the performance of anti-EGFR and enhances cancer progression in breast tissues. The biosensor was designed by a dual amplification approach involving amplification refractory mutation system (ARMS) PCR and multi-enzyme labelled Fe_2O_3/Au nanoparticles (Figure 8). ARMS is utilized for the identification of mutations like a change in a single nucleotide and deletion. ARMS being highly productive amplification method were utilized for incorporating thiol groups and biotin labels to the target allele producing biotin-tagged amplicons. In the fabrication step, thiolated amplicons were incorporated onto Fe_2O_3/Au nanocomposite by strong Au-S binding and streptavidin-alkaline phosphatases (Sa-ALPs) was captured through biotin-streptavidin binding. Ascorbic acid being product of enzymatic reaction was used for the detection of mutant alleles. In the last step, the oxidation peaks of ascorbic acid were measured on SPCE which can detect as low as 0.8% BRAF V600E mutation in the presence of wild-type background. The diagnostic technique is highly sensitive compared to DNA sequencing as well as agarose gel electrophoresis.

Furthermore, another enzymatic approach based on a multiplexed biochip in a 2-electrode electrochemical device for detection of matrix metalloproteinases (MMP2 and MMP7) has been developed [128]. MMP2 and MMP7 are used to monitor initiation, progression and transition from other cancers and considered as a useful biomarker for ovarian and colorectal cancer detection. Microfluidic chip consists of a glass substrate with Au as working and Pt as a reference electrode having layers of photoresist–polydimethylsiloxane over it. MMP specific peptide is deposited on Au electrode surface and work through enzymatic hydrolysis process as MMP (contains hydrolyzing enzyme-zinc dependent endopeptidases) were injected into the biochip. The MMP molecules were measured simultaneously by EIS method with a detection limit of 0.5pg/mL in a detection range from 0.1–400 ng/mL and 0.001–100 ng/mL for MMP2 and MMP7, respectively in just 1hour. The developed polymer microfluidic chip has been found to be low cost, portable, require less time and sample hence could be useful for point of care testing multiplex biochip for many other forms of cancer. Moreover, the device could be applied for other biomarkers having hydrolyzing activity available with them.

Figure 8: Pictorial illustration of the developmental steps involved in electrochemical

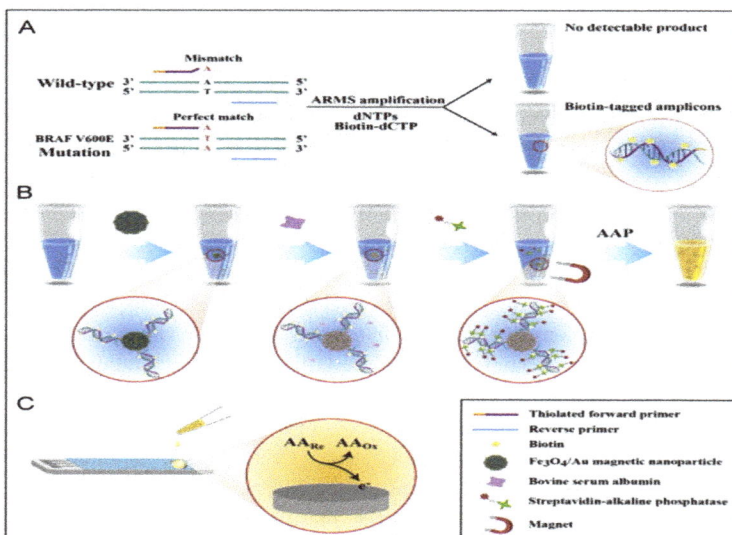

detection of BRAF V600 mutation in colorectal cancer (A) ARMS amplification producing biotin-tagged amplicons; (B) Fabrication of the amplicons onto Fe_2O_3/Au nanocomposite; (C) Oxidation peaks of ascorbic acid at SPCE surface for electrochemical detection.

5.2   Antibody-based biosensor

An electrochemical immunosensor for the diagnosis of CEA has been reported using sol-gel chemistry [129]. FTO electrodes were hydroxylated for silanization with APTES making it a suitable platform for the fabrication of CEA-antibodies. CEA-Ab acts as a biorecognition element for linkage to its receptor molecule i.e. anti-CEA. EIS and SWV were used to check the sensitivity of developed CEA based biosensor having detection limit 0.42ng/mL and 0.043ng/mL, respectively. The developed sol-gel chemistry is effective in terms of simplicity, less response time and cost-effectiveness. Another method targeting CEA has been reported using a displacement type immunosensing strategy for the detection of colorectal cancer [130]. The electrochemical signals thus produced gets affected due to the displacement of electron mediator depending upon the amount of CEA present in the sample hence could detect as low as 0.03ng/mL of target molecules. This method exhibited better sensitivity than previously reported method and could be expanded for a wide variety of targets and environmental contaminants which are devoid of glucose.

Autoantibodies are generated against tumor-associated antigens (TAA) by the immune system hence could be useful in disease diagnosis as well as prognosis. In a study, an electrochemical disposable biosensor for detecting p53 specific autoantibodies was developed for the first time by Garranzo-Asensio et al. [131]. P53 is an important biomarker involved in tumour formation which is released into the circulatory fluids after disruption of tumour cells. The fabrication method involves magnetic capturing of MBs (magnetic microcarriers) labelled HaloTag fusion p53 protein onto the surface of screen-printed carbon electrodes (SPCE) and amperometric determination of p53 specific autoantibodies by measuring the cathodic current produced by hydroquinone/H_2O_2 system. The method is a highly effective approach to check the amount of p53 antibodies present during disease progression and therapeutic administrations and could be easily miniaturized. The halo tag immobilized detection platform could be further used for protein-protein and drug-protein interaction. Recently, a label-free disposable electrochemical immunosensor for p53 detection has been fabricated by a spin coating method using tetra armed star-shaped poly(glycidyl methacrylate) (StarPGMA) on ITO surface [132]. Anti-p53 antibodies then covalently bind to the epoxy groups present on the electrode surface. The immunosensor exhibited detection limit of about 7fg/mL in linear range from 0.02-4pg/mL.

Current antibody-based diagnostic assay is highly expensive, associated with multistep preparation and false positive results. So, the use of affinity peptides could be a better option because of their small size and low cost. Moreover, linear or cyclic forms of peptides are easy to modify at molecular levels hence easy to immobilize on the electrode surface. Lim et al. [133] developed a diagnostic method for an adenoma to carcinoma progression that involves the use of chemically modified affinity peptides specific for LRG1 (leucine-rich α-2-glycoprotein 1). Plasma level of LRG1 plays a crucial role in CRC occurrence involved in apoptosis and cell survival. Out of LRG1 BP1-BP4, high-affinity peptide linker BP3 was chosen as a recognition layer fabricated on the gold electrode for LRG1 detection and performance was checked by EIS and CV. The sensing method showed detection limit 0.025ug/mL and could be useful as a miniaturized point of care sensing platform to detect adenoma-carcinoma transition.

Advancement in the sensing methodology was observed in a study where chemokines were detected using their receptor molecules. Chung et al. [134] for the first time designed a biofunctional nanobiosensor for chemokine screening and diagnosis using a biosensing platform. The sensor was designed by applying CXCR2 receptor on the surface of pre-modified GCE with AuNPs followed by electropolymerization of 2,2':5',2"-terthiophene-3' (p-benzoic acid) (TBA) [CXCR2/pTBA/AuNPs/GCE]. Among all the chemokine ligands present (CXCL5, CXCL8 and CXCL13), only CXCL5 showed the highest affinity to the CXCR2 receptor. The biosensor showed LOD 0.078 ± 0.004 ng/mL within 25min. It could be successfully applied for direct clinical analysis of colorectal cancer sample with high sensitivity and selectivity and found to be a better alternative to ELISA method which requires several hours to complete the reaction.

5.3   Nucleic acid-based biosensor

CeO_2/chitosan nanocomposite film as an immobilization matrix has been fabricated for DNA-DNA hybridization detection in colorectal cancer by Feng et al. [135]. After that, the only report has been published by Wang et al. 2014 on nucleic acid-based biosensor for specific detection of colorectal cancer. A multiple signal amplification procedure for the electrochemical detection of K-ras gene has been reported [136]. K-ras gene Mutations in K-ras genes is highly associated with the tumorigenesis, cancer progression and poor prognosis. Here, carboxylic multiwalled carbon nanotubes (MWCNTs) doped nylon 6 (PA6) composite nanofibers (MWCNTs–PA6) were deposited on GCE by electrospinning followed by electropolymerization of thionine to form MWCNT-PA6-PTH further used as immobilization support for the deposition of the ssDNA1 probe. The prepared surface was then used to produce a sandwich assay by hybridizing K-ras gene over it and further modifying it with gold nanoparticles labeled ssDNA2 (ssDNA1/K-ras gene/AuNPs–ssDNA2). Lastly, a network of thiocyanuric acid (TA)/AuNPs resulting in signal enhancement with noticeable sensitivity down to 30fM and is also capable of discrimination with two, one base mismatched sequences (G/C and A/T) upto 54.3 and 51.9%, respectively.

5.4   Biomimetic biosensor

Using aptamer as bioreceptor molecules, researchers also developed an electrochemical biosensor for colorectal cancer detection. An aptasensor for colon cancer detection was demonstrated for CEA detection which is highly expressed on the surface of HCT 116 human cell line [137]. For designing the aptasensor, Au electrode surface was modified with thiol groups using 11-mercaptoundecanoic acid (11-MUA) and further modified with EDC/NHS and KCHA10a aptamer. Applicability of proposed biosensor was checked by flow cytometry, florescence microscopy and electrochemical studies. It was

Biosensors: Materials and Applications   Materials Research Forum LLC
Materials Research Foundations **47** (2019) 157-210   doi: http://dx.doi.org/10.21741/9781644900130-5

found that a more than 7cells can be detected with this method in the linear detection range of 1-100cells. Ahmadzadeh-Raji, Ghafar-Zadeh and Amoabediny [138] again reported a biosensing method for HCT 116 detection in colorectal cancer with LOD 6cells/mL. In another study, an aptamer-cell-aptamer based electrochemical approach was applied for testing CT-26 cells, with an improved limit of detection [139]. A sandwich platform was fabricated by capturing target CT-26 cells onto the surface of modified SPE electrode, aptamer/AuNPs/SBA-15-3-aminopropyltriethoxysilane/SPE (APT/AuNPs/SBA-15-pr-NH2/SPE) and further treating it with aptamer molecule for sensing by CV and EIS method (Figure 9). The developed method could detect as low as 2cells/mL of CT-26 cells by simple, rapid, label free and less expensive fabrication platform.

Figure 9: Fabrication strategy for the detection of CT-26 cells using sandwich type electrochemical aptasensor.

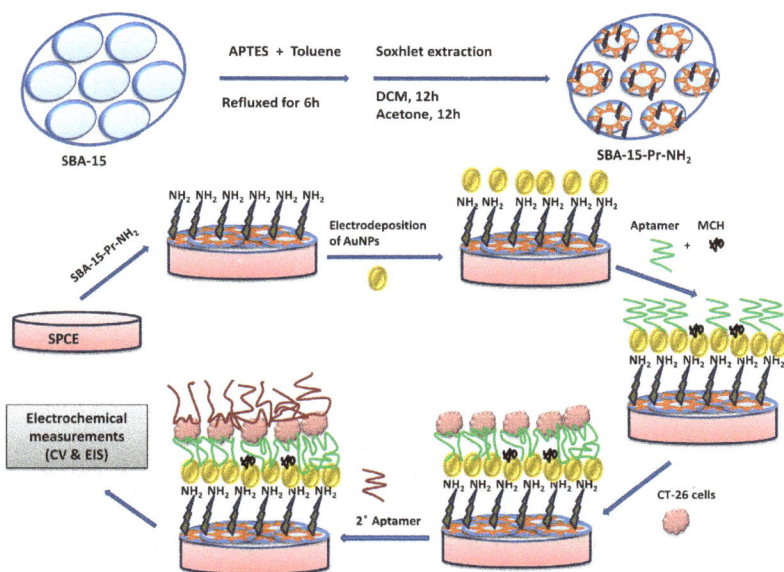

Table 6: List of electrochemical biosensors developed for the determination of colorectal cancer biomarkers.

Biosensing platform	Biomarker	Detection method	Linear range	LOD	Ref.

ssDNA probe/CeO$_2$/CHIT/GCE	DNA	DPV	1.59×10^{-11}- 1.16×10^{-7}mol/L	1×10^{-11}mol/L	[136]
ALP/BSA/DNA/Fe$_2$O$_3$/Au/ SPCE	BRAF V600E	DPV	0.8-50%	0.8%	[128]
TA/AuNPs-ssDNA2/K-ras/ ssDNA1/MWCNT-PA6-PTH	K-ras	DPV	0.1-100pM	30fM	[137]
Hydrolyzation of peptide immobilized on Au electrode by MMP	MMP2, MMP7	EIS	0.05-200ng/mL, 0.0005-50ng/mL	0.5pg/mL	[129]
Au coated with 11MUA/EDC/NHS/Apt/B SA	HCT 116	CV	1-100cells	7cells	[138]
Anti-CEA/BSA/CEA-Ab/APTES/FTO	CEA	EIS, SWV	0.50 - 1.5ng/mL, 0.25 - 1.5ng/mL	0.42ng/m L 0.043ng/ mL	[130]
11MUA/EDC/NHS/Apt/B SA/AuNPs ITOsasa8	CEA	CV	1-25cells/mL	6cells/m L	[139]
Ab$_1$/PAMAM-Au-PBA-ARS/GCE	CEA	SWV	0.01-50ng/mL	0.003ng/ mL	[131]
Apt/AuNPs/SBA-15-pr-NH$_2$/SPE	CT26	CV, EIS	$10 - 1.0 \times 10^5$ cells/mL, $1.0 \times 10^5 - 6 \times 10^6$ cells/mL	2cells/m L	[140]
LRG1B1-B4/Au electrode	LRG1	CV, EIS	0.0025 - 20μg/mL	0.025μg/ mL	[134]
p53 antigen/BSA/Anti-p53-Ab/PGMA coated ITO	P53	EIS	0.02 - 4pg/mL	7fg/mL	[133]
GCE/AuNPs/pTBA/CXCR 2	CXCL5	CV	0.1 - 10ng/mL	0.078±0. 004 ng/mL	[135]

ALP: Alkaline phosphate; TA: Thiocyanuric acid; MWCNT-PA6: Multiwalled carbon nanotubes doped nylon 6 composite nanofibre; PTH: Poly(thionine); 11MUA-11: mercaptoandecanoic acid; EDC: ethyl(dimethylaminopropyl)carbodiimide; NHS: N-hydroxysuccinimide; APTES: (3-Aminopropyl)triethoxysilane; PBA: Phenylboronic acid; ARS: Alizarin Red S; SBA-15-pr-NH$_2$: SBA-15-3-aminopropyltriethoxysilane; PGMA: Poly(glycidylmethacrylate); pTBA: 2,2':5',2"-terthiophene-3' (p-benzoic acid); CXCR2: Chemokine receptor

Conclusion and Future Prospective

A number of electrochemical biosensing platforms for the identification of commonly occurring cancers have been reviewed in this chapter. The main goal of cancer-detecting electrochemical biosensor is to have clinical applications keeping in mind the POCT requirements. Despite significant progress made in the development of various biomolecule based biosensors, there are still some challenges that need to be addressed in terms of selectivity, sensitivity, biocompatibility, longevity and reliability of electrochemical biosensors. However, aptamer and antibodies based biosensing approaches are the preferred choices for the researchers in the area of cancer diagnosis due to improved sensitivity, limit of detection and selectivity. Further, the use of novel nanomaterials such as metal oxides can lead to the development of reliable, sensitive and miniaturized biosensing device for cancer detection. Above all, there is an urgent need for substantial focus on the detection of specific biomarkers related to an organ-specific cancer as there are numerous biomarkers which are specific for more than one type of cancer thereby, limiting their applicability in one step detection of cancer using biosensor. Hence, it is necessary to accelerate the development of biosensors which could detect a cancer-specific biomarker with enhanced sensitivity making itself applicable for clinical care.

References

[1]   L.A. Torre, R.L. Siegel, E.M. Ward, A. Jemal, Global cancer incidence and mortality rates and trends--an update, cancer epidemiol. Biomarkers Prev. 25 (2016) 16–27. https://doi.org/10.1158/1055-9965.EPI-15-0578

[2]   R.L. Siegel, K.D. Miller, A. Jemal, Cancer statistics, CA. Cancer J. Clin. 68 (2018) 7–30. https://doi.org/10.3322/caac.21442

[3]   B. Gupta, N. Kumar, Worldwide incidence, mortality and time trends for cancer of the oesophagus, Eur. J. Cancer Prev. 26 (2017) 107–118. https://doi.org/10.1097/CEJ.0000000000000249

[4]   A. del Sol, R. Balling, L. Hood, D. Galas, Diseases as network perturbations, Curr. Opin. Biotechnol. 21 (2010) 566–571. https://doi.org/10.1016/j.copbio.2010.07.010

[5]   S.K. Chatterjee, B.R. Zetter, Cancer biomarkers: knowing the present and predicting the future, Futur. Oncol. 1 (2005) 37–50. https://doi.org/10.1517/14796694.1.1.37

[6]   S. Kumar, A. Mohan, R. Guleria, Biomarkers in cancer screening, research and detection: present and future: a review, Biomarkers. 11 (2006) 385–405. https://doi.org/10.1080/13547500600775011

[7]   B. Jin, P. Wang, H. Mao, B. Hu, H. Zhang, Z. Cheng, Z. Wu, X. Bian, C. Jia, F. Jing, Q. Jin, J. Zhao, Multi-nanomaterial electrochemical biosensor based on label-

free graphene for detecting cancer biomarkers, Biosens. Bioelectron. 55 (2014) 464–469. https://doi.org/10.1016/j.bios.2013.12.025

[8]   V.S.P.K.S.A. Jayanthi, A.B. Das, U. Saxena, Recent advances in biosensor development for the detection of cancer biomarkers, Biosens. Bioelectron. 91 (2017) 15–23. https://doi.org/10.1016/j.bios.2016.12.014

[9]   B. Bohunicky, S.A. Mousa, Biosensors: The new wave in cancer diagnosis, Nanotechnol. Sci. Appl. 4 (2011) 1–10. https://doi.org/10.2147/NSA.S13465

[10]   R. Ranjan, E.N. Esimbekova, V.A. Kratasyuk, Rapid biosensing tools for cancer biomarkers, Biosens. Bioelectron. 87 (2017) 918–930. https://doi.org/10.1016/j.bios.2016.09.061

[11]   S. Kanchi, M.I. Sabela, P.S. Mdluli, Inamuddin, K. Bisetty, Smartphone based bioanalytical and diagnosis applications: A review., Biosens. Bioelectron. 102 (2018) 136–149. https://doi.org/10.1016/j.bios.2017.11.021

[12]   L. Wang, Q. Xiong, F. Xiao, H. Duan, 2D nanomaterials based electrochemical biosensors for cancer diagnosis, Biosens. Bioelectron. 89 (2017) 136–151. https://doi.org/10.1016/j.bios.2016.06.011

[13]   S.N. Topkaya, M. Azimzadeh, M. Ozsoz, Electrochemical biosensors for cancer biomarkers detection: recent advances and challenges, Electroanalysis. 28 (2016) 1402–1419. https://doi.org/10.1002/elan.201501174

[14]   N. Sohrabi, A. Valizadeh, S.M. Farkhani, A. Akbarzadeh, Basics of DNA biosensors and cancer diagnosis, Artif. Cells, Nanomedicine, Biotechnol. 44 (2016) 654–663. https://doi.org/10.3109/21691401.2014.976707

[15]   M. Choudhary, A. Singh, S. Kaur, K. Arora, Enhancing lung cancer diagnosis: electrochemical simultaneous bianalyte immunosensing using carbon nanotubes–chitosan nanocomposite, Appl. Biochem. Biotechnol. 174 (2014) 1188–1200. https://doi.org/10.1007/s12010-014-1020-1

[16]   K.K. Hussain, N.G. Gurudatt, T.A. Mir, Y.-B. Shim, Amperometric sensing of HIF1α expressed in cancer cells and the effect of hypoxic mimicking agents, Biosens. Bioelectron. 83 (2016) 312–318. https://doi.org/10.1016/j.bios.2016.04.068

[17]   P.J. Marangos, D.E. Schmechel, Neuron Specific Enolase, A clinically useful marker for neurons and neuroendocrine cells, Annu. Rev. Neurosci. 10 (1987) 269–295. https://doi.org/10.1146/annurev.ne.10.030187.001413

[18]   H. Wang, H. Han, Z. Ma, Conductive hydrogel composed of 1,3,5-benzenetricarboxylic acid and Fe3+ used as enhanced electrochemical

immunosensing substrate for tumor biomarker, Bioelectrochemistry. 114 (2017) 48–53. https://doi.org/10.1016/j.bioelechem.2016.12.006

[19]   Z. Wei, J. Zhang, A. Zhang, Y. Wang, X. Cai, Electrochemical detecting lung cancer-associated antigen based on graphene-gold nanocomposite, molecules. 22 (2017) 392. https://doi.org/10.3390/molecules22030392

[20]   Y. Dai, C.C. Liu, Detection of 17 β-estradiol in environmental samples and for health care using a single-use, cost effective biosensor based on differential pulse voltammetry (DPV), Biosensors. 7 (2017) 15–27. https://doi.org/10.3390/bios7020015

[21]   E.B. Aydın, M.K. Sezgintürk, A sensitive and disposable electrochemical immunosensor for detection of SOX2, a biomarker of cancer, Talanta. 172 (2017) 162–170. https://doi.org/10.1016/j.talanta.2017.05.048

[22]   Y. Fan, J. Liu, Y. Wang, J. Luo, H. Xu, S. Xu, X. Cai, A wireless point-of-care testing system for the detection of neuron-specific enolase with microfluidic paper-based analytical devices, Biosens. Bioelectron. 95 (2017) 60–66. https://doi.org/10.1016/j.bios.2017.04.003

[23]   Q. Tian, Y. Wang, R. Deng, L. Lin, Y. Liu, J. Li, Carbon nanotube enhanced label-free detection of microRNAs based on hairpin probe triggered solid-phase rolling-circle amplification, Nanoscale. 7 (2015) 987–993. https://doi.org/10.1039/C4NR05243A

[24]   S. Liu, W. Su, Z. Li, X. Ding, Electrochemical detection of lung cancer specific microRNAs using 3D DNA origami nanostructures, Biosens. Bioelectron. 71 (2015) 57–61. https://doi.org/10.1016/j.bios.2015.04.006

[25]   X.-W. Xu, X.-H. Weng, C.-L. Wang, W.-W. Lin, A.-L. Liu, W. Chen, X.-H. Lin, Detection EGFR exon 19 status of lung cancer patients by DNA electrochemical biosensor, Biosens. Bioelectron. 80 (2016) 411–417. https://doi.org/10.1016/j.bios.2016.02.009

[26]   S. Su, Y. Wu, D. Zhu, J. Chao, X. Liu, Y. Wan, Y. Su, X. Zuo, C. Fan, L. Wang, On-Electrode synthesis of shape-controlled hierarchical flower-like gold nanostructures for efficient interfacial DNA assembly and sensitive electrochemical sensing of microRNA, Small. 12 (2016) 3794–3801. https://doi.org/10.1002/smll.201601066

[27]   D. Voccia, M. Sosnowska, F. Bettazzi, G. Roscigno, E. Fratini, V. De Franciscis, G. Condorelli, R. Chitta, F. D'Souza, W. Kutner, I. Palchetti, Direct determination of small RNAs using a biotinylated polythiophene impedimetric genosensor,

Biosens. Bioelectron. 87 (2017) 1012–1019.
https://doi.org/10.1016/j.bios.2016.09.058

[28] B. Bo, T. Zhang, Y. Jiang, H. Cui, P. Miao, Triple signal amplification strategy for ultrasensitive determination of miRNA based on duplex specific nuclease and bridge DNA–gold nanoparticles, Anal. Chem. 90 (2018) 2395–2400. https://doi.org/10.1021/acs.analchem.7b05447

[29] S. Liu, W. Su, Y. Li, L. Zhang, X. Ding, Manufacturing of an electrochemical biosensing platform based on hybrid DNA hydrogel: Taking lung cancer-specific miR-21 as an example, Biosens. Bioelectron. 103 (2018) 1–5. https://doi.org/10.1016/j.bios.2017.12.021

[30] M. Amouzadeh Tabrizi, M. Shamsipur, L. Farzin, A high sensitive electrochemical aptasensor for the determination of VEGF 165 in serum of lung cancer patient, Biosens. Bioelectron. 74 (2015) 764–769. https://doi.org/10.1016/j.bios.2015.07.032

[31] G.S. Zamay, T.N. Zamay, O.S. Kolovskaya, A. V. Krat, Y.E. Glazyrin, A. V. Dubinina, A.S. Zamay, Development of a biosensor for electrochemical detection of tumor-associated proteins in blood plasma of cancer patients by aptamers, Dokl. Biochem. Biophys. 466 (2016) 70–73. https://doi.org/10.1134/S1607672916010208

[32] M. Ahmed, L.G. Carrascosa, A.A. Ibn Sina, E.M. Zarate, D. Korbie, K. Ru, M.J.A. Shiddiky, P. Mainwaring, M. Trau, Detection of aberrant protein phosphorylation in cancer using direct gold-protein affinity interactions, Biosens. Bioelectron. 91 (2017) 8–14. https://doi.org/10.1016/j.bios.2016.12.012

[33] D. Pu, H. Liang, F. Wei, D. Akin, Z. Feng, Q. Yan, Y. Li, Y. Zhen, L. Xu, G. Dong, H. Wan, J. Dong, X. Qiu, C. Qin, D. Zhu, X. Wang, T. Sun, W. Zhang, C. Li, X. Tang, Y. Qiao, D.T.W. Wong, Q. Zhou, Evaluation of a novel saliva-based epidermal growth factor receptor mutation detection for lung cancer: A pilot study, Thorac. Cancer. 7 (2016) 428–436. https://doi.org/10.1111/1759-7714.12350

[34] M.H. Akhtar, T.A. Mir, N.G. Gurudatt, S. Chung, Y.-B. Shim, Sensitive NADH detection in a tumorigenic cell line using a nano-biosensor based on the organic complex formation, Biosens. Bioelectron. 85 (2016) 488–495. https://doi.org/10.1016/j.bios.2016.05.045

[35] J. Ferlay, I. Soerjomataram, R. Dikshit, S. Eser, C. Mathers, M. Rebelo, D.M. Parkin, D. Forman, F. Bray, Cancer incidence and mortality worldwide: Sources, methods and major patterns in GLOBOCAN 2012, Int. J. Cancer. 136 (2015) E359–E386. https://doi.org/10.1002/ijc.29210

Biosensors: Materials and Applications
Materials Research Foundations **47** (2019) 157-210

Materials Research Forum LLC
doi: http://dx.doi.org/10.21741/9781644900130-5

[36] L.A. Torre, F. Bray, R.L. Siegel, J. Ferlay, J. Lortet-Tieulent, A. Jemal, Global
      cancer statistics, 2012, CA. Cancer J. Clin. 65 (2015) 87–108.
      https://doi.org/10.3322/caac.21262

[37] S. Mittal, H. Kaur, N. Gautam, A.K. Mantha, Biosensors for breast cancer
      diagnosis: A review of bioreceptors, biotransducers and signal amplification
      strategies, Biosens. Bioelectron. 88 (2017) 217–231.
      https://doi.org/10.1016/j.bios.2016.08.028

[38] J. Zhao, Y. Yan, L. Zhu, X. Li, G. Li, An amperometric biosensor for the detection
      of hydrogen peroxide released from human breast cancer cells, Biosens.
      Bioelectron. 41 (2013) 815–819. https://doi.org/10.1016/j.bios.2012.10.019

[39] H. Fan, Y. Zhang, D. Wu, H. Ma, X. Li, Y. Li, H. Wang, H. Li, B. Du, Q. Wei,
      Construction of label-free electrochemical immunosensor on mesoporous carbon
      nanospheres for breast cancer susceptibility gene, Anal. Chim. Acta. 770 (2013)
      62–67. https://doi.org/10.1016/j.aca.2013.01.066

[40] S. Rauf, G.K. Mishra, J. Azhar, R.K. Mishra, K.Y. Goud, M.A.H. Nawaz, J.L.
      Marty, A. Hayat, Carboxylic group riched graphene oxide based disposable
      electrochemical immunosensor for cancer biomarker detection, Anal. Biochem.
      545 (2018) 13–19. https://doi.org/10.1016/j.ab.2018.01.007

[41] H. Li, J. He, S. Li, A.P.F. Turner, Electrochemical immunosensor with N-doped
      graphene-modified electrode for label-free detection of the breast cancer
      biomarker CA 15-3, Biosens. Bioelectron. 43 (2013) 25–29.
      https://doi.org/10.1016/j.bios.2012.11.037

[42] M. Emami, M. Shamsipur, R. Saber, R. Irajirad, An electrochemical
      immunosensor for detection of a breast cancer biomarker based on antiHER2–iron
      oxide nanoparticle bioconjugates, Analyst. 139 (2014) 2858–2866.
      https://doi.org/10.1039/C4AN00183D

[43] R.C.B. Marques, S. Viswanathan, H.P.A. Nouws, C. Delerue-Matos, M.B.
      González-García, Electrochemical immunosensor for the analysis of the breast
      cancer biomarker HER2 ECD, Talanta. 129 (2014) 594–599.
      https://doi.org/10.1016/j.talanta.2014.06.035

[44] M. Shamsipur, M. Emami, L. Farzin, R. Saber, A sandwich-type electrochemical
      immunosensor based on in situ silver deposition for determination of serum level
      of HER2 in breast cancer patients, Biosens. Bioelectron. 103 (2018) 54–61.
      https://doi.org/10.1016/j.bios.2017.12.022

[45] X. Ren, T. Yan, S. Zhang, X. Zhang, P. Gao, D. Wu, B. Du, Q. Wei, Ultrasensitive
      dual amplification sandwich immunosensor for breast cancer susceptibility gene

based on sheet materials, Analyst. 139 (2014) 3061–3068.
https://doi.org/10.1039/C4AN00099D

[46] M. Amouzadeh Tabrizi, M. Shamsipur, R. Saber, S. Sarkar, N. Zolfaghari, An ultrasensitive sandwich-type electrochemical immunosensor for the determination of SKBR-3 breast cancer cell using rGO-TPA/FeHCF nano labeled Anti-HCT as a signal tag, Sensors Actuators B Chem. 243 (2017) 823–830. https://doi.org/10.1016/j.snb.2016.12.061

[47] F. Mouffouk, S. Aouabdi, E. Al-Hetlani, H. Serrai, T. Alrefae, L. Leo Chen, New generation of electrochemical immunoassay based on polymeric nanoparticles for early detection of breast cancer, Int. J. Nanomedicine. Volume 12 (2017) 3037–3047. https://doi.org/10.2147/IJN.S127086

[48] S. Carvajal, S.N. Fera, A.L. Jones, T.A. Baldo, I.M. Mosa, J.F. Rusling, C.E. Krause, Disposable inkjet-printed electrochemical platform for detection of clinically relevant HER-2 breast cancer biomarker, Biosens. Bioelectron. 104 (2018) 158–162. https://doi.org/10.1016/j.bios.2018.01.003

[49] A. Benvidi, A.D. Firouzabadi, S.M. Moshtaghiun, M. Mazloum-Ardakani, M.D. Tezerjani, Ultrasensitive DNA sensor based on gold nanoparticles/reduced graphene oxide/glassy carbon electrode, Anal. Biochem. 484 (2015) 24–30. https://doi.org/10.1016/j.ab.2015.05.009

[50] A.A. Saeed, J.L.A. Sánchez, C.K. O'Sullivan, M.N. Abbas, DNA biosensors based on gold nanoparticles-modified graphene oxide for the detection of breast cancer biomarkers for early diagnosis, Bioelectrochemistry. 118 (2017) 91–99. https://doi.org/10.1016/j.bioelechem.2017.07.002

[51] Y. Wang, X. Huang, H. Li, L. Guo, Sensitive impedimetric DNA biosensor based on (Nb,V) codoped TiO2 for breast cancer susceptible gene detection, Mater. Sci. Eng. C. 77 (2017) 867–873. https://doi.org/10.1016/j.msec.2017.03.260

[52] A. Benvidi, S. Jahanbani, Self-assembled monolayer of SH-DNA strand on a magnetic bar carbon paste electrode modified with Fe_3O_4 @Ag nanoparticles for detection of breast cancer mutation, J. Electroanal. Chem. 768 (2016) 47–54. https://doi.org/10.1016/j.jelechem.2016.02.038

[53] L. Chen, X. Liu, C. Chen, Impedimetric biosensor modified with hydrophilic material of tannic acid/polyethylene glycol and dopamine-assisted deposition for detection of breast cancer-related BRCA1 gene, J. Electroanal. Chem. 791 (2017) 204–210. https://doi.org/10.1016/j.jelechem.2017.03.001

[54] W. Wang, X. Fan, S. Xu, J.J. Davis, X. Luo, Low fouling label-free DNA sensor based on polyethylene glycols decorated with gold nanoparticles for the detection

Biosensors: Materials and Applications   Materials Research Forum LLC
Materials Research Foundations **47** (2019) 157-210   doi: http://dx.doi.org/10.21741/9781644900130-5

of breast cancer biomarkers, Biosens. Bioelectron. 71 (2015) 51–56.
https://doi.org/10.1016/j.bios.2015.04.018

[55] M. Cui, Y. Wang, H. Wang, Y. Wu, X. Luo, A label-free electrochemical DNA
      biosensor for breast cancer marker BRCA1 based on self-assembled antifouling
      peptide monolayer, Sensors Actuators B Chem. 244 (2017) 742–749.
      https://doi.org/10.1016/j.snb.2017.01.060

[56] T. Kilic, S.N. Topkaya, D. Ozkan Ariksoysal, M. Ozsoz, P. Ballar, Y. Erac, O.
      Gozen, Electrochemical based detection of microRNA, mir21 in breast cancer
      cells, Biosens. Bioelectron. 38 (2012) 195–201.
      https://doi.org/10.1016/j.bios.2012.05.031

[57] C.-Y. Hong, X. Chen, T. Liu, J. Li, H.-H. Yang, J.-H. Chen, G.-N. Chen,
      Ultrasensitive electrochemical detection of cancer-associated circulating
      microRNA in serum samples based on DNA concatamers, Biosens. Bioelectron.
      50 (2013) 132–136. https://doi.org/10.1016/j.bios.2013.06.040

[58] A.R. Cardoso, F.T.C. Moreira, R. Fernandes, M.G.F. Sales, Novel and simple
      electrochemical biosensor monitoring attomolar levels of miRNA-155 in breast
      cancer, Biosens. Bioelectron. 80 (2016) 621–630.
      https://doi.org/10.1016/j.bios.2016.02.035

[59] R.M. Torrente-Rodríguez, S. Campuzano, E. López-Hernández, V.R.-V. Montiel,
      R. Barderas, R. Granados, J.M. Sánchez-Puelles, J.M. Pingarrón, Simultaneous
      detection of two breast cancer-related miRNAs in tumor tissues using p19-based
      disposable amperometric magnetobiosensing platforms, Biosens. Bioelectron. 66
      (2015) 385–391. https://doi.org/10.1016/j.bios.2014.11.047

[60] H.-A. Rafiee-Pour, M. Behpour, M. Keshavarz, A novel label-free electrochemical
      miRNA biosensor using methylene blue as redox indicator: application to breast
      cancer biomarker miRNA-21, Biosens. Bioelectron. 77 (2016) 202–207.
      https://doi.org/10.1016/j.bios.2015.09.025

[61] L. Ribovski, V. Zucolotto, B.C. Janegitz, A label-free electrochemical DNA
      sensor to identify breast cancer susceptibility, Microchem. J. 133 (2017) 37–42.
      https://doi.org/10.1016/j.microc.2017.03.011

[62] L. Tian, K. Qian, J. Qi, Q. Liu, C. Yao, W. Song, Y. Wang, Gold nanoparticles
      superlattices assembly for electrochemical biosensor detection of microRNA-21,
      Biosens. Bioelectron. 99 (2018) 564–570.
      https://doi.org/10.1016/j.bios.2017.08.035

[63] J.A. Ribeiro, C.M. Pereira, A.F. Silva, M.G.F. Sales, Disposable electrochemical
      detection of breast cancer tumour marker CA 15-3 using poly(Toluidine Blue) as

imprinted polymer receptor, Biosens. Bioelectron. 109 (2018) 246–254. https://doi.org/10.1016/j.bios.2018.03.011

[64] C. V. Uliana, C.R. Peverari, A.S. Afonso, M.R. Cominetti, R.C. Faria, Fully disposable microfluidic electrochemical device for detection of estrogen receptor alpha breast cancer biomarker, Biosens. Bioelectron. 99 (2018) 156–162. https://doi.org/10.1016/j.bios.2017.07.043

[65] Q. Sheng, N. Cheng, W. Bai, J. Zheng, Ultrasensitive electrochemical detection of breast cancer cells based on DNA-rolling-circle-amplification-directed enzyme-catalyzed polymerization, Chem. Commun. 51 (2015) 2114–2117. https://doi.org/10.1039/C4CC08954E

[66] S. Cai, M. Chen, M. Liu, W. He, Z. Liu, D. Wu, Y. Xia, H. Yang, J. Chen, A signal amplification electrochemical aptasensor for the detection of breast cancer cell via free-running DNA walker, Biosens. Bioelectron. 85 (2016) 184–189. https://doi.org/10.1016/j.bios.2016.05.003

[67] M. Yan, G. Sun, F. Liu, J. Lu, J. Yu, X. Song, An aptasensor for sensitive detection of human breast cancer cells by using porous GO/Au composites and porous PtFe alloy as effective sensing platform and signal amplification labels, Anal. Chim. Acta. 798 (2013) 33–39. https://doi.org/10.1016/j.aca.2013.08.046

[68] M. Su, H. Liu, L. Ge, Y. Wang, S. Ge, J. Yu, M. Yan, Aptamer-Based electrochemiluminescent detection of MCF-7 cancer cells based on carbon quantum dots coated mesoporous silica nanoparticles, Electrochim. Acta. 146 (2014) 262–269. https://doi.org/10.1016/j.electacta.2014.08.129

[69] X. Chen, Q. Zhang, C. Qian, N. Hao, L. Xu, C. Yao, Electrochemical aptasensor for mucin 1 based on dual signal amplification of poly(o-phenylenediamine) carrier and functionalized carbon nanotubes tracing tag, Biosens. Bioelectron. 64 (2015) 485–492. https://doi.org/10.1016/j.bios.2014.09.052

[70] X. Zhu, J. Yang, M. Liu, Y. Wu, Z. Shen, G. Li, Sensitive detection of human breast cancer cells based on aptamer–cell–aptamer sandwich architecture, Anal. Chim. Acta. 764 (2013) 59–63. https://doi.org/10.1016/j.aca.2012.12.024

[71] K. Wang, M.-Q. He, F.-H. Zhai, R.-H. He, Y.-L. Yu, A novel electrochemical biosensor based on polyadenine modified aptamer for label-free and ultrasensitive detection of human breast cancer cells, Talanta. 166 (2017) 87–92. https://doi.org/10.1016/j.talanta.2017.01.052

[72] L. Tian, J. Qi, K. Qian, O. Oderinde, Q. Liu, C. Yao, W. Song, Y. Wang, Copper (II) oxide nanozyme based electrochemical cytosensor for high sensitive detection

of circulating tumor cells in breast cancer, J. Electroanal. Chem. 812 (2018) 1–9. https://doi.org/10.1016/j.jelechem.2017.12.012

[73] T.S.C.R. Rebelo, C.M. Pereira, M.G.F. Sales, J.P. Noronha, J. Costa-Rodrigues, F. Silva, M.H. Fernandes, Sarcosine oxidase composite screen-printed electrode for sarcosine determination in biological samples, Anal. Chim. Acta. 850 (2014) 26–32. https://doi.org/10.1016/j.aca.2014.08.005

[74] V. Narwal, P. Kumar, P. Joon, C.S. Pundir, Fabrication of an amperometric sarcosine biosensor based on sarcosine oxidase/chitosan/CuNPs/c-MWCNT/Au electrode for detection of prostate cancer, Enzyme Microb. Technol. 113 (2018) 44–51. https://doi.org/10.1016/j.enzmictec.2018.02.010

[75] A. Salimi, B. Kavosi, F. Fathi, R. Hallaj, Highly sensitive immunosensing of prostate-specific antigen based on ionic liquid–carbon nanotubes modified electrode: Application as cancer biomarker for prostatebiopsies, Biosens. Bioelectron. 42 (2013) 439–446. https://doi.org/10.1016/j.bios.2012.10.053

[76] H.V. Tran, B. Piro, S. Reisberg, L. Huy Nguyen, T. Dung Nguyen, H.T. Duc, M.C. Pham, An electrochemical ELISA-like immunosensor for miRNAs detection based on screen-printed gold electrodes modified with reduced graphene oxide and carbon nanotubes, Biosens. Bioelectron. 62 (2014) 25–30. https://doi.org/10.1016/j.bios.2014.06.014

[77] B. Zhang, B. Liu, G. Chen, D. Tang, Redox and catalysis 'all-in-one' infinite coordination polymer for electrochemical immunosensor of tumor markers, Biosens. Bioelectron. 64 (2015) 6–12. https://doi.org/10.1016/j.bios.2014.08.024

[78] W. Hong, S. Lee, E. Jae Kim, M. Lee, Y. Cho, A reusable electrochemical immunosensor fabricated using a temperature-responsive polymer for cancer biomarker proteins, Biosens. Bioelectron. 78 (2016) 181–186. https://doi.org/10.1016/j.bios.2015.11.040

[79] G.G. Gutiérrez-Zúñiga, J.L. Hernández-López, Sensitivity improvement of a sandwich-type ELISA immunosensor for the detection of different prostate-specific antigen isoforms in human serum using electrochemical impedance spectroscopy and an ordered and hierarchically organized interfacial supramolecu, Anal. Chim. Acta. 902 (2016) 97–106. https://doi.org/10.1016/j.aca.2015.10.042

[80] L. Li, J. Xu, X. Zheng, C. Ma, X. Song, S. Ge, J. Yu, M. Yan, Growth of gold-manganese oxide nanostructures on a 3D origami device for glucose-oxidase label based electrochemical immunosensor, Biosens. Bioelectron. 61 (2014) 76–82. https://doi.org/10.1016/j.bios.2014.05.012

[81]   M. Pal, R. Khan, Graphene oxide layer decorated gold nanoparticles based immunosensor for the detection of prostate cancer risk factor, Anal. Biochem. 536 (2017) 51–58. https://doi.org/10.1016/j.ab.2017.08.001

[82]   M.S. Khan, K. Dighe, Z. Wang, I. Srivastava, E. Daza, A.S. Schwartz-Dual, J. Ghannam, S.K. Misra, D. Pan, Detection of prostate specific antigen (PSA) in human saliva using an ultra-sensitive nanocomposite of graphene nanoplatelets with diblock- co -polymers and Au electrodes, Analyst. 143 (2018) 1094–1103. https://doi.org/10.1039/C7AN01932G

[83]   N. Blel, N. Fourati, M. Souiri, C. Zerrouki, A. Omezzine, A. Bouslama, A. Othmane, Ultrasensitive electrochemical sensors for psa detection: related surface functionalization strategies, Curr. Top. Med. Chem. 17 (2017). https://doi.org/10.2174/1568026617666170821152757

[84]   H.D. Jang, S.K. Kim, H. Chang, J.-W. Choi, 3D label-free prostate specific antigen (PSA) immunosensor based on graphene–gold composites, Biosens. Bioelectron. 63 (2015) 546–551. https://doi.org/10.1016/j.bios.2014.08.008

[85]   G. Sun, L. Zhang, Y. Zhang, H. Yang, C. Ma, S. Ge, M. Yan, J. Yu, X. Song, Multiplexed enzyme-free electrochemical immunosensor based on ZnO nanorods modified reduced graphene oxide-paper electrode and silver deposition-induced signal amplification strategy, Biosens. Bioelectron. 71 (2015) 30–36. https://doi.org/10.1016/j.bios.2015.04.007

[86]   H.V. Tran, B. Piro, S. Reisberg, L.D. Tran, H.T. Duc, M.C. Pham, Label-free and reagentless electrochemical detection of microRNAs using a conducting polymer nanostructured by carbon nanotubes: Application to prostate cancer biomarker miR-141, Biosens. Bioelectron. 49 (2013) 164–169. https://doi.org/10.1016/j.bios.2013.05.007

[87]   M. Sharafeldin, G.W. Bishop, S. Bhakta, A. El-Sawy, S.L. Suib, J.F. Rusling, Fe 3 O 4 nanoparticles on graphene oxide sheets for isolation and ultrasensitive amperometric detection of cancer biomarker proteins, Biosens. Bioelectron. 91 (2017) 359–366. https://doi.org/10.1016/j.bios.2016.12.052

[88]   L.-H. Pan, S.-H. Kuo, T.-Y. Lin, C.-W. Lin, P.-Y. Fang, H.-W. Yang, An electrochemical biosensor to simultaneously detect VEGF and PSA for early prostate cancer diagnosis based on graphene oxide/ssDNA/PLLA nanoparticles, Biosens. Bioelectron. 89 (2017) 598–605. https://doi.org/10.1016/j.bios.2016.01.077

[89]   M.-S. Wu, Z. Liu, H.-W. Shi, H.-Y. Chen, J.-J. Xu, Visual Electrochemiluminescence Detection of Cancer Biomarkers on a Closed Bipolar

Electrode Array Chip, Anal. Chem. 87 (2015) 530–537.
https://doi.org/10.1021/ac502989f

[90] N. Gao, B. Ling, Z. Gao, L. Wang, H. Chen, Near-infrared-emitting NaYF4:Yb,Tm/Mn upconverting nanoparticle/gold nanorod electrochemiluminescence resonance energy transfer system for sensitive prostate-specific antigen detection, Anal. Bioanal. Chem. 409 (2017) 2675–2683. https://doi.org/10.1007/s00216-017-0212-2

[91] H. Ma, X. Li, T. Yan, Y. Li, Y. Zhang, D. Wu, Q. Wei, B. Du, Electrochemiluminescent immunosensing of prostate-specific antigen based on silver nanoparticles-doped Pb (II) metal-organic framework, Biosens. Bioelectron. 79 (2016) 379–385. https://doi.org/10.1016/j.bios.2015.12.080

[92] T.S.C.R. Rebelo, J.P. Noronha, M. Galésio, H. Santos, M. Diniz, M.G.F. Sales, M.H. Fernandes, J. Costa-Rodrigues, Testing the variability of PSA expression by different human prostate cancer cell lines by means of a new potentiometric device employing molecularly antibody assembled on graphene surface, Mater. Sci. Eng. C. 59 (2016) 1069–1078. https://doi.org/10.1016/j.msec.2015.11.032

[93] P.M.S. Silva, A.L.R. Lima, B.V.M. Silva, L.C.B.B. Coelho, R.F. Dutra, M.T.S. Correia, Cratylia mollis lectin nanoelectrode for differential diagnostic of prostate cancer and benign prostatic hyperplasia based on label-free detection, Biosens. Bioelectron. 85 (2016) 171–177. https://doi.org/10.1016/j.bios.2016.05.004

[94] D. Pihíková, Š. Belicky, P. Kasák, T. Bertok, J. Tkac, Sensitive detection and glycoprofiling of a prostate specific antigen using impedimetric assays, Analyst. 141 (2016) 1044–1051. https://doi.org/10.1039/C5AN02322J

[95] D. Pihikova, P. Kasak, P. Kubanikova, R. Sokol, J. Tkac, Aberrant sialylation of a prostate-specific antigen: Electrochemical label-free glycoprofiling in prostate cancer serum samples, Anal. Chim. Acta. 934 (2016) 72–79. https://doi.org/10.1016/j.aca.2016.06.043

[96] D. Pihikova, Z. Pakanova, M. Nemcovic, P. Barath, S. Belicky, T. Bertok, P. Kasak, J. Mucha, J. Tkac, Sweet characterisation of prostate specific antigen using electrochemical lectin-based immunosensor assay and MALDI TOF/TOF analysis: Focus on sialic acid, Proteomics. 16 (2016) 3085–3095. https://doi.org/10.1002/pmic.201500463

[97] K. Ren, J. Wu, F. Yan, H. Ju, Ratiometric electrochemical proximity assay for sensitive one-step protein detection, Sci. Rep. 4 (2015) 4360. https://doi.org/10.1038/srep04360

[98] K. Ren, J. Wu, F. Yan, Y. Zhang, H. Ju, Immunoreaction-triggered DNA assembly for one-step sensitive ratiometric electrochemical biosensing of protein biomarker, Biosens. Bioelectron. 66 (2015) 345–349. https://doi.org/10.1016/j.bios.2014.11.046

[99] S. Lee, H. Jo, J. Her, H.Y. Lee, C. Ban, Ultrasensitive electrochemical detection of engrailed-2 based on homeodomain-specific DNA probe recognition for the diagnosis of prostate cancer, Biosens. Bioelectron. 66 (2015) 32–38. https://doi.org/10.1016/j.bios.2014.11.003

[100] A. Saheb, S. Patterson, M. Josowicz, Probing for DNA methylation with a voltammetric DNA detector, Analyst. 139 (2014) 786–792. https://doi.org/10.1039/C3AN02154H

[101] Y. Chang, Y. Zhuo, Y. Chai, R. Yuan, Host–guest recognition-assisted electrochemical release: its reusable sensing application based on DNA cross configuration-fueled target cycling and strand displacement reaction amplification, Anal. Chem. 89 (2017) 8266–8272. https://doi.org/10.1021/acs.analchem.7b01272

[102] P. Jolly, V. Tamboli, R.L. Harniman, P. Estrela, C.J. Allender, J.L. Bowen, Aptamer–MIP hybrid receptor for highly sensitive electrochemical detection of prostate specific antigen, Biosens. Bioelectron. 75 (2016) 188–195. https://doi.org/10.1016/j.bios.2015.08.043

[103] A. Rahi, N. Sattarahmady, H. Heli, Label-free electrochemical aptasensing of the human prostate-specific antigen using gold nanospears, Talanta. 156–157 (2016) 218–224. https://doi.org/10.1016/j.talanta.2016.05.029

[104] M. Souada, B. Piro, S. Reisberg, G. Anquetin, V. Noël, M.C. Pham, Label-free electrochemical detection of prostate-specific antigen based on nucleic acid aptamer, Biosens. Bioelectron. 68 (2015) 49–54. https://doi.org/10.1016/j.bios.2014.12.033

[105] E. Heydari-Bafrooei, N.S. Shamszadeh, Electrochemical bioassay development for ultrasensitive aptasensing of prostate specific antigen, Biosens. Bioelectron. 91 (2017) 284–292. https://doi.org/10.1016/j.bios.2016.12.048

[106] G. Sun, Y. Zhang, Q. Kong, X. Zheng, J. Yu, X. Song, CuO-induced signal amplification strategy for multiplexed photoelectrochemical immunosensing using CdS sensitized ZnO nanotubes arrays as photoactive material and AuPd alloy nanoparticles as electron sink, Biosens. Bioelectron. 66 (2015) 565–571. https://doi.org/10.1016/j.bios.2014.12.020

[107] Z. Yang, B. Kasprzyk-Hordern, S. Goggins, C.G. Frost, P. Estrela, A novel immobilization strategy for electrochemical detection of cancer biomarkers: DNA-

directed immobilization of aptamer sensors for sensitive detection of prostate specific antigens, Analyst. 140 (2015) 2628–2633. https://doi.org/10.1039/C4AN02277G

[108] I. Tzouvadaki, P. Jolly, X. Lu, S. Ingebrandt, G. de Micheli, P. Estrela, S. Carrara, Label-free ultrasensitive memristive aptasensor, Nano Lett. 16 (2016) 4472–4476. https://doi.org/10.1021/acs.nanolett.6b01648

[109] K. Settu, J.-T. Liu, C.-J. Chen, J.-Z. Tsai, Development of carbon–graphene-based aptamer biosensor for EN2 protein detection, Anal. Biochem. 534 (2017) 99–107. https://doi.org/10.1016/j.ab.2017.07.012

[110] B.P. Crulhas, A.E. Karpik, F.K. Delella, G.R. Castro, V.A. Pedrosa, Electrochemical aptamer-based biosensor developed to monitor PSA and VEGF released by prostate cancer cells, Anal. Bioanal. Chem. 409 (2017) 6771–6780. https://doi.org/10.1007/s00216-017-0630-1

[111] K. Mohan, K.C. Donavan, J.A. Arter, R.M. Penner, G.A. Weiss, Sub-nanomolar Detection of prostate-specific membrane antigen in synthetic urine by synergistic, dual-ligand phage, J. Am. Chem. Soc. 135 (2013) 7761–7767. https://doi.org/10.1021/ja4028082

[112] M.S. Chiriacò, E. Primiceri, A. Montanaro, F. de Feo, L. Leone, R. Rinaldi, G. Maruccio, On-chip screening for prostate cancer: an EIS microfluidic platform for contemporary detection of free and total PSA, Analyst. 138 (2013) 5404. https://doi.org/10.1039/c3an00911d

[113] Y. Uludag, F. Narter, E. Sağlam, G. Köktürk, M.Y. Gök, M. Akgün, S. Barut, S. Budak, An integrated lab-on-a-chip-based electrochemical biosensor for rapid and sensitive detection of cancer biomarkers, Anal. Bioanal. Chem. 408 (2016) 7775–7783. https://doi.org/10.1007/s00216-016-9879-z

[114] R.M. Mohamadi, I. Ivanov, J. Stojcic, R.K. Nam, E.H. Sargent, S.O. Kelley, Sample-to-answer isolation and mrna profiling of circulating tumor cells, Anal. Chem. 87 (2015) 6258–6264. https://doi.org/10.1021/acs.analchem.5b01019

[115] M. Moscovici, A. Bhimji, S.O. Kelley, Rapid and specific electrochemical detection of prostate cancer cells using an aperture sensor array, Lab Chip. 13 (2013) 940. https://doi.org/10.1039/c2lc41049d

[116] T. Xu, Y. Song, W. Gao, T. Wu, L.-P. Xu, X. Zhang, S. Wang, Superwettable electrochemical biosensor toward detection of cancer biomarkers, ACS Sensors. 3 (2018) 72–78. https://doi.org/10.1021/acssensors.7b00868

[117] F. Haggar, R. Boushey, Colorectal cancer epidemiology: incidence, mortality, survival, and risk factors, Clin. Colon Rectal Surg. 22 (2009) 191–197. https://doi.org/10.1055/s-0029-1242458

[118] G. Zarkavelis, Current and future biomarkers in colorectal cancer, Ann. Gastroenterol. (2017). https://doi.org/10.20524/aog.2017.0191

[119] G. Lech, Colorectal cancer tumour markers and biomarkers: Recent therapeutic advances, World J. Gastroenterol. 22 (2016) 1745. https://doi.org/10.3748/wjg.v22.i5.1745

[120] T. Tanaka, M. Tanaka, T. Tanaka, R. Ishigamori, Biomarkers for colorectal cancer, Int. J. Mol. Sci. 11 (2010) 3209–3225. https://doi.org/10.3390/ijms11093209

[121] X. Gu, Z. She, T. Ma, S. Tian, H.-B. Kraatz, Electrochemical detection of carcinoembryonic antigen, Biosens. Bioelectron. 102 (2018) 610–616. https://doi.org/10.1016/j.bios.2017.12.014

[122] H. Shu, W. Wen, H. Xiong, X. Zhang, S. Wang, Novel electrochemical aptamer biosensor based on gold nanoparticles signal amplification for the detection of carcinoembryonic antigen, Electrochem. Commun. 37 (2013) 15–19. https://doi.org/10.1016/j.elecom.2013.09.018

[123] H. Zeng, D.A.Y. Agyapong, C. Li, R. Zhao, H. Yang, C. Wu, Y. Jiang, Y. Liu, A carcinoembryonic antigen optoelectronic immunosensor based on thiol-derivative-nanogold labeled anti-CEA antibody nanomaterial and gold modified ITO, Sensors Actuators B Chem. 221 (2015) 22–27. https://doi.org/10.1016/j.snb.2015.06.062

[124] L. Zhao, C. Li, H. Qi, Q. Gao, C. Zhang, Electrochemical lectin-based biosensor array for detection and discrimination of carcinoembryonic antigen using dual amplification of gold nanoparticles and horseradish peroxidase, Sensors Actuators B Chem. 235 (2016) 575–582. https://doi.org/10.1016/j.snb.2016.05.136

[125] Y. Li, Y. Chen, D. Deng, L. Luo, H. He, Z. Wang, Water-dispersible graphene/amphiphilic pyrene derivative nanocomposite: High AuNPs loading capacity for CEA electrochemical immunosensing, Sensors Actuators B Chem. 248 (2017) 966–972. https://doi.org/10.1016/j.snb.2017.02.138

[126] S.X. Lee, H.N. Lim, I. Ibrahim, A. Jamil, A. Pandikumar, N.M. Huang, Horseradish peroxidase-labeled silver/reduced graphene oxide thin film-modified screen-printed electrode for detection of carcinoembryonic antigen, Biosens. Bioelectron. 89 (2017) 673–680. https://doi.org/10.1016/j.bios.2015.12.030

[127] B. Situ, N. Cao, B. Li, Q. Liu, L. Lin, Z. Dai, X. Zou, Z. Cai, Q. Wang, X. Yan, L. Zheng, Sensitive electrochemical analysis of BRAF V600E mutation based on an amplification-refractory mutation system coupled with multienzyme functionalized Fe3O4/Au nanoparticles, Biosens. Bioelectron. 43 (2013) 257–263. https://doi.org/10.1016/j.bios.2012.12.021

[128] S.Y. Hwang, I.J. Seo, S.Y. Lee, Y. Ahn, Microfluidic multiplex biochip based on a point-of-care electrochemical detection system for matrix metalloproteinases, J. Electroanal. Chem. 756 (2015) 118–123. https://doi.org/10.1016/j.jelechem.2015.08.015

[129] L. Truta, M. Sales, Sol-gel chemistry in biosensing devices of electrical transduction: application to CEA cancer biomarker, Curr. Top. Med. Chem. 15 (2015) 256–261. https://doi.org/10.2174/1568026614666141229113318

[130] B. Zhang, C. Ding, Displacement-type amperometric immunosensing platform for sensitive determination of tumour markers, Biosens. Bioelectron. 82 (2016) 112–118. https://doi.org/10.1016/j.bios.2016.03.053

[131] M. Garranzo-Asensio, A. Guzman-Aranguez, C. Povés, M.J. Fernández-Aceñero, R.M. Torrente-Rodríguez, V. Ruiz-Valdepeñas Montiel, G. Domínguez, L.S. Frutos, N. Rodríguez, M. Villalba, J.M. Pingarrón, S. Campuzano, R. Barderas, Toward liquid biopsy: determination of the humoral immune response in cancer patients using halotag fusion protein-modified electrochemical bioplatforms, Anal. Chem. 88 (2016) 12339–12345. https://doi.org/10.1021/acs.analchem.6b03526

[132] M. Aydın, E.B. Aydın, M.K. Sezgintürk, A disposable immunosensor using ITO based electrode modified by a star-shaped polymer for analysis of tumor suppressor protein p53 in human serum, Biosens. Bioelectron. 107 (2018) 1–9. https://doi.org/10.1016/j.bios.2018.02.017

[133] J.M. Lim, M.Y. Ryu, J.W. Yun, T.J. Park, J.P. Park, Electrochemical peptide sensor for diagnosing adenoma-carcinoma transition in colon cancer, Biosens. Bioelectron. 98 (2017) 330–337. https://doi.org/10.1016/j.bios.2017.07.013

[134] S. Chung, P. Chandra, J.P. Koo, Y.-B. Shim, Development of a bifunctional nanobiosensor for screening and detection of chemokine ligand in colorectal cancer cell line, Biosens. Bioelectron. 100 (2018) 396–403. https://doi.org/10.1016/j.bios.2017.09.031

[135] K.-J. Feng, Y.-H. Yang, Z.-J. Wang, J.-H. Jiang, G.-L. Shen, R.-Q. Yu, A nano-porous CeO_2/Chitosan composite film as the immobilization matrix for colorectal cancer DNA sequence-selective electrochemical biosensor, Talanta. 70 (2006) 561–565. https://doi.org/10.1016/j.talanta.2006.01.009

[136] X. Wang, G. Shu, C. Gao, Y. Yang, Q. Xu, M. Tang, Electrochemical biosensor based on functional composite nanofibers for detection of K-ras gene via multiple signal amplification strategy, Anal. Biochem. 466 (2014) 51–58. https://doi.org/10.1016/j.ab.2014.08.023

[137] M. Raji, G. Amoabediny, P. Tajik, M. Hosseini, E. Ghafar-Zadeh, An Apta-Biosensor for Colon Cancer Diagnostics, Sensors. 15 (2015) 22291–22303. https://doi.org/10.3390/s150922291

[138] M. Ahmadzadeh-Raji, E. Ghafar-Zadeh, G. Amoabediny, An optically-transparent aptamer-based detection system for colon cancer applications using gold nanoparticles electrodeposited on indium tin oxide, Sensors. 16 (2016) 1071. https://doi.org/10.3390/s16071071

[139] A.B. Hashkavayi, J.B. Raoof, R. Ojani, S. Kavoosian, Ultrasensitive electrochemical aptasensor based on sandwich architecture for selective label-free detection of colorectal cancer (CT26) cells, Biosens. Bioelectron. 92 (2017) 630–637. https://doi.org/10.1016/j.bios.2016.10.042

Biosensors: Materials and Applications   Materials Research Forum LLC
Materials Research Foundations **47** (2019) 211-240   doi: http://dx.doi.org/10.21741/9781644900130-6

Chapter 6

Role of Nanoparticles in Combating Infections

Kanwal Rehman[1], Muhammad Sajid Hamid Akash[2*]

[1]Institute of Pharmacy, Physiology and Pharmacology, University of Agriculture, Faisalabad, Pakistan

[2]Department of Pharmaceutical Chemistry, Faculty of Pharmaceutical Sciences, Government College University Faisalabad, Pakistan

*sajidakash@gmail.com / sajidakash@gcuf.edu.pk

Abstract

Infections still remain a serious threat to human beings globally. Bacterial infections are one of the major causes of deaths worldwide and increase the overall medical costs, mortality and morbidity due to the development of antimicrobial resistance by means of different mechanisms. Complications related to infectious diseases are being increased day by day, more particularly in developing countries due to limited resources and less availability of the choice of antimicrobial agents. The excessive use of antimicrobial agents creates drug resistance into the pathogens, which demand additional efforts to develop advanced and more effective therapeutic strategies to prevent and treat such kind of infections. So, there was an urgent need to develop new strategies and therapies to cure microbial infections. In response to this need, pharmaceutical scientists are trying their best to design such materials that enhance the biological activity and pharmacokinetic properties of antimicrobial agents against microbial infections. Currently, the best available strategy that can be implemented to prevent the development of drug resistance in pathogens is nanoparticle-based delivery of antimicrobial agents in order to enhance the effectiveness of antimicrobial agents against infections. Nanoparticles have experimentally been proven to have a wide range of antimicrobial potential against various pathogens and an ideal carrier system for intracellular delivery of antimicrobial agents. In this chapter, we have comprehensively summarized the role of polymeric-based nanoparticles for combating the infectious diseases.

Keywords

Antimicrobial Agents, Infections, Polymers, Nanoparticles, Delivery System

Contents

Biosensors: Materials and Applications   Materials Research Forum LLC
Materials Research Foundations **47** (2019) 211-240   doi: http://dx.doi.org/10.21741/9781644900130-6

1.   Introduction

Combating infection means to take action to reduce, destroy, or prevent microbial agent which is harmful to the body. At the start of the 20th century, bacterial and fungal-based infectious diseases were the most prominent causes of death worldwide [1] and discovery of antibiotics provided the innovations for the treatment of these infections. In the late 1940s, the commercial production of first penicillin antibiotic was begun. Since that time, infections were cured with antibiotics and claimed to be a great success until 1970-1980s. But later on, it became a big challenge for scientists due to the development of antimicrobial resistance [2]. A schematic representation of the development of antibiotics and the development of resistance has been briefly described in figure 1 [3]. Nowadays, multidrug resistant (MDR) bacteria are becoming a serious threat to human health because they produce many types of infections every year [4,5]. After that in the 21st century, the development and modifications of antibiotic agents have played a critical role in combating chronic infectious diseases [6,7].

Development of antimicrobial resistance in pathogens and decreased intracellular penetration of antimicrobial agents have escalated the dire need of such types of strategies that have the ability to address these challenges [8–10]. It has been estimated that the rate of fatality due to the development of antimicrobial resistance is predicted to increase into the millions in 2050 while, the development of strategies for the treatment of antibiotic-resistant microbial agents has been very slow [11]. Antibiotic-resistant microbial agents often exist as biofilms that are protected by the layer of extracellular polymeric matrices. The formation of biofilm in antibiotic-resistant microbial agents is a major problem that makes them impervious to the conventional antimicrobial agents [12,13]. Recently, it has been experimentally evidenced that polymeric-based nanoparticles have the potential to break down the formation of biofilm and cross the extracellular membrane of microbial agents [11].

In this chapter, we have briefly described the various challenges that are being faced during the treatment of infectious diseases and the use of nanotechnology for efficient delivery of antibiotics. Moreover, we have also comprehensively summarized the role of polymeric-based nanoparticles for combating the infectious diseases and recent advances for the intracellular delivery of antimicrobial agents loaded with polymeric-based nanoparticles.

2.   Challenges for the treatment of microbial infections

Despite the wide success of antibiotics against different types of microbes and pathogens, the treatment of bacterial infection still faces various challenges, particularly the

Biosensors: Materials and Applications   Materials Research Forum LLC
Materials Research Foundations **47** (2019) 211-240   doi: http://dx.doi.org/10.21741/9781644900130-6

development of antibiotic resistance and decreased penetration of antibiotics across the extracellular membrane. Despite of this, there are certain limitations with the use of antimicrobial agents which are: short half-life, high toxicity, poor bioavailability after oral administration, less stability at physiological conditions, poor penetration across the cell membrane and microbial resistance to these antibiotics [14–17]. Numerous mechanisms and pathways are involved to develop antibiotic resistance within the microbe that has been schematically described in figure 2 [18].

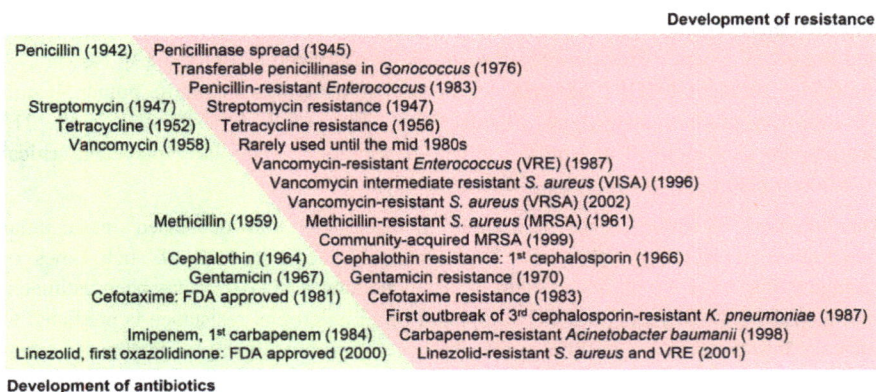

Development of resistance

Penicillin (1942)   Penicillinase spread (1945)
                    Transferable penicillinase in *Gonococcus* (1976)
                    Penicillin-resistant *Enterococcus* (1983)
Streptomycin (1947)   Streptomycin resistance (1947)
Tetracycline (1952)   Tetracycline resistance (1956)
Vancomycin (1958)   Rarely used until the mid 1980s
                    Vancomycin-resistant *Enterococcus* (VRE) (1987)
                    Vancomycin intermediate resistant *S. aureus* (VISA) (1996)
                    Vancomycin-resistant *S. aureus* (VRSA) (2002)
Methicillin (1959)   Methicillin-resistant *S. aureus* (MRSA) (1961)
                    Community-acquired MRSA (1999)
Cephalothin (1964)   Cephalothin resistance: 1st cephalosporin (1966)
Gentamicin (1967)   Gentamicin resistance (1970)
Cefotaxime: FDA approved (1981)   Cefotaxime resistance (1983)
                    First outbreak of 3rd cephalosporin-resistant *K. pneumoniae* (1987)
Imipenem, 1st carbapenem (1984)   Carbapenem-resistant *Acinetobacter baumanii* (1998)
Linezolid, first oxazolidinone: FDA approved (2000)   Linezolid-resistant *S. aureus* and VRE (2001)

Development of antibiotics

Figure 1: History of antimicrobial agent development vs. subsequent acquaintance of resistance by microorganisms. Adopted from Ref. [3].

3.   Role of nanotechnology in therapeutic delivery of antimicrobial agents

Nanotechnology is an important scientific field that has significant potential to generate innovative and new material, especially in future clinical applications and overcomes many hurdles that are difficult to solve with other traditional methods. It involves the study, synthesis, design and characterization of nanoscale materials that have at least one dimension on the nanometer scale ($\sim$ 1–100 nm). From the last few years, the application of nanotechnology has been growing rapidly in developing new material in the field of biological sciences [19–23]. In such a scenario, nanotechnology provides an attractive, effective and safe platform for the rapid development and modifications of conventional antimicrobial agents for the treatment of fungal and bacterial infections. Nanotechnology-based delivery systems have improved pharmaceutical characteristics, biodistribution, sustained release and bioavailability of therapeutic agents and minimize the adverse

Biosensors: Materials and Applications   Materials Research Forum LLC
Materials Research Foundations **47** (2019) 211-240   doi: http://dx.doi.org/10.21741/9781644900130-6

effects of antimicrobial agents which are associated with them [24–27]. Nanomaterials show distinct physical and chemical properties than that of bulk material due to unique properties (small size and large surface area to volume ratio). Due to these features, nanomaterials have a tendency to enter into the bacterial cell and increase the therapeutic activity of antimicrobial agent against antimicrobial-drug resistant microorganisms with less toxicity [28–35]. Moreover, nanoparticulate systems offer some extra advantages, such as solubilization of hydrophobic compound, improvement in bioavailability, stability, ability to cross the blood-brain barrier, protection from physical, chemical and biological degradation and prolonged therapeutic potential of antimicrobial agents. Moreover, nano-particulate systems improve the ability to cross the biological barriers to reach the targeted infectious cells for treatment [36–38].

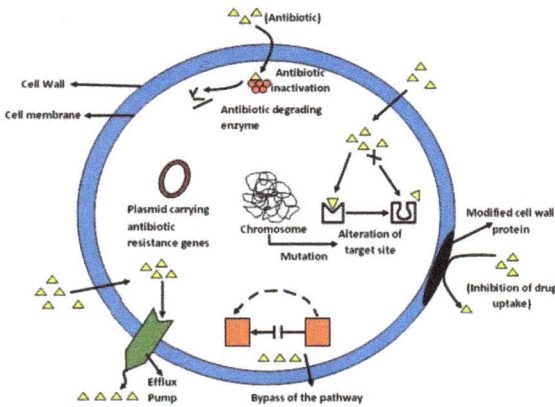

Figure 2: Mechanisms of development of antibiotic resistance microbe. Adopted from Ref. [18].

Around 1965, the first nanotechnology was used to enhance the therapeutic effect of a drug. From the last several years, nanotechnology has delivered many valuable therapeutic agents to their targeted sites [39–43]. Therapeutic agents are encapsulated or loaded into the nanoparticles in different ways including liposomes, polymeric nanoparticles, micelles, etc. [43,44]. There are many advantages of nanotechnology notably it has the ability to improve the drug solubility, control and sustain the release, precise drug targeting, prolong systemic circulation for the treatment of variety diseases

Biosensors: Materials and Applications
Materials Research Foundations **47** (2019) 211-240

Materials Research Forum LLC
doi: http://dx.doi.org/10.21741/9781644900130-6

[45–48]. Among the various strategies that are being investigated, polymeric-based nanoparticles have been extensively instigated to enhance the therapeutic effectiveness of existing antibiotics [32,42,49]. In the proceeding sections, we have briefly focused on the role of nanoparticles in combating microbial infections.

4.   Polymeric-based nanoparticles as an antimicrobial agents

It has been well documented that antibiotics exhibit a wide range of success against microbial infections, but still the development of resistance against antibiotic is a major challenge and this drawback has now been overcome by the use of polymeric-based nanoparticles. These nanoparticles have exclusively been investigated to increase the therapeutic potentials of antibiotics and increase the sensitivity of MDR-bacteria [18,50]. These nanoparticles exhibit antibacterial potentials via the involvement of various mechanisms that have been schematically elaborated in figure 3 [18].

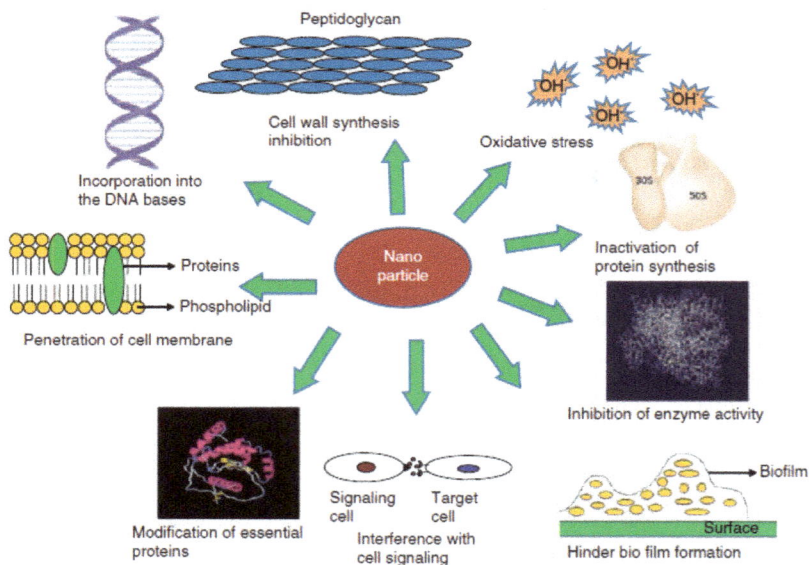

Figure 3: Schematic representation of the mode of action of antibacterial activity of nanoparticles. Adopted from Ref. [18].

Biosensors: Materials and Applications   Materials Research Forum LLC
Materials Research Foundations **47** (2019) 211-240   doi: http://dx.doi.org/10.21741/9781644900130-6

4.1   Metallic nanoparticles as antibacterial agents

It has been well documented that metals have long been used for wound healing, prevent infection and treatment of burns. Metallic nanoparticles have been used for the treatment of different chronic diseases [51,52]. It has been evidenced that metallic ions such as $AgNO_3$ penetrate into the bacterial cell wall and cross extracellular membrane via interaction with sulphur containing proteins. Various mechanisms are involved to penetrate the nanoparticles inside the bacterial cell (Figure 4). Once it enters into the bacterial cell, it damages bacterial DNA and respiratory enzymes which become the cause of cell death of the bacteria. The tiny size (<10 nm) and large surface area of metallic compounds are more effective than the larger size of metallic compounds, because small size metallic particles are easily penetrated into the bacterial cell wall and membrane [53–67]. Metallic nanoparticles easily interact with DNA bases (part of the bacterial cell membrane which have sulfur and phosphorous) and induce the cell death [68]. In the following sub-sections, we have discussed the role of various metallic-based nanoparticles that may have exhibited their antimicrobial properties against various types of pathogens.

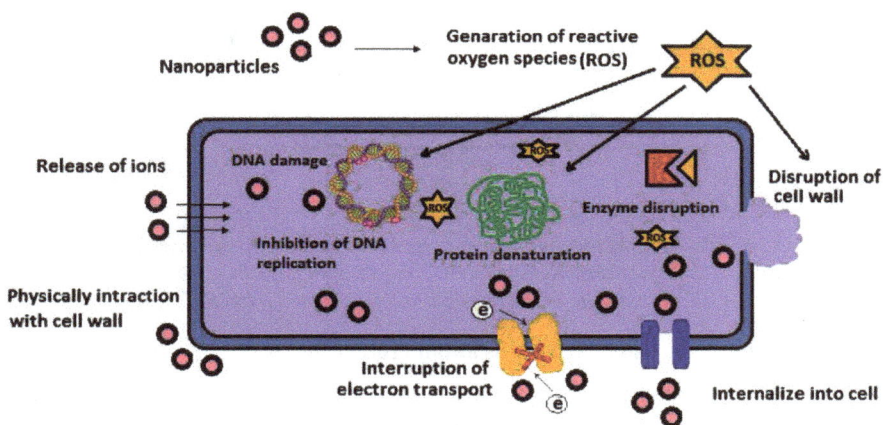

Figure 4: Schematic representation of various mechanisms of antimicrobial activity of the metal nanoparticles. Adopted from Ref. [52].

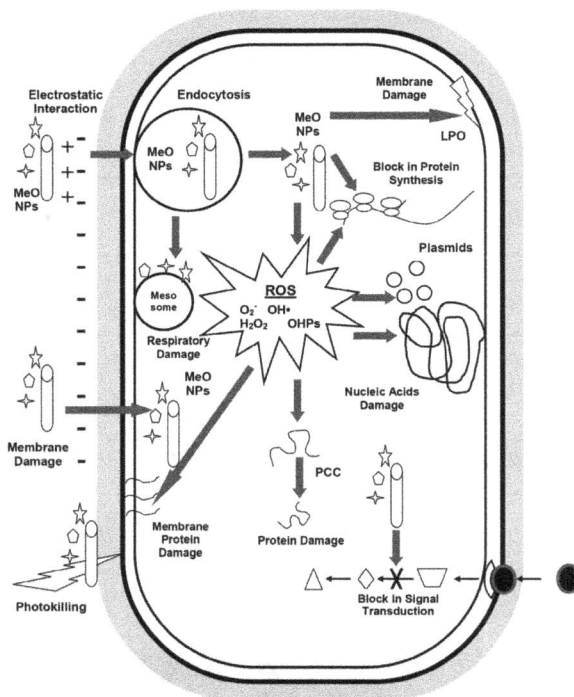

Figure 5: Overview of the antimicrobial mechanism of action of metallic oxide nanoparticles. Abbreviations: ROS; reactive oxygen species, LPO; lipid peroxidation, MeO-NPs; metallic oxide nanoparticles. Adopted from Ref. [84].

4.1.1   Metallic oxides nanoparticles as antimicrobial agents

It has been reported that different metals (for example zinc, gold, copper, nickel, titanium, cadmium, iron and magnesium) and their oxides nanoparticles exhibit antimicrobial effects by reducing the extent of infections. Furthermore, these metal oxides nanoparticles have shown reasonable stability and safety profile when they are in small particle size [26,69–71]. These nanoparticles have strong antibacterial potential and less toxicity when compared to the same material with large particle size. Moreover, the positive charge of the metal particle helps to bind the nanoparticle with a negatively charged bacterial cell wall (Figure 5) which is an ideal feature of these nanoparticles

Biosensors: Materials and Applications   Materials Research Forum LLC
Materials Research Foundations **47** (2019) 211-240   doi: http://dx.doi.org/10.21741/9781644900130-6

[70,72–80]. It has been evidenced from experimental studies that many metallic oxides nanoparticles such as Fe_3O_4 showed antibacterial activity when they were in nano size and do not exhibit antibacterial activity when having large particle size [81–83]. Once entered into the bacterial cell, metallic oxides nanoparticles show antibacterial activity due to the generation of ROS which damages the bacterial cell walls (Figure 5). Nanoparticles have the strong capability to destroy the cell membrane which ultimately becomes the cause of cell death.

4.1.2 Silver nanoparticles as antimicrobial agents

Silver nanoparticles show strong antibacterial activity as compared to its salts due to large surface area which has a strong affinity and better contact with pathogens such as *Aeromonas hydrophila* [85]. Penicillin conjugated silver nanoparticles have shown to exhibit significant antibacterial activity against different gram-positive and gram-negative resistant microbial strains [86]. Silver nanoparticles alone have some shortcomings that have been overcome by conjugating the silver nanoparticles with polysaccharides, chitin and pluronic [87–89]. During wound healing, this polysaccharide-, chitin- and pluronic-based silver nanoparticles play a potential role to combat the bacterial growth by destroying the stability of microbial cell membranes and also by increasing the bioavailability of the drug. It has also been found that pluronic-coated silver nanoparticles are very effective against antibiotic-resistant *Staphylococcus aureus* strains [89].

4.1.3 Zinc oxide nanoparticles as antimicrobial agents

The antibacterial activity of ZnO-NPs against several microbial agents has attained considerable interest due to the increased specific surface area of ZnO-NPs [90–93]. ZnO-NPs exhibit antimicrobial activity via involvement of generation of ROS which leads to the cell wall damage due to ZnO-localized interaction, enhanced membrane permeability, internalization of ZnO-NPs and uptake of toxic dissolved zinc ions [94].

4.1.4 Gold nanoparticles as antimicrobial agents

Gold nanoparticles (Au-NPs) have a wide range of pharmacological properties and they have exclusively been used for gene, radiation and cancer therapy, and numerous medical diagnosis purposes [95–97]. Development of antibiotic resistance in pathogens is one of the major challenge and antimicrobial activity of Au-NPs against several pathogens such as *P. aeruginosa, E.coli, K. pneumoniae, Bacillus subtillis, Staphylococcus aureus, Corynebacterium pseudotuberculosis,* and *S. choleraesius*, have been extensively studied [98–101]. It has been found that Au-NPs exhibit antimicrobial activity via disruption of cell wall and membrane of the pathogen and generation of ROS [102]. Similarly, antimicrobial peptide-based Au-NPs [11] and gold nanorods [103] have also been shown to have potent antimicrobial activity against several antibiotic-resistant microbial agents.

4.1.5 Copper oxide nanoparticles as antimicrobial agents

Another important metallic compound which exhibits antimicrobial properties is copper. Copper oxide (CuO) has also antifungal and antibacterial properties. It is much cheaper as compared to silver and is very effective against a wide range of microbes that induce hospital-acquired infections [104–107]. CuO-NPs exhibit better antibacterial activity via generation of ROS in gram-negative bacteria as compared to that in gram-positive bacteria [107]. The potential antimicrobial effects of CuO-NPs may help to develop a sustainable particulate system for the inhibition of microbial infections. In recent studies, it has been extensively reported that CuO-NPs have potential to control the infections induced by various types of bacterial and fungal species [108–110].

4.1.6 Metal-halogen complex-based nanoparticles as antimicrobial agents

It has been evidenced that different halogens such as chlorine, bromine, iodine and fluorine, etc. have antibacterial activity, but these halogens may have also some toxic effects when they are used directly. To avoid this drawback, scientists have developed metal halogen complexes like Mg-halogen nanoparticles for the treatment of microbial infections. Mg-halogens have the ability to damage the microbial cell membrane. The antimicrobial capability and holding capacity of Mg for halogens can be increased by converting it into nanoparticles [111,112].

4.2   Chitosan-based nanoparticles as antimicrobial agents

Chitosan is a natural polysaccharide and has antimicrobial properties due to its poly-cationic character which has the ability to interact with negatively charged bacterial cell walls [113]. Chitosan-based nanoparticles have been extensively studied for the evaluation of antimicrobial activities against various types of gram-positive and gram-negative bacteria [114]. By the help of this interaction, it inhibits the synthesis of mRNA and DNA. Chitosan-based nanoparticles have found to be more effective against a wide range of bacteria such as *E.coli, Klebsiella pneumonia* and *Pseudomonas aeruginosa* [115]. The antimicrobial activity of chitosan-based nanoparticles is based on the size of the nanoparticle. Nano-sized chitosan has high antibacterial activity as compared to that of the large particle size of chitosan because, at nano-scale, chitosan has a large surface to volume ratio, higher charge density and interaction affinity with microbes [116]. Chitosan-loaded silver nanoparticles exhibit a larger zone of inhibition as compared to that of simple silver nanoparticles. It has been also shown that hydrogel-based chitosan-g-poly (glutamic acid) polyelectrolyte complex showed antimicrobial activity against *E.coli* and *S. aureus* [117]. Similarly, chitosan-dextran hydrogel complex also demonstrated efficient antimicrobial activity against *Streptomyces pyogenes, E.coli and S. aureus* [118].

5.   Polymeric-based nanoparticles as microbial diagnostic agents

Nanoparticles have been extensively investigated for the diagnosis of antibiotic-resistant pathogens (Table 1). The physicochemical properties of nanoparticles are responsible for accurate, sensitive, cost-effective and rapid diagnosis of antibiotic-resistant pathogens [119]. In table 1, we have briefly summarized the role of various nanoparticles for the diagnosis of various pathogens along with their limitation of detection. The studies mentioned in table 1 indicate that Ag- and Au-based nanoparticles have been extensively studied due to their their unique physicochemical properties for specific and selective identification of proteins and DNA/RNA sequences associated with the presence of various types of pathogens [120].

6.   Recent advances in intracellular delivery of nanoparticle-based antibiotics

The incidences of antimicrobial resistance against various pathogens are being increased day by day and delivery of nanoparticle-based antibiotics can improve the cellular penetration, retention and distribution of antibiotics to overcome the MDR developed by a microbial agent against certain antibiotic [18]. In the following sub-sections, we have briefly described various nanoparticle-based delivery of various antibiotics for the treatment of infectious diseases.

6.1   Amphotericin B

Amphotericin B is an antifungal drug that is being widely used for the treatment of various types of fungal infections such as cryptococcosis, coccidioidomycosis, blastomycosis, candidiasis, aspergillosis, phaeohyphomycosis, hyalohyphomycosis, leishmaniasis, extracutaneous sporotrichosis and mucormycosis. Microbial resistance developed against amphotericin B limits the therapeutic applications against fungal infections [143,144]. The development of antimicrobial resistance against amphotericin B has been encountered using polymeric-based nanoparticle delivery of amphotericin B with PLGA [145,146]. PLGA-based nanoparticles significantly increased the oral bioavailability along with increased antifungal activity and reduced toxicity.

Table 1: Nanoparticles as microbial diagnostic agents. Adapted from Ref. [83].

Nanoparticles	Type of pathogen to be detected	Detection limit	Ref.
Ag-NPs	MRSA, *Listeria* spp., *E. coli* & *P. aeruginosa*	10^3 CFU/mL	[121]
	SRV	0.05 pg/mL	[120]
	Salmonella spp.	50–100 CFU/mL	[122]
	Aviar Influenza Virus H7	1.6 pg/mL	[123]
Au-coated-NPs	*Legionella* spp.	-	[124]
Ag-Au-NPs	*S. epidermidis, B. megaterium, E. coli* & *Salmonella enteric*	102 CFU/mL	[125]
Au-NPs	TB	0.75 µg 2 h	[126]
	Bacillus. Anthracis	-	[127]
	MRSA	66 pg/µL (<105 CFU/mL)	[126]
	S. enterica	37 fM	[128]
	Acinetobacter baumani	0.8125 ng/µL	[129]
	MTBC and *Plasmodium*		[130]
	MDRTB		[131]
	BIOFILMS: Amycolatopsis azurea, B. licheniformic, B. megaterium, E. coli, P. aeruginosa	-	[132]
	E. coli	5x104 CFU/mL 25 min	[133]
	Influenza Virus H5N1	<105 copies of transcribed RNA; 2.5 h	[134]
	E. coli 0157:H7	103 CFU/mL	[135]
	SIFILIS	0.98pg/mL	[136]
	Influenza Virus	10 pg/mL	[137]
Aptamer-conjugated-Au-NPs	*C. difficile*	1 nM	[138]
	Salmonella	3 CFU/mL	[139]
	S. aureus	10 CFU/mL 1.5 h	[140]
	TB	30 µg/mL 2 h	[141]
Ag-Au core shell NPs	HPV	0.05 pmol/µL	[142]

Abbreviations. Ag-NPs: silver nanoparticles, Au-coated-NPs: gold-coated nanoparticles, Ag-Au-NPs: silver-gold nanoparticles, MRSA: Methicillin-resistant *S. aureus*, TB: Tuberculosis, HPV: Human Papilloma Virus.

Biosensors: Materials and Applications   Materials Research Forum LLC
Materials Research Foundations **47** (2019) 211-240   doi: http://dx.doi.org/10.21741/9781644900130-6

6.2   Aminoglycosides

Aminoglycosides notably gentamicin are widely used for the treatment of various types of microbial infections. The antimicrobial activity of aminoglycosides is reduced in the acidic endosomal pH due to the slow penetration of aminoglycosides across the cell membrane and remain confined within the endosomal compartment [147,148]. Despite this, aminoglycosides also face the development of antimicrobial resistance against various types of pathogens [149]. This problem has been overcome by the use of a number of polymeric-based nanoparticles [150,151]. Encapsulation of aminoglycosides with polymeric-based nanoparticles has significantly attained the considerable interest as they have the ability to increase the intracellular delivery and cellular uptake of encapsulated antibiotics [152].

6.3   Beta-lactam antibiotics

Beta-lactam antibiotics are one of the most important antibiotics and are being used as the first choice for the treatment of numerous types of infections. Due to less penetration into the cells and lysosomal membranes, they are not preferably prescribed for the treatment of intracellular infections. This problem has been overcome by the encapsulation of these antibiotics with various types of biodegradable polymeric-based nanoparticles notably polyethylcyanoacrylate, chitosan and polyisohexylcyanoacrylate nanoparticles [153,154]. These polymeric-based nanoparticles have experimentally proven to have the ability to increase the intracellular penetration and retention time of the beta-lactam antibiotics [152].

6.4   Tetracycline antibiotics

Tetracyclines also have poor cellular penetration and this limitation has been overcome by the use of biodegradable polymeric-based nanoparticles [152]. These nanoparticles have the ability to increase the intracellular penetration of tetracyclines into the pathogens [155,156].

6.5   Fluoroquinolone antibiotics

These antibiotics permeate into the phagocytes, but when the extracellular concentration falls down, this flux out rapidly via organic anion carriers [156]. This drawback has been overcome by the use of polymeric-based nanoparticles which have the ability to increase the intracellular delivery and longevity of fluoroquinolones against various type of intracellular pathogens such as *Francisella tularensis*, *Chlamydia pneumoniae*, *Salmonella*, *L. monocytogenes* and *Mycobacterium tuberculosis* [157].

6.6   Macrolide antibiotics

Despite of efficient penetration across the cellular membrane, macrolides are poorly accumulated into the cells and their potential activity decrease in acidic pH [158]. This problem has been overcome by encapsulating the macrolides with polymeric-based nanoparticles [152].

6.7   Cephalosporins

Cephalosporins belong to the class of β-lactam antibiotics and have a wide range of antibacterial activities. Cephalosporins also face MDR against various types of pathogens [149] and this problem has been overcome by using the nanoparticles for delivery of cephalosporins [154,159]. They increased the therapeutic effect of antibiotics than that of the non-modified drug against the pathogen.

6.8   Nanoparticle-based antibacterial vaccination

The concept of vaccination against different infectious diseases began in the late 18th century after the discovery of vaccination against smallpox by Edward Jenner. Vaccines are made to improve the host's immune response against MDR bacterial and fungal pathogens. They initiate a powerful immune response and protective immunity in the host's body against infectious disease [160,161]. In vaccination, a specific immunity is introduced in the recipient against a specific pathogen. It has been evidenced from experimental studies that nanomaterials may exhibit the intrinsic immunomodulatory effects by acting as adjuvants or immune potentiators [162]. Nanoparticles-based antibacterial vaccines have been developed from the prevention of bacterial infections [162–164]. The results have demonstrated that the antibacterial vaccine in nano form has a great impact on the prevention and provide excellent outcomes compared to the bulk antimicrobial agents in treating infections [165].

Conclusion

The use of nanotechnology in medicine helps us to create a novel carrier system that has the ability to improve the pharmaceutical potential of conventional antimicrobial agents. It has been concluded from the brief review of the current literature cited in this chapter that polymeric-based nanoparticle drug delivery approaches are being applied to prolong the antimicrobial activity and prevent MDR against the pathogens. Nanoparticles have been found as potential drug carrier systems to enhance the therapeutic potential of antimicrobial agents. Experimental studies exhibit that antimicrobial agents loaded in nanoparticles showed better pharmaceutical potential than that of its potent agent. The up-to-date knowledge presented here will be helpful for the pharmaceutical researchers who are trying to find a solution against MDR pathogens.

References

[1]   M.L. Cohen, Changing patterns of infectious disease, Nature. 406 (2000) 762–767. https://doi.org/10.1038/35021206

[2]   G. Taubes, The bacteria fight back, Science (80-.). 321 (2008) 356–361. https://doi.org/10.1126/science.321.5887.356

[3]   A.J. Huh, Y.J. Kwon, "Nanoantibiotics": A new paradigm for treating infectious diseases using nanomaterials in the antibiotics resistant era, J. Control. Release. 156 (2011) 128–145. https://doi.org/10.1016/j.jconrel.2011.07.002

[4]   S.B. Levy, B. Marshall, Antibacterial resistance worldwide: causes, challenges and responses, Nat. Med. 10 (2004) S122–S129. https://doi.org/10.1038/nm1145

[5]   G.L. French, The continuing crisis in antibiotic resistance, Int. J. Antimicrob. Agents. 36 (2010) S3–S7. https://doi.org/10.1016/S0924-8579(10)70003-0

[6]   A.L. Demain, Antibiotics: Natural products essential to human health, Med. Res. Rev. 29 (2009) 821–842. https://doi.org/10.1002/med.20154

[7]   A.S. Pina, A. Hussain, A.C.A. Roque, An historical overview of drug discovery, in: Methods Mol. Biol., 2010: pp. 3–12. https://doi.org/10.1007/978-1-60761-244-5_1

[8]   V. Aloush, S. Navon-Venezia, Y. Seigman-Igra, S. Cabili, Y. Carmeli, Multidrug-resistant Pseudomonas aeruginosa: risk factors and clinical impact, Antimicrob. Agents Chemother. 50 (2006) 43–48. https://doi.org/10.1128/AAC.50.1.43-48.2006

[9]   V. Manchanda, S. Sanchaita, N. Singh, Multidrug resistant acinetobacter., J. Glob. Infect. Dis. 2 (2010) 291–304. https://doi.org/10.4103/0974-777X.68538

[10] F. Guilhelmelli, N. Vilela, P. Albuquerque, L. da S. Derengowski, I. Silva-Pereira, C.M. Kyaw, Antibiotic development challenges: the various mechanisms of action of antimicrobial peptides and of bacterial resistance, Front. Microbiol. 4 (2013) 353. https://doi.org/10.3389/fmicb.2013.00353

[11] U. Rajchakit, V. Sarojini, Recent developments in antimicrobial-peptide-conjugated gold nanoparticles, Bioconjug. Chem. 28 (2017) 2673–2686. https://doi.org/10.1021/acs.bioconjchem.7b00368

[12] N. Høiby, T. Bjarnsholt, M. Givskov, S. Molin, O. Ciofu, Antibiotic resistance of bacterial biofilms, Int. J. Antimicrob. Agents. 35 (2010) 322–332. https://doi.org/10.1016/j.ijantimicag.2009.12.011

[13] J.A. Otter, K. Vickery, J.T. Walker, E. deLancey Pulcini, P. Stoodley, S.D. Goldenberg, J.A.G. Salkeld, J. Chewins, S. Yezli, J.D. Edgeworth, Surface-attached cells, biofilms and biocide susceptibility: implications for hospital cleaning and disinfection, J. Hosp. Infect. 89 (2015) 16–27. https://doi.org/10.1016/j.jhin.2014.09.008

[14] M.J. Hajipour, K.M. Fromm, A. Akbar Ashkarran, D. Jimenez de Aberasturi, I.R. de Larramendi, T. Rojo, V. Serpooshan, W.J. Parak, M. Mahmoudi, Antibacterial properties of nanoparticles, Trends Biotechnol. 30 (2012) 499–511. https://doi.org/10.1016/j.tibtech.2012.06.004

[15] P.A. Patel, V.B. Patravale, AmbiOnp: solid lipid nanoparticles of amphotericin B for oral administration., J. Biomed. Nanotechnol. 7 (2011) 632–9. http://www.ncbi.nlm.nih.gov/pubmed/22195480 (accessed January 2, 2019)

[16] M.B. Chaudhari, P.P. Desai, P.A. Patel, V.B. Patravale, Solid lipid nanoparticles of amphotericin B (AmbiOnp): in vitro and in vivo assessment towards safe and effective oral treatment module, Drug Deliv. Transl. Res. 6 (2015) 354–64. https://doi.org/10.1007/s13346-015-0267-6

[17] D.M. Casa, T.C.M.M. Carraro, L.E.A. de Camargo, L.F. Dalmolin, N.M. Khalil, R.M. Mainardes, Poly(L-lactide) nanoparticles reduce amphotericin b cytotoxicity and maintain its in vitro antifungal activity., J. Nanosci. Nanotechnol. 15 (2015) 848–54. http://www.ncbi.nlm.nih.gov/pubmed/26328449 (accessed January 2, 2019)

[18] R. Singh, M.S. Smitha, S.P. Singh, The role of nanotechnology in combating multi-drug resistant bacteria., J. Nanosci. Nanotechnol. 14 (2014) 4745–56. http://www.ncbi.nlm.nih.gov/pubmed/24757944 (accessed January 2, 2019)

[19] C. Noguera, Physics and chemistry at oxide surfaces, Cambridge University Press, 1996

[20] H.H. Kung, Transition metal oxides : surface chemistry and catalysis, Elsevier, 1989.

[21] J.M. Vohs, The surface science of metal oxides. By V. E. Henrich and P. A. Cox, Cambridge University Press, Cambridge, U.K., 1994, 464 pp. hardcover $99.95; paperback $39.95, AIChE J. 44 (1998) 502–503. https://doi.org/10.1002/aic.690440230

[22]   J.A. Rodríguez, M. Fernández Garcia, Synthesis, properties, and applications of
       oxide nanomaterials, Wiley-Interscience, 2007

[23]   M. Fernández-García, A. Martínez-Arias, J.C. Hanson, J.A. Rodriguez,
       Nanostructured oxides in chemistry: characterization and properties, (2004).
       https://doi.org/10.1021/CR030032F

[24]   S.D. Puckett, E. Taylor, T. Raimondo, T.J. Webster, The relationship between the
       nanostructure of titanium surfaces and bacterial attachment, Biomaterials. 31
       (2010) 706–713. https://doi.org/10.1016/j.biomaterials.2009.09.081

[25]   K.M. Tarquinio, N.K. Kothurkar, D.Y. Goswami, R.C. Sanders, A.L. Zaritsky,
       A.M. LeVine, Bactericidal effects of silver plus titanium dioxide-coated
       endotracheal tubes on Pseudomonas aeruginosa and Staphylococcus aureus., Int. J.
       Nanomedicine. 5 (2010) 177–83. http://www.ncbi.nlm.nih.gov/pubmed/20463933
       (accessed January 2, 2019)

[26]   N. Tran, A. Mir, D. Mallik, A. Sinha, S. Nayar, T.J. Webster, Bactericidal effect of
       iron oxide nanoparticles on Staphylococcus aureus., Int. J. Nanomedicine. 5
       (2010) 277–83. http://www.ncbi.nlm.nih.gov/pubmed/20463943 (accessed January
       2, 2019)

[27]   P.A. Tran, T.J. Webster, Selenium nanoparticles inhibit Staphylococcus aureus
       growth., Int. J. Nanomedicine. 6 (2011) 1553–8.
       https://doi.org/10.2147/IJN.S21729

[28]   S.M. Moghimi, A.C. Hunter, T.L. Andresen, Factors controlling nanoparticle
       pharmacokinetics: an integrated analysis and perspective, Annu. Rev. Pharmacol.
       Toxicol. 52 (2012) 481–503. https://doi.org/10.1146/annurev-pharmtox-010611-
       134623

[29]   A.E. Nel, L. Mädler, D. Velegol, T. Xia, E.M. V. Hoek, P. Somasundaran, F.
       Klaessig, V. Castranova, M. Thompson, Understanding biophysicochemical
       interactions at the nano–bio interface, Nat. Mater. 8 (2009) 543–557.
       https://doi.org/10.1038/nmat2442

[30]   N. Khlebtsov, L. Dykman, Biodistribution and toxicity of engineered gold
       nanoparticles: a review of in vitro and in vivo studies., Chem. Soc. Rev. 40 (2011)
       1647–71. https://doi.org/10.1039/c0cs00018c

[31]   M.-C. Bowman, T.E. Ballard, C.J. Ackerson, D.L. Feldheim, D.M. Margolis, C.
       Melander, Inhibition of HIV Fusion with Multivalent Gold Nanoparticles, J. Am.
       Chem. Soc. 130 (2008) 6896–6897. https://doi.org/10.1021/ja710321g

[32]   O.M. Koo, I. Rubinstein, H. Onyuksel, Role of nanotechnology in targeted drug delivery and imaging: a concise review, Nanomedicine Nanotechnology, Biol. Med. 1 (2005) 193–212. https://doi.org/10.1016/j.nano.2005.06.004

[33]   R.T. Sadikot, I. Rubinstein, Long-acting, multi-targeted nanomedicine: addressing unmet medical need in acute lung injury, J. Biomed. Nanotechnol. 5 (2009) 614–619. https://doi.org/10.1166/jbn.2009.1078

[34]   R.T. Sadikot, Peptide Nanomedicines for treatment of acute lung injury, in: Methods Enzymol., 2012: pp. 315–324. https://doi.org/10.1016/B978-0-12-391860-4.00016-1

[35]   K.S. Brandenburg, I. Rubinstein, R.T. Sadikot, H. Önyüksel, Polymyxin B self-associated with phospholipid nanomicelles, Pharm. Dev. Technol. 17 (2012) 654–660. https://doi.org/10.3109/10837450.2011.572893

[36]   P. Couvreur, C. Vauthier, Nanotechnology: intelligent design to treat complex disease, Pharm. Res. 23 (2006) 1417–1450. https://doi.org/10.1007/s11095-006-0284-8

[37]   P.R. Lockman, R.J. Mumper, M.A. Khan, D.D. Allen, Nanoparticle technology for drug delivery across the blood-brain barrier, Drug Dev. Ind. Pharm. 28 (2002) 1–13. https://doi.org/10.1081/DDC-120001481

[38]   T.K.M. Mbela ', J.H. Poupaert ', P. Dumont, Poly(diethylmethylidene malonate) nanoparticles as primaquine delivery system to liver, 1992

[39]   M.S.H. Akash, K. Rehman, Recent progress in biomedical applications of Pluronic (PF127): Pharmaceutical perspectives, J. Control. Release. 209 (2015) 120–138. https://doi.org/10.1016/j.jconrel.2015.04.032

[40]   M.S.H. Akash, K. Rehman, S. Chen, Pluronic F127-based thermosensitive gels for delivery of therapeutic proteins and peptides, Polym. Rev. 54 (2014) 573–597. https://doi.org/10.1080/15583724.2014.927885

[41]   M.S.H. Akash, K. Rehman, S. Chen, Natural and synthetic polymers as drug carriers for delivery of therapeutic proteins, Polym. Rev. 55 (2015) 371–406. https://doi.org/10.1080/15583724.2014.995806

[42]   M.S.H. Akash, K. Rehman, S. Chen, Polymeric-based particulate systems for delivery of therapeutic proteins, Pharm. Dev. Technol. 21 (2016) 367–378. https://doi.org/10.3109/10837450.2014.999785

[43]   M.S.H. Akash, K. Rehman, M. Tariq, S. Chen, Development of therapeutic proteins: Advances and challenges. Turk J Biol. 2015;39(3):343-58

Biosensors: Materials and Applications   Materials Research Forum LLC
Materials Research Foundations **47** (2019) 211-240   doi: http://dx.doi.org/10.21741/9781644900130-6

[44]   R.A. Petros, J.M. DeSimone, Strategies in the design of nanoparticles for
       therapeutic applications, Nat. Rev. Drug Discov. 9 (2010) 615–627.
       https://doi.org/10.1038/nrd2591

[45]   M.E. Davis, Z. Chen, D.M. Shin, Nanoparticle therapeutics: an emerging treatment
       modality for cancer, Nat. Rev. Drug Discov. 7 (2008) 771–782.
       https://doi.org/10.1038/nrd2614

[46]   L. Zhang, D. Pornpattananangku, C.-M.J. Hu, C.-M. Huang, Development of
       nanoparticles for antimicrobial drug delivery., Curr. Med. Chem. 17 (2010) 585–
       94

[47]   M.S.H. Akash, K. Rehman, S. Chen, IL-1Ra and its delivery strategies: inserting
       the association in perspective, Pharm. Res. 30 (2013) 2951–2966.
       https://doi.org/10.1007/s11095-013-1118-0

[48]   M.S.H. Akash, K. Rehman, N. Li, J.-Q. Gao, H. Sun, S. Chen, Sustained delivery
       of IL-1Ra from Pluronic F127-based thermosensitive gel prolongs its therapeutic
       potentials, Pharm. Res. 29 (2012) 3475–3485. https://doi.org/10.1007/s11095-012-
       0843-0

[49]   H.M. Mansour, Y.-S. Rhee, X. Wu, Nanomedicine in pulmonary delivery., Int. J.
       Nanomedicine. 4 (2009) 299–319.
       http://www.ncbi.nlm.nih.gov/pubmed/20054434 (accessed January 2, 2019)

[50]   W. Gao, S. Thamphiwatana, P. Angsantikul, L. Zhang, Nanoparticle approaches
       against bacterial infections, Wiley Interdiscip. Rev. Nanomedicine
       Nanobiotechnology. 6 (2014) 532–547. https://doi.org/10.1002/wnan.1282

[51]   R.Y. Pelgrift, A.J. Friedman, Nanotechnology as a therapeutic tool to combat
       microbial resistance, Adv. Drug Deliv. Rev. 65 (2013) 1803–1815.
       https://doi.org/10.1016/j.addr.2013.07.011

[52]   S.M. Dizaj, F. Lotfipour, M. Barzegar-Jalali, M.H. Zarrintan, K. Adibkia,
       Antimicrobial activity of the metals and metal oxide nanoparticles, Mater. Sci.
       Eng. C. 44 (2014) 278–284. https://doi.org/10.1016/j.msec.2014.08.031

[53]   B. Fadeel, Nanosafety: towards safer design of nanomedicines, J. Intern. Med. 274
       (2013) 578–580. https://doi.org/10.1111/joim.12137

[54]   M. Banerjee, S. Mallick, A. Paul, A. Chattopadhyay, S.S. Ghosh, Heightened
       reactive oxygen species generation in the antimicrobial activity of a three
       component iodinated chitosan–silver nanoparticle composite, Langmuir. 26 (2010)
       5901–5908. https://doi.org/10.1021/la9038528

[55]   F. Martinez-Gutierrez, P.L. Olive, A. Banuelos, E. Orrantia, N. Nino, E.M.
       Sanchez, F. Ruiz, H. Bach, Y. Av-Gay, Synthesis, characterization, and evaluation

of antimicrobial and cytotoxic effect of silver and titanium nanoparticles, Nanomedicine Nanotechnology, Biol. Med. 6 (2010) 681–688. https://doi.org/10.1016/j.nano.2010.02.001

[56] S. Ghosh, R. Kaushik, K. Nagalakshmi, S.L. Hoti, G.A. Menezes, B.N. Harish, H.N. Vasan, Antimicrobial activity of highly stable silver nanoparticles embedded in agar-agar matrix as a thin film., Carbohydr. Res. 345 (2010) 2220–7. https://doi.org/10.1016/j.carres.2010.08.001

[57] S. Shrivastava, T. Bera, A. Roy, G. Singh, P. Ramachandrarao, D. Dash, Characterization of enhanced antibacterial effects of novel silver nanoparticles, Nanotechnology. 18 (2007) 225103. https://doi.org/10.1088/0957-4484/18/22/225103

[58] A.R. Shahverdi, A. Fakhimi, H.R. Shahverdi, S. Minaian, Synthesis and effect of silver nanoparticles on the antibacterial activity of different antibiotics against Staphylococcus aureus and Escherichia coli, Nanomedicine Nanotechnology, Biol. Med. 3 (2007) 168–171. https://doi.org/10.1016/j.nano.2007.02.001

[59] J.S. Kim, E. Kuk, K.N. Yu, J.-H. Kim, S.J. Park, H.J. Lee, S.H. Kim, Y.K. Park, Y.H. Park, C.-Y. Hwang, Y.-K. Kim, Y.-S. Lee, D.H. Jeong, M.-H. Cho, Antimicrobial effects of silver nanoparticles, Nanomedicine Nanotechnology, Biol. Med. 3 (2007) 95–101. https://doi.org/10.1016/j.nano.2006.12.001

[60] Y. Ma, T. Zhou, C. Zhao, Preparation of chitosan-nylon-6 blended membranes containing silver ions as antibacterial materials., Carbohydr. Res. 343 (2008) 230–7. https://doi.org/10.1016/j.carres.2007.11.006

[61] P. Sanpui, A. Murugadoss, P. Prasad, S. Ghosh, A. Chattopadhyay, The antibacterial properties of a novel chitosan–Ag-nanoparticle composite, Int. J. Food Microbiol. 124 (2008) 142–146. https://doi.org/10.1016/j.ijfoodmicro.2008.03.004

[62] L. Qi, Z. Xu, X. Jiang, C. Hu, X. Zou, Preparation and antibacterial activity of chitosan nanoparticles, (2004). https://doi.org/10.1016/j.carres.2004.09.007

[63] S. Chadwick, C. Kriegel, M. Amiji, Nanotechnology solutions for mucosal immunization, Adv. Drug Deliv. Rev. 62 (2010) 394–407. https://doi.org/10.1016/j.addr.2009.11.012

[64] A.J. Friedman, G. Han, M.S. Navati, M. Chacko, L. Gunther, A. Alfieri, J.M. Friedman, Sustained release nitric oxide releasing nanoparticles: Characterization of a novel delivery platform based on nitrite containing hydrogel/glass composites, Nitric Oxide. 19 (2008) 12–20. https://doi.org/10.1016/j.niox.2008.04.003

Biosensors: Materials and Applications   Materials Research Forum LLC
Materials Research Foundations **47** (2019) 211-240   doi: http://dx.doi.org/10.21741/9781644900130-6

[65]   M. Potara, E. Jakab, A. Damert, O. Popescu, V. Canpean, S. Astilean, Synergistic antibacterial activity of chitosan–silver nanocomposites on *Staphylococcus aureus*, Nanotechnology. 22 (2011) 135101. https://doi.org/10.1088/0957-4484/22/13/135101

[66]   M. Rai, A. Yadav, A. Gade, Silver nanoparticles as a new generation of antimicrobials, Biotechnol. Adv. 27 (2009) 76–83. https://doi.org/10.1016/J.BIOTECHADV.2008.09.002

[67]   S. Pal, Y.K. Tak, J.M. Song, Does the antibacterial activity of silver nanoparticles depend on the shape of the nanoparticle? A study of the Gram-negative bacterium Escherichia coli., Appl. Environ. Microbiol. 73 (2007) 1712–20. https://doi.org/10.1128/AEM.02218-06

[68]   Q.L. Feng, J. Wu, G.Q. Chen, F.Z. Cui, T.N. Kim, J.O. Kim, A mechanistic study of the antibacterial effect of silver ions on Escherichia coli and Staphylococcus aureus., J. Biomed. Mater. Res. 52 (2000) 662–8

[69]   G. Applerot, A. Lipovsky, R. Dror, N. Perkas, Y. Nitzan, R. Lubart, A. Gedanken, Enhanced antibacterial activity of nanocrystalline ZnO due to increased ROS-mediated cell injury, Adv. Funct. Mater. 19 (2009) 842–852. https://doi.org/10.1002/adfm.200801081

[70]   N. Jones, B. Ray, K.T. Ranjit, A.C. Manna, Antibacterial activity of ZnO nanoparticle suspensions on a broad spectrum of microorganisms, FEMS Microbiol. Lett. 279 (2008) 71–76. https://doi.org/10.1111/j.1574-6968.2007.01012.x

[71]   E. Taylor, T.J. Webster, Reducing infections through nanotechnology and nanoparticles., Int. J. Nanomedicine. 6 (2011) 1463–73. https://doi.org/10.2147/IJN.S22021

[72]   A. Nanda, M. Saravanan, Biosynthesis of silver nanoparticles from Staphylococcus aureus and its antimicrobial activity against MRSA and MRSE, Nanomedicine Nanotechnology, Biol. Med. 5 (2009) 452–456. https://doi.org/10.1016/j.nano.2009.01.012

[73]   D.S. Balaji, S. Basavaraja, R. Deshpande, D.B. Mahesh, B.K. Prabhakar, A. Venkataraman, Extracellular biosynthesis of functionalized silver nanoparticles by strains of Cladosporium cladosporioides fungus, Colloids Surfaces B Biointerfaces. 68 (2009) 88–92. https://doi.org/10.1016/j.colsurfb.2008.09.022

[74]   S. Basavaraja, S.D. Balaji, A. Lagashetty, A.H. Rajasab, A. Venkataraman, Extracellular biosynthesis of silver nanoparticles using the fungus Fusarium

Biosensors: Materials and Applications   Materials Research Forum LLC
Materials Research Foundations **47** (2019) 211-240   doi: http://dx.doi.org/10.21741/9781644900130-6

semitectum, Mater. Res. Bull. 43 (2008) 1164–1170.
https://doi.org/10.1016/J.MATERRESBULL.2007.06.020

[75]   J.D. Holmes, P.R. Smith, R. Evans-Gowing, D.J. Richardson, D.A. Russell, J.R.
       Sodeau, Energy-dispersive X-ray analysis of the extracellular cadmium sulfide
       crystallites of Klebsiella aerogenes, Arch. Microbiol. 163 (1995) 143–147.
       https://doi.org/10.1007/BF00381789

[76]   M. Saravanan, A. Nanda, Extracellular synthesis of silver bionanoparticles from
       Aspergillus clavatus and its antimicrobial activity against MRSA and MRSE,
       Colloids Surfaces B Biointerfaces. 77 (2010) 214–218.
       https://doi.org/10.1016/j.colsurfb.2010.01.026

[77]   K. Chaloupka, Y. Malam, A.M. Seifalian, Nanosilver as a new generation of
       nanoproduct in biomedical applications, Trends Biotechnol. 28 (2010) 580–588.
       https://doi.org/10.1016/j.tibtech.2010.07.006

[78]   N.G. Durmus, E.N. Taylor, K.M. Kummer, T.J. Webster, Enhanced efficacy of
       superparamagnetic iron oxide nanoparticles against antibiotic-resistant biofilms in
       the presence of metabolites, Adv. Mater. 25 (2013) 5706–5713.
       https://doi.org/10.1002/adma.201302627

[79]   G.D. Savi, M.M. da Silva Paula, J.C. Possato, T. Barichello, D. Castagnaro, V.M.
       Scussel, Biological activity of gold nanoparticles towards filamentous pathogenic
       fungi, J. Nano Res. 20 (2012) 11–20.
       https://doi.org/10.4028/www.scientific.net/JNanoR.20.11

[80]   L. Brunet, D.Y. Lyon, E.M. Hotze, P.J.J. Alvarez, M.R. Wiesner, Comparative
       Photoactivity and antibacterial properties of C 60 fullerenes and titanium dioxide
       nanoparticles, Environ. Sci. Technol. 43 (2009) 4355–4360.
       https://doi.org/10.1021/es803093t

[81]   M. Premanathan, K. Karthikeyan, K. Jeyasubramanian, G. Manivannan, Selective
       toxicity of ZnO nanoparticles toward Gram-positive bacteria and cancer cells by
       apoptosis through lipid peroxidation, Nanomedicine Nanotechnology, Biol. Med. 7
       (2011) 184–192. https://doi.org/10.1016/j.nano.2010.10.001

[82]   J. Ma, Z. Xiong, T. David Waite, W.J. Ng, X.S. Zhao, Enhanced inactivation of
       bacteria with silver-modified mesoporous TiO_2 under weak ultraviolet irradiation,
       Microporous Mesoporous Mater. 144 (2011) 97–104.
       https://doi.org/10.1016/j.micromeso.2011.03.040

[83]   E. Torres-Sangiao, A. Holban, M. Gestal, Advanced nanobiomaterials: vaccines,
       diagnosis and treatment of infectious diseases, Molecules. 21 (2016) 867.
       https://doi.org/10.3390/molecules21070867

Biosensors: Materials and Applications   Materials Research Forum LLC
Materials Research Foundations **47** (2019) 211-240   doi: http://dx.doi.org/10.21741/9781644900130-6

[84]   A. Raghunath, E. Perumal, Metal oxide nanoparticles as antimicrobial agents: a
       promise for the future, Int. J. Antimicrob. Agents. 49 (2017) 137–152.
       https://doi.org/10.1016/j.ijantimicag.2016.11.011

[85]   B. Sarkar, A. Mahanty, S.P. Netam, S. Mishra, N. Pradhan, M. Samanta,
       Inhibitory role of silver nanoparticles against important fish pathogen, Aeromonas
       hydrophila, Int. J. Nanomater. Biostructures. 2 (2012) 70–74.
       http://www.urpjournals.com (accessed January 3, 2019)

[86]   V. Ahmed, J. Kumar, M. Kumar, M.B. Chauhan, M. Vij, M. Ganguli, N.S.
       Chauhan, Synthesis, characterization of penicillin G capped silver nanoconjugates
       to combat β-lactamase resistance in infectious microorganism., J. Biotechnol. 163
       (2013) 419–24. https://doi.org/10.1016/j.jbiotec.2012.12.002

[87]   A. Travan, C. Pelillo, I. Donati, E. Marsich, M. Benincasa, T. Scarpa, S. Semeraro,
       G. Turco, R. Gennaro, S. Paoletti, Non-cytotoxic silver nanoparticle-
       polysaccharide nanocomposites with antimicrobial activity, Biomacromolecules.
       10 (2009) 1429–1435. https://doi.org/10.1021/bm900039x

[88]   K. Madhumathi, P.T. Sudheesh Kumar, S. Abhilash, V. Sreeja, H. Tamura, K.
       Manzoor, S. V. Nair, R. Jayakumar, Development of novel chitin/nanosilver
       composite scaffolds for wound dressing applications, J. Mater. Sci. Mater. Med.
       21 (2010) 807–813. https://doi.org/10.1007/s10856-009-3877-z

[89]   B. Marta, E. Jakab, M. Potara, T. Simon, F. Imre-Lucaci, L. Barbu-Tudoran, O.
       Popescu, S. Astilean, Pluronic-coated silver nanoprisms: Synthesis,
       characterization and their antibacterial activity, Colloids Surfaces A Physicochem.
       Eng. Asp. 441 (2014) 77–83. https://doi.org/10.1016/J.COLSURFA.2013.08.076

[90]   R. Dobrucka, J. Długaszewska, Biosynthesis and antibacterial activity of ZnO
       nanoparticles using Trifolium pratense flower extract, Saudi J. Biol. Sci. 23 (2016)
       517–523. https://doi.org/10.1016/j.sjbs.2015.05.016

[91]   L.S. Reddy, M.M. Nisha, M. Joice, P.N. Shilpa, Antimicrobial activity of zinc
       oxide (ZnO) nanoparticle against *Klebsiella pneumoniae*, Pharm. Biol. 52 (2014)
       1388–1397. https://doi.org/10.3109/13880209.2014.893001

[92]   Y. Liu, L. He, A. Mustapha, H. Li, Z.Q. Hu, M. Lin, Antibacterial activities of
       zinc oxide nanoparticles against Escherichia coli O157:H7, J. Appl. Microbiol.
       107 (2009) 1193–1201. https://doi.org/10.1111/j.1365-2672.2009.04303.x

[93]   Y. Xie, Y. He, P.L. Irwin, T. Jin, X. Shi, Antibacterial activity and mechanism of
       action of zinc oxide nanoparticles against Campylobacter jejuni., Appl. Environ.
       Microbiol. 77 (2011) 2325–31. https://doi.org/10.1128/AEM.02149-10

Biosensors: Materials and Applications   Materials Research Forum LLC
Materials Research Foundations 47 (2019) 211-240   doi: http://dx.doi.org/10.21741/9781644900130-6

[94] A. Sirelkhatim, S. Mahmud, A. Seeni, N.H.M. Kaus, L.C. Ann, S.K.M. Bakhori, H. Hasan, D. Mohamad, Review on Zinc oxide nanoparticles: antibacterial activity and toxicity mechanism, Nano-Micro Lett. 7 (2015) 219–242. https://doi.org/10.1007/s40820-015-0040-x

[95] A. Giasuddin, K. Jhuma, A.M. Haq, Use of gold nanoparticles in diagnostics, surgery and medicine: a review, Bangladesh J. Med. Biochem. 5 (2013) 56–60. https://doi.org/10.3329/bjmb.v5i2.13346

[96] S. Jain, D.G. Hirst, J.M. O'Sullivan, Gold nanoparticles as novel agents for cancer therapy, Br. J. Radiol. 85 (2012) 101–113. https://doi.org/10.1259/bjr/59448833

[97] A. Mesbahi, A review on gold nanoparticles radiosensitization effect in radiation therapy of cancer., Reports Pract. Oncol. Radiother. J. Gt. Cancer Cent. Pozn. Polish Soc. Radiat. Oncol. 15 (2010) 176–80. https://doi.org/10.1016/j.rpor.2010.09.001

[98] Jacob D. Gibson, and Bishnu P. Khanal, E.R. Zubarev, Paclitaxel-functionalized gold nanoparticles, JACS (2007). https://doi.org/10.1021/JA075181K

[99] M.M. Mohamed, S.A. Fouad, H.A. Elshoky, G.M. Mohammed, T.A. Salaheldin, Antibacterial effect of gold nanoparticles against Corynebacterium pseudotuberculosis, Int. J. Vet. Sci. Med. 5 (2017) 23–29. https://doi.org/10.1016/J.IJVSM.2017.02.003

[100] S. Shamaila, N. Zafar, S. Riaz, R. Sharif, J. Nazir, S. Naseem, Gold nanoparticles: an efficient antimicrobial agent against enteric bacterial human pathogen., Nanomater. (Basel, Switzerland). 6 (2016). https://doi.org/10.3390/nano6040071

[101] J.A. Lemire, J.J. Harrison, R.J. Turner, Antimicrobial activity of metals: mechanisms, molecular targets and applications, Nat. Rev. Microbiol. 11 (2013) 371–384. https://doi.org/10.1038/nrmicro3028

[102] M. Seong, D.G. Lee, Reactive oxygen species-independent apoptotic pathway by gold nanoparticles in Candida albicans., Microbiol. Res. 207 (2018) 33–40. https://doi.org/10.1016/j.micres.2017.11.003

[103] N. Mahmoud, A. Alkilany, E. Khalil, A. Al-Bakri, Antibacterial activity of gold nanorods against *Staphylococcus aureus* and *Propionibacterium acnes*: misinterpretations and artifacts, Int. J. Nanomedicine. Volume 12 (2017) 7311–7322. https://doi.org/10.2147/IJN.S145531

[104] J.P. Ruparelia, A.K. Chatterjee, S.P. Duttagupta, S. Mukherji, Strain specificity in antimicrobial activity of silver and copper nanoparticles, Acta Biomater. 4 (2008) 707–716. https://doi.org/10.1016/j.actbio.2007.11.006

[105] G. Ren, D. Hu, E.W.C. Cheng, M.A. Vargas-Reus, P. Reip, R.P. Allaker, Characterisation of copper oxide nanoparticles for antimicrobial applications, Int. J. Antimicrob. Agents. 33 (2009) 587–590. https://doi.org/10.1016/j.ijantimicag.2008.12.004

[106] M. Ahamed, H.A. Alhadlaq, M.A.M. Khan, P. Karuppiah, N.A. Al-Dhabi, Synthesis, characterization, and antimicrobial activity of copper oxide nanoparticles, J. Nanomater. 2014 (2014) 1–4. https://doi.org/10.1155/2014/637858

[107] G. Applerot, J. Lellouche, A. Lipovsky, Y. Nitzan, R. Lubart, A. Gedanken, E. Banin, Understanding the antibacterial mechanism of CuO Nanoparticles: revealing the route of induced oxidative stress, Small. 8 (2012) 3326–3337. https://doi.org/10.1002/smll.201200772

[108] H.R. Ashjari, M.S.S. Dorraji, V. Fakhrzadeh, H. Eslami, M.H. Rasoulifard, M. Rastgouy-Houjaghan, P. Gholizadeh, H.S. Kafil, Starch-based polyurethane/CuO nanocomposite foam: Antibacterial effects for infection control., Int. J. Biol. Macromol. 111 (2018) 1076–1082. https://doi.org/10.1016/j.ijbiomac.2018.01.137

[109] D. Devipriya, S.M. Roopan, Cissus quadrangularis mediated ecofriendly synthesis of copper oxide nanoparticles and its antifungal studies against Aspergillus niger, Aspergillus flavus, Mater. Sci. Eng. C. 80 (2017) 38–44. https://doi.org/10.1016/j.msec.2017.05.130

[110] Q. Maqbool, S. Iftikhar, M. Nazar, F. Abbas, A. Saleem, T. Hussain, R. Kausar, S. Anwaar, N. Jabeen, Green fabricated CuO nanobullets via Olea europaea leaf extract shows auspicious antimicrobial potential, IET Nanobiotechnology. 11 (2017) 463–468. https://doi.org/10.1049/iet-nbt.2016.0125

[111] P.K. Stoimenov, R.L. Klinger, G.L. Marchin, K.J. Klabunde, Metal oxide nanoparticles as bactericidal agents, (2002). https://doi.org/10.1021/LA0202374

[112] J. Lellouche, E. Kahana, S. Elias, A. Gedanken, E. Banin, Antibiofilm activity of nanosized magnesium fluoride, Biomaterials. 30 (2009) 5969–5978. https://doi.org/10.1016/J.BIOMATERIALS.2009.07.037

[113] A. Landriscina, J. Rosen, A.J. Friedman, Biodegradable chitosan nanoparticles in drug delivery for infectious disease., Nanomedicine (Lond). 10 (2015) 1609–19. https://doi.org/10.2217/nnm.15.7

[114] V. Saharan, A. Mehrotra, R. Khatik, P. Rawal, S.S. Sharma, A. Pal, Synthesis of chitosan based nanoparticles and their in vitro evaluation against phytopathogenic fungi, Int. J. Biol. Macromol. 62 (2013) 677–683. https://doi.org/10.1016/j.ijbiomac.2013.10.012

[115] B. Jamil, H. Habib, S. Abbasi, H. Nasir, A. Rahman, A. Rehman, H. Bokhari, M. Imran, Cefazolin loaded chitosan nanoparticles to cure multi drug resistant Gram-negative pathogens., Carbohydr. Polym. 136 (2016) 682–91. https://doi.org/10.1016/j.carbpol.2015.09.078

[116] K. Blecher, A. Nasir, A. Friedman, The growing role of nanotechnology in combating infectious disease, Virulence. 2 (2011) 395–401. https://doi.org/10.4161/viru.2.5.17035

[117] C.T. Tsao, C.H. Chang, Y.Y. Lin, M.F. Wu, J.-L. Wang, J.L. Han, K.H. Hsieh, Antibacterial activity and biocompatibility of a chitosan-gamma-poly(glutamic acid) polyelectrolyte complex hydrogel., Carbohydr. Res. 345 (2010) 1774–80. https://doi.org/10.1016/j.carres.2010.06.002

[118] M.A. Aziz, J.D. Cabral, H.J.L. Brooks, S.C. Moratti, L.R. Hanton, Antimicrobial properties of a chitosan dextran-based hydrogel for surgical use., Antimicrob. Agents Chemother. 56 (2012) 280–7. https://doi.org/10.1128/AAC.05463-11

[119] M. Qasim, D.-J. Lim, H. Park, D. Na, Nanotechnology for diagnosis and treatment of infectious diseases, J. Nanosci. Nanotechnol. 14 (2014) 7374–7387. https://doi.org/10.1166/jnn.2014.9578

[120] L. Zhan, S.J. Zhen, X.Y. Wan, P.F. Gao, C.Z. Huang, A sensitive surface-enhanced Raman scattering enzyme-catalyzed immunoassay of respiratory syncytial virus., Talanta. 148 (2016) 308–12. https://doi.org/10.1016/j.talanta.2015.10.081

[121] L. Chen, N. Mungroo, L. Daikuara, S. Neethirajan, Label-free NIR-SERS discrimination and detection of foodborne bacteria by in situ synthesis of Ag colloids, J. Nanobiotechnology. 13 (2015) 45. https://doi.org/10.1186/s12951-015-0106-4

[122] Z. Wang, N. Duan, J. Li, J. Ye, S. Ma, G. Le, Ultrasensitive chemiluminescent immunoassay of Salmonella with silver enhancement of nanogold labels, Luminescence. 26 (2011) 136–141. https://doi.org/10.1002/bio.1196

[123] J. Huang, Z. Xie, Z. Xie, S. Luo, L. Xie, L. Huang, Q. Fan, Y. Zhang, S. Wang, T. Zeng, Silver nanoparticles coated graphene electrochemical sensor for the ultrasensitive analysis of avian influenza virus H7, Anal. Chim. Acta. 913 (2016) 121–127. https://doi.org/10.1016/j.aca.2016.01.050

[124] J. Li, T. Qin, X.X. Jia, A.H. Deng, X. Zhang, W.H. Fan, S.D. Huo, T.Y. Wen, W.J. Liu, Rapid Identification of Legionella Pathogenicity by Surface-Enhanced Raman Spectroscopy., Biomed. Environ. Sci. 28 (2015) 437–44. https://doi.org/10.3967/bes2015.061

Biosensors: Materials and Applications   Materials Research Forum LLC
Materials Research Foundations **47** (2019) 211-240   doi: http://dx.doi.org/10.21741/9781644900130-6

[125] A. Sivanesan, E. Witkowska, W. Adamkiewicz, Ł. Dziewit, A. Kamińska, J. Waluk, Nanostructured silver–gold bimetallic SERS substrates for selective identification of bacteria in human blood, Analyst. 139 (2014) 1037. https://doi.org/10.1039/c3an01924a

[126] B. Veigas, A.R. Fernandes, P. V Baptista, AuNPs for identification of molecular signatures of resistance., Front. Microbiol. 5 (2014) 455. https://doi.org/10.3389/fmicb.2014.00455

[127] H. Deng, X. Zhang, A. Kumar, G. Zou, X. Zhang, X.-J. Liang, Long genomic DNA amplicons adsorption onto unmodified gold nanoparticles for colorimetric detection of Bacillus anthracis, Chem. Commun. 49 (2013) 51–53. https://doi.org/10.1039/C2CC37037A

[128] K. Kalidasan, J.L. Neo, M. Uttamchandani, Direct visual detection of Salmonella genomic DNA using gold nanoparticles, Mol. Biosyst. 9 (2013) 618. https://doi.org/10.1039/c3mb25527a

[129] M.A.F. Khalil, H.M.E. Azzazy, A.S. Attia, A.G.M. Hashem, A sensitive colorimetric assay for identification of *Acinetobacter baumannii* using unmodified gold nanoparticles, J. Appl. Microbiol. 117 (2014) 465–471. https://doi.org/10.1111/jam.12546

[130] B. Veigas, P. Pedrosa, F.F. Carlos, L. Mancio-Silva, A.R. Grosso, E. Fortunato, M.M. Mota, P. V Baptista, One nanoprobe, two pathogens: gold nanoprobes multiplexing for point-of-care., J. Nanobiotechnology. 13 (2015) 48. https://doi.org/10.1186/s12951-015-0109-1

[131] P. Pedrosa, Gold nanoprobes for multi loci assessment of multi-drug resistant tuberculosis, Tuberculosis. 94 (2014)

[132] X. Li, H. Kong, R. Mout, K. Saha, D.F. Moyano, S.M. Robinson, S. Rana, X. Zhang, M.A. Riley, V.M. Rotello, Rapid identification of bacterial biofilms and biofilm wound models using a multichannel nanosensor, ACS Nano. 8 (2014) 12014–12019. https://doi.org/10.1021/nn505753s

[133] C. Pöhlmann, I. Dieser, M. Sprinzl, A lateral flow assay for identification of Escherichia coli by ribosomal RNA hybridisation, Analyst. 139 (2014) 1063. https://doi.org/10.1039/c3an02059b

[134] J. Zhao, S. Tang, J. Storhoff, S. Marla, Y.P. Bao, X. Wang, E.Y. Wong, V. Ragupathy, Z. Ye, I.K. Hewlett, Multiplexed, rapid detection of H5N1 using a PCR-free nanoparticle-based genomic microarray assay, BMC Biotechnol. 10 (2010) 74. https://doi.org/10.1186/1472-6750-10-74

[135] M. Chen, Z. Yu, D. Liu, T. Peng, K. Liu, S. Wang, Y. Xiong, H. Wei, H. Xu, W. Lai, Dual gold nanoparticle lateflow immunoassay for sensitive detection of Escherichia coli O157:H7, Anal. Chim. Acta. 876 (2015) 71–76. https://doi.org/10.1016/j.aca.2015.03.023

[136] X.-M. Nie, R. Huang, C.-X. Dong, L.-J. Tang, R. Gui, J.-H. Jiang, Plasmonic ELISA for the ultrasensitive detection of Treponema pallidum., Biosens. Bioelectron. 58 (2014) 314–9. https://doi.org/10.1016/j.bios.2014.03.007

[137] S.R. Ahmed, J. Kim, T. Suzuki, J. Lee, E.Y. Park, Detection of influenza virus using peroxidase-mimic of gold nanoparticles., Biotechnol. Bioeng. 113 (2016) 2298–303. https://doi.org/10.1002/bit.25982

[138] P. Luo, Y. Liu, Y. Xia, H. Xu, G. Xie, Aptamer biosensor for sensitive detection of toxin A of Clostridium difficile using gold nanoparticles synthesized by Bacillus stearothermophilus, Biosens. Bioelectron. 54 (2014) 217–221. https://doi.org/10.1016/j.bios.2013.11.013

[139] X. Ma, Y. Jiang, F. Jia, Y. Yu, J. Chen, Z. Wang, An aptamer-based electrochemical biosensor for the detection of Salmonella., J. Microbiol. Methods. 98 (2014) 94–8. https://doi.org/10.1016/j.mimet.2014.01.003

[140] Y.-C. Chang, C.-Y. Yang, R.-L. Sun, Y.-F. Cheng, W.-C. Kao, P.-C. Yang, Rapid single cell detection of Staphylococcus aureus by aptamer-conjugated gold nanoparticles, Sci. Rep. 3 (2013) 1863. https://doi.org/10.1038/srep01863

[141] B. Veigas, J.M. Jacob, M.N. Costa, D.S. Santos, M. Viveiros, J. Inácio, R. Martins, P. Barquinha, E. Fortunato, P.V. Baptista, Gold on paper–paper platform for Au-nanoprobe TB detection, Lab Chip. 12 (2012) 4802. https://doi.org/10.1039/c2lc40739f

[142] X.Z. Li, S. Kim, W. Cho, S.-Y. Lee, Optical detection of nanoparticle-enhanced human papillomavirus genotyping microarrays, Biomed. Opt. Express. 4 (2013) 187. https://doi.org/10.1364/BOE.4.000187

[143] D. Ellis, Amphotericin B: spectrum and resistance, J. Antimicrob. Chemother. 49 (2002) 7–10. https://doi.org/10.1093/jac/49.suppl_1.7

[144] B. Purkait, A. Kumar, N. Nandi, A.H. Sardar, S. Das, S. Kumar, K. Pandey, V. Ravidas, M. Kumar, T. De, D. Singh, P. Das, Mechanism of amphotericin B resistance in clinical isolates of Leishmania donovani., Antimicrob. Agents Chemother. 56 (2012) 1031–41. https://doi.org/10.1128/AAC.00030-11

[145] J.L. Italia, M.M. Yahya, D. Singh, M.N. V. Ravi Kumar, Biodegradable nanoparticles improve oral bioavailability of amphotericin b and show reduced

Biosensors: Materials and Applications   Materials Research Forum LLC
Materials Research Foundations **47** (2019) 211-240   doi: http://dx.doi.org/10.21741/9781644900130-6

nephrotoxicity compared to intravenous fungizone®, Pharm. Res. 26 (2009) 1324–1331. https://doi.org/10.1007/s11095-009-9841-2

[146] X. Tang, H. Zhu, L. Sun, W. Hou, S. Cai, R. Zhang, F. Liu, Enhanced antifungal effects of amphotericin B-TPGS-b-(PCL-ran-PGA) nanoparticles in vitro and in vivo., Int. J. Nanomedicine. 9 (2014) 5403–13. https://doi.org/10.2147/IJN.S71623

[147] A.M. Abraham, A. Walubo, The effect of surface charge on the disposition of liposome-encapsulated gentamicin to the rat liver, brain, lungs and kidneys after intraperitoneal administration, Int. J. Antimicrob. Agents. 25 (2005) 392–397. https://doi.org/10.1016/j.ijantimicag.2005.01.018

[148] R. Schiffelers, G. Storm, I. Bakker-Woudenberg, Liposome-encapsulated aminoglycosides in pre-clinical and clinical studies, J. Antimicrob. Chemother. 48 (2001) 333–344. https://doi.org/10.1093/jac/48.3.333

[149] M. Bala, V. Singh, A. Bhargava, M. Kakran, N.C. Joshi, R. Bhatnagar, Gentamicin Susceptibility among a Sample of Multidrug-Resistant Neisseria gonorrhoeae Isolates in India., Antimicrob. Agents Chemother. 60 (2016) 7518–7521. https://doi.org/10.1128/AAC.01907-16

[150] C. Scott, D.J. Abdelghany, R.J. Quinn, B.F. Ingram, R.F. Gilmore, C.C. Donnelly, C.J. Taggart, Gentamicin-loaded nanoparticles show improved antimicrobial effects towards Pseudomonas aeruginosa infection, Int. J. Nanomedicine. 7 (2012) 4053. https://doi.org/10.2147/IJN.S34341

[151] U. Posadowska, M. Brzychczy-Włoch, A. Drożdż, M. Krok-Borkowicz, M. Włodarczyk-Biegun, P. Dobrzyński, W. Chrzanowski, E. Pamuła, Injectable hybrid delivery system composed of gellan gum, nanoparticles and gentamicin for the localized treatment of bone infections, Expert Opin. Drug Deliv. 13 (2016) 613–620. https://doi.org/10.1517/17425247.2016.1146673

[152] S. Xie, Y. Tao, Y. Pan, W. Qu, G. Cheng, L. Huang, D. Chen, X. Wang, Z. Liu, Z. Yuan, Biodegradable nanoparticles for intracellular delivery of antimicrobial agents, J. Control. Release. 187 (2014) 101–117. https://doi.org/10.1016/j.jconrel.2014.05.034

[153] E. Yazar, A.L. Bas, Y.O. Birdane, K. Yapar, M. Elmas, B. Tras, Determination of intracellular (neutrophil and monocyte) concentrations of free and liposome encapsulated ampicillin in sheep, Vet. Med. (Praha). (2006). https://doi.org/10.17221/5517-VETMED

[154] N.M. Zaki, M.M. Hafez, Enhanced antibacterial effect of ceftriaxone sodium-loaded chitosan nanoparticles against intracellular *Salmonella typhimurium*, AAPS PharmSciTech. 13 (2012) 411–421. https://doi.org/10.1208/s12249-012-9758-7

Biosensors: Materials and Applications   Materials Research Forum LLC
Materials Research Foundations **47** (2019) 211-240   doi: http://dx.doi.org/10.21741/9781644900130-6

[155] M.N. Seleem, N. Jain, N. Pothayee, A. Ranjan, J.S. Riffle, N. Sriranganathan, Targeting *Brucella melitensis* with polymeric nanoparticles containing streptomycin and doxycycline, FEMS Microbiol. Lett. 294 (2009) 24–31. https://doi.org/10.1111/j.1574-6968.2009.01530.x

[156] H. Pinto-Alphandary, A. Andremont, P. Couvreur, Targeted delivery of antibiotics using liposomes and nanoparticles: research and applications., Int. J. Antimicrob. Agents. 13 (2000) 155–68. http://www.ncbi.nlm.nih.gov/pubmed/10724019 (accessed January 2, 2019)

[157] S. Chono, T. Tanino, T. Seki, K. Morimoto, Efficient drug delivery to alveolar macrophages and lung epithelial lining fluid following pulmonary administration of liposomal ciprofloxacin in rats with pneumonia and estimation of its antibacterial effects, Drug Dev. Ind. Pharm. 34 (2008) 1090–1096. https://doi.org/10.1080/03639040801958421

[158] N. Mor, J. Vanderkolk, N. Mezo, L. Heifets, Effects of clarithromycin and rifabutin alone and in combination on intracellular and extracellular replication of Mycobacterium avium., Antimicrob. Agents Chemother. 38 (1994) 2738–42. http://www.ncbi.nlm.nih.gov/pubmed/7695255 (accessed January 2, 2019)

[159] B. Garnaik, P.N. Chaudhari, K.K. Mohite, B.S. Selukar, S.S. Nande, S.P. Parwe, Synthesis of ciprofloxacin-conjugated poly (L-lactic acid) polymer for nanofiber fabrication and antibacterial evaluation, Int. J. Nanomedicine. 9 (2014) 1463. https://doi.org/10.2147/IJN.S54971

[160] S.A. Plotkin, Vaccines: past, present and future, Nat. Med. 10 (2005) S5–S11. https://doi.org/10.1038/nm1209

[161] R.N. Germain, Vaccines and the future of human immunology, Immunity. 33 (2010) 441–450. https://doi.org/10.1016/J.IMMUNI.2010.09.014

[162] D.J. Irvine, M.C. Hanson, K. Rakhra, T. Tokatlian, Synthetic nanoparticles for vaccines and immunotherapy, Chem. Rev. 115 (2015) 11109–11146. https://doi.org/10.1021/acs.chemrev.5b00109

[163] A. Salvador, M. Igartua, R.M. Hernández, J.L. Pedraz, An overview on the field of micro- and nanotechnologies for synthetic peptide-based vaccines, J. Drug Deliv. 2011 (2011) 1–18. https://doi.org/10.1155/2011/181646

[164] S.M. Moghimi, A.C. Hunter, J.C. Murray, Nanomedicine: current status and future prospects, FASEB J. 19 (2005) 311–330. https://doi.org/10.1096/fj.04-2747rev

[165] A.S. Fauci, D.M. Morens, The perpetual challenge of infectious diseases, N. Engl. J. Med. 366 (2012) 454–461. https://doi.org/10.1056/NEJMra1108296

Biosensors: Materials and Applications   Materials Research Forum LLC
Materials Research Foundations **47** (2019) 241-288   doi: http://dx.doi.org/10.21741/9781644900130-7

Chapter 7

Theranostic Application of Nanoparticulated Systems: Present and Future Prospects

Rout George Kerry[1], Sabuj Sahoo[1], Gitishree Das[2], Jayanta Kumar Patra[2*]

[1]P. G. Department of Biotechnology, Utkal Univesity, Vani Vihar, Bhubaneswar-751004, Odisha, India

[2]Research Institute of Biotechnology & Medical Converged Science, Dongguk University-Seoul, Ilsandong-gu, Gyeonggi-do 10326, Republic of Korea

* jkpatra.cet@gmail.com

Abstract

Nano particulated systems are biocompatible materials or devices, engineered with a purpose to deliverer desired bioactive compounds to a targeted location without inducing any secondary reactions or side-effects. The diversified ability of this bioengineered molecule to breach the biological barriers to reach the targeted location in the biological system uplifts its other versatile nature of active distribution. Furthermore, its negligible toxicity and biodegradability has resulted in making it a unique candidate for its purpose as nanoparticulated system. These nano-based systems are currently exploited in conjugation to a heterogeneous array of bioactive natural phytochemicals or synthetic compounds as a therapy against various diseases or disorders. Some diseases or disorders include obesity, diabetes, liver fibrosis, cardiovascular disorders, neurodegenerative disorders, cancers of various forms and microbial infections. Despite of the ability of these nano-based systems to be a novel therapy against a number of diseases and disorders their utilization and commercialization is restrained. This procrastination could be relinquished if pertinent mechanisms of their molecular interactions are properly acknowledged. Henceforth the objective of the present paper is to provide an overview of the types of nano carriers employed in diversified nano particulated systems based on their theranostic application, beneficial as well as deleterious impacts, present status and future prospects.

Keywords

Nanocarriers, Bioconjugation, Biodistribution, Biocompatible, Diagnostics, Drug Delivery, Functionalize

Biosensors: Materials and Applications   Materials Research Forum LLC
Materials Research Foundations **47** (2019) 241-288   doi: http://dx.doi.org/10.21741/9781644900130-7

Contents

1.   Introduction

During the early eighties from the works of Michael Faraday and Richard Feynman the concept of nanoparticles (NPs) is basically derived [1]. Concurrently, manipulation at the atomic level and production of nanoscale objects originated from the development of a polymer drug-conjugate and liposome, pioneering research of Jatzkewitz and Bangham respectively [2,3]. Further, Eric Drexler revolutionized the concept of molecular manipulations of NPs with his paper and a book entitled as "Molecular Engineering: An Approach to the Development of General Capabilities for Molecular Manipulation" and "engines of creation: the coming era of nanotechnology" respectively [4]. A unique platform is provided by nanotechnology where it is not so stringent to modify or develop nano particulated systems from metallic or organic compounds having fruitful application in the field of biomedical therapy. The physical, chemical and biological properties such as size, shape, stability, biocompatibility, biodegradability, poly dispersity, optical properties, surface Plasmon resonance are some of the major attributions of these nano particulated systems. Currently, these properties are highly explored in various clinical and biomedical applications such as contrasting agents in MRI imaging, cell labeling, anti-microbial, biomarkers and as drug-delivery systems for diagnosis or theranostic purposes [5,6].

Biosensors: Materials and Applications   Materials Research Forum LLC
Materials Research Foundations **47** (2019) 241-288   doi: http://dx.doi.org/10.21741/9781644900130-7

Out of all the possible nanotechnology based applications, the targeted drug-delivery systems are highly exploited in a search of finding a multifunctional therapy against various diseases and disorders such as obesity, diabetes, liver fibrosis, cardiovascular disorders, neuro degenerative disorders, cancers of various forms and microbial infections [7-10]. NPs such as chitosan conjugated γ-poly (glutamic acid) (γ-PGA) along with anti-diabetic peptides; insulin and the exendin-4 were proven to be effective in diabetes [11]. Nanotechnology-based nanomedicine such as PEGylated liposomal doxorubicin, cyclodextrin-containing camptothecin, polymeric micelle containing paclitaxel, magnesium oxide nanoparticles, PEG-irinotecan (NKTR 102), lipid NPs containing small interfering RNAs (siRNA), block copolymer vaccine containing peptides are being evaluated for their targeted anti-cancers activity for the treatment of ovary, pancreas, lung, gastro esophageal, breast, liver and skin cancer [12].

In the past few years naturally occurring nanosized vesicles secreted by monocytes and macrophages named exosomes are being used as an exosomal-based delivery system for a potent antioxidant, to treat Parkinson's disease (PD) [13]. Nanoparticulated systems made of zen, encapsulating quercetin, which through oral absorption contributes notably in improving bioavailability of bioactive compounds such as flavonoids as a possible therapy against Alzheimer's disease (AD) [9,14]. Despite the great beneficiary efficiency of this orally administered nanotechnology based drug delivery system, it suffers some drawbacks such as it tends to show gastrointestinal side effects and lack of brain targeting [15]. Therefore, polymeric NPs such as PEGylated poly [α,β-(N-2-hydroxyethyl)-d,l-aspartamide], galantamine loaded poly (lactic-co-glycolic acid) (PLGA), liposomes containing rivastigmine HCl along with phosphatidylcholine; dihexadecyl phosphate; cholesterol; glycerol and efavirenz-loaded poly (ethylene oxide)/ poly (propylene oxide) micelles were brought into light which are administered by intravenous, intranasal or by subcutaneous an alternative to oral uptake [15-17].

Dendrimers, liposomes, solid lipid NPs (SLN) and nanopolymers such as poly (lactic-co-glycolic) acid loaded with specific natural or synthetic bioactive compounds are being investigated for their theranostic application in the case of chronic lung diseases such as asthma, tuberculosis hypertensions, etc. [18]. It is evidenced that nano encapsulated antibiotics such asrifampicin, isoniazid, moxifloxacin and streptomycine were found to be more effective than conventional free antibiotics against tuberculosis [19]. AIDS, which threatens to cause a great plague in the present generation, could also be possibly treated with the help of nanotechnology based therapy. The effectiveness of antiretroviral (ARV) drug depends on its ability to penetrate through BBB and blood-cerebrospinal fluid barrier (BCSFB) which most of the conventional medicine fails to breach. Therefore, an alternative to these medicines are nano medicines which could efficiently penetrate BBB and BCSFB with negligible toxicity [20]. Likewise there is various nanotechnology based experimental therapy for diabetes, obesity liver fibrosis which are described below along with drug targets and the demerits as well as future prospects.

2. Types of nanocarriers

Nano particulated systems are basically nano carriers involving multiple types of organic or inorganic biocompatible molecules which together with the therapeutic compounds range approximately within 1000 nm [21]. Here the agent with therapeutic efficiency could be either fabricated on the surface or encapsulated within the nano particulated system [22,23]. These nano-based systems could be segregated in to various types based on the composition or source of NPs used for the synthesis of nanocarriers which further could be engineered into nano particulated systems.

These nanocarriers could be synthesized either from inorganic, organic or both of the NPs. Inorganic NPs including gold (Au), silver (Ag), Ag-Au, copper (Cu), zinc (Zn)/ (ZnO), iron (Fe), iron oxide (Fe_2O_3), silica based NPs (porous, non-porous, mesoporous), carbon based NPs (quantum dots, carbon cnanotubes, fullerenes, graphene, nanodiamonds), titanium based NPs, etc. in the form of nanoshells and nanocages are commonly been synthesized [24,25]. The synthesis of NPs from various biological sources (plants and microbial) and their considerable potential in biomedical application in conjugation with other nanocarrierss resulting in a hybrid nanoparticulated system is a significant achievement of the present time. Inorganic compounds such as iron (III) oxide (Fe_2O_3) are paramagnetic in nature and occur naturally as magnetite mineral. These super paramagnetic iron oxide NPs are slowly been given priority for various clinical and biomedical applications as a novel entity because of their ultrafine size, magnetic properties and biocompatibility [26-31]. Ag NPs are still charming the current scientists in every phase of scientific advancement to explore new dimension for their utilization specifically biomedical research [32]. Current utilization of these MNPs is growing day by day starting from paints to very sophisticated *in vitro* and *in vivo* DDS. AuNPs are also extensively exploited in different domains of science diversified domains; biomedical is one of those fields where the particles have shown its effectiveness [33]. Silica NPs on the other hand is an efficient nanocarrier in the drug delivery system due to its specific size, volume, distribution and high surface tolerability to silanol functionalization are some of the other modifiable features [34]. Carbon based nanocarriers such as carbon nanotubes, fullerenes, nanodiamonds (NDs), carbon nanohorns, carbon nanodots, graphenes and its derivatives are also widely applied in the field of nanomedicine or as drug delivery vehicles along with their ability to coronate with different biomolecules and bioimaging has been largely exploited in recent times [35].

But the most efficient nanocarriers in biomedical science are organic NPs which includes polymeric nanoparticles (PNP), lipid based nanoparticles, dendrimers (DEN), liquid crystal (LC) systems, niosomes (NIO) microemulsions (MEs), nanomicelles (NAs) [21]. PNPs are chain of CH_2 based polymers having an appearance criss-crossed matrix or scaffold. By self-emulsion polymerization or by induced polymerization could result in the formation of a colloidal polymeric NPs simultaneously encapsulating a therapeutic agent [36]. Solid-lipid-NPs and nanostructured lipids are lipid based, extensively used nanocarriers in the biomedical field. These nanocarriers basically consist of a mixture of one or more lipid, surfactants and water which are previously known to be biocompatible and biodegradable in nature [37]. Another nano-sized highly symmetrical/ ordered, branched macromolecules with well-defined, homogenous, and monodispersed structure are dendrimers [38]. These hyper branched macromolecules with cautiously customized architecture and the functionalizable end-groups further provide an additional opportunity for their user dependent modulation of physicochemical or biological activities [38]. Niosomes were primarily used in cosmetic industry but presently are being used in biomedical sciences as a nano carrier for the purpose of a drug delivery system. These nano-systems are thermodynamically stable polyhedral, multilamellar or unilamellar nanostructures results from self-assembly of amphiphilic surfactant which are non-ionic in an aqueous medium with hydrating mixture of cholesterol [39]. These systems are equipped to carry a diversified array of drugs either natural or synthetic origin irrespective of their nature that is amphiphilic, hydrophilic, lipophilic with further increasing the bioavailability and ameliorated side-effects [40,41]. Similarly, isotropic liquid mixtures of oil, water and surfactant, in combination with a co-surfactant gives rise to clear, thermodynamically stable micro/nanosystems called micro/nanoemulsions. These systems in combination with various antibiotics such as ciprofloxacin, cephalosporin, caftriaxone, cefotaxime, etc. are also highly explored for their biomedical potency [42].

3.   Targeted delivery and control release

The nanoparticulated systems possess varying degree of magnetic, thermal, optical and electrical properties which is basically due to their high surface area and limited quantum mechanical effects [43]. These properties orchestrate a predominant role in targeted drug delivery. In general, nanocarriers loaded with a drug, could be delivered to the targeted site through passive, active or physical targeting methods [43,44].

Materials Research Forum LLC
doi: http://dx.doi.org/10.21741/9781644900130-7

Enhanced permeability and retention effect (EPR) are the working principle of passive targeting that makes targeted cells to selectively absorb nanoparticulated systems [45]. NPs properties such as particle size, shape and surface charge greatly influences EPR effect resulting in the modulation of circulation time, penetration speed and intracellular internalization of the NPs [46,47]. Active targeting alternatively involves ligands such as antibodies, proteins and peptides functionalized on the surface of nano formulated particles, which at targeted site interacts selectively with the appropriate over expressed receptors [48-50]. The efficiency of the method could be substantially improved if multiple ligands or diversity of a single ligands having high binding affinity for the targeted receptors used [43]. Lastly, when external sources or fields such as photothermal and magnetic hyperthermia therapy are used to guide NPs to the targeted site and control release, then the targeting method could be categorized as physical targeting [51].

The most dynamic and irreplaceable application displayed by these nanoparticulated systems is their ability of controled release. In the current century of nanomedicine, it is actually this phenomenon which is extensively explored for development of advanced therapeutics. The phenomenon of controled release in the nanoparticulated system is stimuli-responsive in nature. This, simply means that the controled release of the nanoparticulated systems could be triggered either by intracellular (pH, ATP, Glutathione, enzyme, glucose and H_2O_2) or exogenous (temperature, light, magnetic field, ultrasound and electericity) stimuli [52].

4. Merits of nanotechnology based therapeutics

Nanotechnology based therapy have numerous merits which are simple reflection of their tuned physical properties (discussed later) in the biomedical science. For instance in diseases such as obesity, diabetes, cardiovascular diseases (CVD), liver fibrosis, cancer, neurodegenerative diseases and microbial infections nanotechnology based therapy shows promising outcomes (**Figure 1,Table 1**).

Obesity

Obesity is at the centre of metabolic disorders such as diabetes, fatty liver disease and other array of additional health problems such as cardiovascular diseases (CVD), atherosclerosis, degenerative disorders and certain type of cancers [78]. Obesity is also linked to elevated secretion of pro-inflammatory, inflammatory, diabetogenic and atherogenic adipokines/ cytokines, mRNA changes and protein profile, malformed extracellular vesicles containing mRNAs, micro RNAs, certain proteins as well as fibrosis and deregulated extra cellular modelling which results in systemic inflammation, insulin resistance and metabolic disorder [79]. The growing epidemic of obesity could be addressed by the emerging nano technological approaches specifically in the food industries. As a powerful public health tool, approaches based on nanotechnology plays a distinctive and chief role in food production which could be beneficial for providing low-calorie foods [80]. From different prospect obesity associated disorders could be addressed by targeting adipose tissue expansion and transformation from an energy storage status that is white adipose tissue (WAT) into an energy expenditure status that is brown-like adipose tissue (BAT) [81].

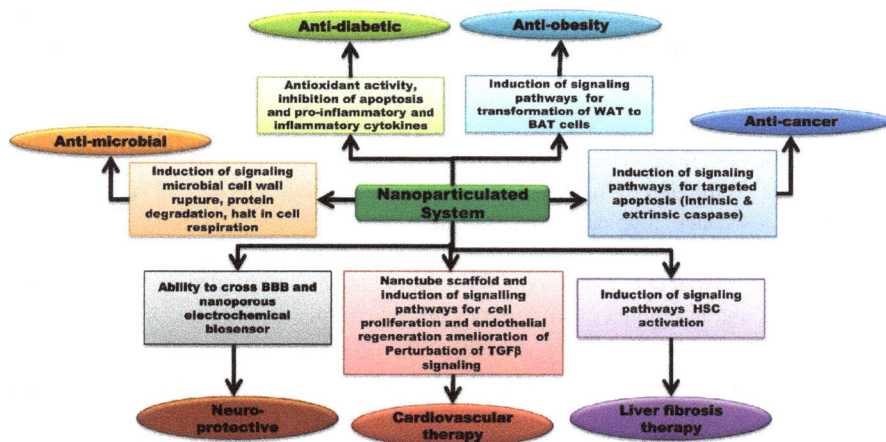

Figure 1: Strategies of nanoparticulated system in biomedical application. HSC-hepatic stellate cell, WAT-white adipose tissue, BAT-brown-like adipose tissue, BBB-blood barain barrier, TGFβ-transforming growth factor β

Currently, the main focus of nanotechnology is development of biodegradable non-toxic drug delivery nanosystem, utilizing nano-emulsions, surfactant micelles, emulsion bilayers and reverse micelles [82]. Xue et al. [81] developed a peptide-functionalized NPs nanoparticulated system to deliver either Peroxisome Proliferator-Activated Receptor gamma (PPARgamma) activator rosiglitazone (Rosi) or prostaglandin E2 analog (16, 16-dimethyl PGE2) to adipose tissue vasculature. Through self-assembly of end-to-end linkage between poly (lactic-coglycolic acid)-b-poly (ethylene glycol) (PLGA-b-PEG) and endothelial-targeted peptide the biodegradable nano particulated system was developed. Transformation of WAT to BAT is induced Rosi realised from the

Biosensors: Materials and Applications   Materials Research Forum LLC
Materials Research Foundations **47** (2019) 241-288   doi: http://dx.doi.org/10.21741/9781644900130-7

nanosystem, which further facilitates enhances delivery at the targeted site [81]. Similarly, there are many natural compounds such as barberine, butein, capsaicin, and fucoxanthin that influence obesity by BAT activation/ browning of WAT and uncoupling protein 1 activation in WAT via either AMPK/Pgc-1α activation, Prdm4 induction or TrpV1 activation [83-87]. Likewise, there are some synthetic compounds that influence obesity by BAT activation/ browning of WAT such as Ppary agonist, Notch inhibitors salsalate, β3-AR agonists and BAY 41-8543 by the means of Prdm16 stabilization Notch pathway inhibition, Pka pathway, β-adrenergic receptor activation or cGMP-dependent pathway [52]. These obesity modulating compounds could be loaded into desired polymeric or liposome based nanoparticulated system and evaluated for their anti-obesity activity as the research in these area is scanty.

Diabetes Mellitus

Both Type-I and II are broad categories of diabetes mellitus. Type-1 diabetes is an autoimmune destruction of insulin-producing pancreatic beta cells and the later is caused by insulin resistance coupled with failure of beta cell to compensate [88]. In type-II diabetes insulin regulates the blood glucose level by signalling the cells to absorb sugar from the bloodstream but the uptake of sugar from blood is interrupted by the destruction of beta cells and is brought about by both the environmental and genetic factors under stress environment [89,90].

Various stress triggered auto-antigens are produced by beta cells in type-1 diabetes which are later recognized by the auto-antibodies and auto-reactive T-cells that precisely lead to autoimmune destruction [91,92]. The auto-reactive T-cells, monocytes and dendritic cells are activated by the pro-inflammatory cytokines, which are in turn activated by their transcription factors NFκB [93]. IL-2 and INF-βsecreted by the activated CD4$^+$ "Th1" T-cells, activates CD8$^+$ T-cells and other immunologically related macrophages. Together these inflammatory cytokines (IL-β, TNF-α and ROS) secreted by the activated macrophage contributes to the destruction of beta-cells [94,95]. In type-II diabetes (insulin resistance) the insulin receptor substrate (IRS) family present on insulin-responsive cells are themselves phosphorylated when insulin binds to them, initiating downstream signalling events [96]. Inhibitions of these downstream signalling pathways are the major mechanism through which inflammatory signalling leads to insulin resistance [88]. The reoccurrence of may lead to variety of complications such as neuropathy, retinopathy and nephropathy both in case of type-1 and II diabetes [78,97].

The present medication such as metformin, sulphonylureas and antihyperglycaemic agents such as glinides, thiazolidinediones, α-glucosidase inhibitors, dipeptidyl peptidase-4 inhibitors and sodium-glucose cotransporter-2 inhibitors shows heterogeneous side effects, low delivery rate, retention time, biodistribution and biodegradability after prolonged administration [98]. Thus, the demand for development of an alternative medication which could compensate the lacuna of conventional drugs is of urgent need, which could be precisely addressed by the exploitation of nanotechnology.

Ameliorated toxicity or side-effects of synthetic drugs and efficient delivery of natural therapeutic by nanoparticulated system has become an emerging focus of current studies [99]. Development of anti-diabetic therapy are being explored using PNPs, polymeric micelles, ceramic NPs, liposomes, DENs and functionalized SiNPs [87,99]. For example insulin loaded N-trimethyl chitosan chloride functionalized poly lactic-co-glycoside NP was formulated by Sheng et al. [100] for oral delivery, which was efficient in oral insulin absorption and breaching the multiple barriers. Shi et al. [101] developed polyethylene glycol-poly (lactic-cohlycolic acid) NPs functionalized with Fc recptors for exenatide oral delivery. It is an effective therapy against type-II diabetes, as exenatide (39-aminoside peptide) is similar to glucagone like peptide-1 in its glucoregulatory action. Similarly, there are other nanotechnology based therapeutics available which shows promising results some of which have been depicted in Table-1.

Cardiovascular Diseases

Visceral obesity, diabetes mellitus and hypertension comes under a cluster of cardiovascular risk factors which begins a sequence of cardiovascular events leading to cardiovascular disease continuum [102]. It has been stated that delayed intervention of these factors will inexorable progress to atherosclerosis/ cerebrovascular diseases (CVD), coronary heart disease, rheumatic and congenital heart diseases, etc. which may further lead to death [102,103]. Within the field of cardiovascular engineering and regenerative medicine, investigation and translation to find a remedy to ameliorate the impact of CVDs, is basically rooted to advancement in biomedical science [56]. Nano particulated systems plays a central role in improving the therapeutic potential of conventional/natural drugs or agents in targeted drug delivery, diagnosis, repair and regeneration of the cardiac tissue, simultaneously without hampering homeostasis of the system [103,104].

The present diversity in application of nanomaterials are ranging from protein level, that is few nanometres to cellular levels that is less than a micrometers, where the nanosystems mimic the cellular extracellular matrix, microenvironment as well as tissue structure hierarchy [105,106]. Polymeric PLGA NP encapsulating pitavastatin without PEGylation was developed by Duivenvoorden et al. to augment the anti-inflammatory effect of statin in coronary artery disease mediated by monocyte or macrophage [107]. Ahadian et al. [108] culture mouse embryoid bodies in the microwells, the surface of which was fabricated gelatine methacryloyl hydrogel-aligned carbon nanotube scaffold.

They found that gelatin methacryloyl hydrogel electrophoretically aligned to carbon nanotube enhance the cardiac differentiation of the mouse embryoid bodies. Later in another approach they introduced carbon nanotube into poly (octamethylene maleate) 1,2,4-butanetricarboxylate (124 polymer) and an elastomeric scaffold for cardiac tissue engineering, which provided electrical conductivity and structural integrity to the nanopolymer (124 polymer) for tissue regeneration [109]. The progress in the nanotechnological strategies to treat various cardiovascular diseases is still improving and much emphasis and enthusiasms must be raised towards the development of advanced theranostic to combat the present need.

Liver fibrosis

Developments of cirrhosis progressing from hepatic fibrosis are the key features displaced by almost every chronic liver disease. The pathophysiology of the extracellular matrix (ECM) deposition and metabolism in liver fibrosis have also gained deeper understanding due to the current advancement of science and technology [110]. Both the proportion of ECM and type varies in fibrotic liver in comparison to normal liver. It also has been found that, in liver fibrosis there is an increased deposition of malformed ECM [111]. At the molecular level, it has been documented that, increased deposition ECM proteins such as fibrillar collagens, fibronectin splice variants, proteoglycans; basement membrane proteins and moderately understood modifications in matrix proteins including glycosylation, glycosaminoglycan side chain sulfation and intermolecular together leads to cross-linking which leads to the development of fibrotic scars [111,112]. Steatohepatitis in particularly non-alcoholic steatohepatitis (NASH) and steatosis are involved in non-alcoholic fatty liver disease (NAFLD) of which steatosis considered as the first step of NAFLD, in the long run may progress to fibrosis and cirrhosis [113,114]. Lastly, metabolic syndromes such as high blood pressure and insulin resistance contribute vastly to the development of fibrosis [113,115].

Further, hepatic stellate cell (HSC) as a major source of ECM deposition has been lately discovered from deeper insight into liver fibrogenesis [116]. And the activation of HSC at the molecular level is again mediated by several signalling pathways and upregulation of several proteins such as interstitial collagen, α-smooth muscle actin, proteoglycans and metalloproteinase [116,117]. Cytokine signals such as TGF-β, PDGH, etc. for activation and maintenance of matrix deposition, IL-10 for reversing this process or promote matrix degradation and subsequently. Along with these signalling pathways there are other potential therapeutic targets such as glucose regulation, leptin, endothelin, angiotensin II and a probable role of PPARγ [110,118].

Biosensors: Materials and Applications   Materials Research Forum LLC
Materials Research Foundations **47** (2019) 241-288   doi: http://dx.doi.org/10.21741/9781644900130-7

Number of conventional approaches has been made for treating liver diseases interferon γ, angiotensin II antagonists and IL10 have unsatisfactory clinical trials results despite the promising results in preclinical trials [119-121]. The major drawback unveiled by the convention therapeutic formulation was lack of targeted delivery [122]. This drawback can be easily achieved by nanoparticulated drug delivery systems which have an advantage over conventional therapeutic formulations that includes tuned, controlled, biodegradable, and biocompatible as well as target specific delivery with low side-effects [117,123]. For example an ultrasound microbubble-cationic nano-liposome complex encapsulating HGF expressing vector, was developed by Zhang et al. [124] with an aim of treating liver fibrosis in a bile duct ligation rat model and understanding its relationship with diffusion-weight MRI parameters. Thomas et al. [125] developed hyaluronic acid micelles carrying losartan an angiotensin type 1 receptor blocker to attenuate HSC activation in a C3H/HeN mice subject and murine fibroblast cells (NIH3T3), human hepatic stellate cells (hHSC) and hepatocyte cell line (FL83B). Based on α-smooth muscle actin expression in activation of HSC cell, the efficiency of NPs was determined. Likewise there are many other examples where nano-based approaches are evaluated for their therapeutic values or as a standard medication against liver fibrosis (Table 1).

Cancer

Cancer is a drastic and complex form of diseases causing a significant number of mortalities globally. It is characterized by the uncontrolled growth and spread of abnormal (unresponsive to cellular signalling) cells within or out of the tissue, resulting in accumulation, local damage and inflammation [126,127]. Cancer otherwise termed as malignant tumour, can be eliminated by heterogeneous cytotoxic approaches such as anti-cancer drugs, γ-irradiation, suicide genes or immunotherapy through apoptosis [128]. Morphological, biochemical changes including cell shrinkage, neuclear DNA damage and membrane blabbing are the common characteristics of apoptosis. These specific alterations within the cells can be induced by multiple stress-inducible molecules such as INK, MAPK/ERK, NFκB, ceramide, and other pro-inflammatory, inflammatory cytokines as well as compound such as granzyme B, released by cytotoxic T cells or NK cells, may directly activate downstream apoptosis [129-131].

The current forms of treatment for cancer includes surgeries, radiation therapy, and/or systemic therapies such as chemotherapy, hormonal therapy, neoadjuvant therapy, immune therapy, gene therapy and targeted therapy which varies based on type and stage of cancer; tumour characteristics; and the patient's age, health, and preferences [126,132]. Despite the availability of a number of treatments, many forms of cancers still remains untreatable, either due to delayed identification, diagnosis, sophistication or high

Biosensors: Materials and Applications   Materials Research Forum LLC
Materials Research Foundations 47 (2019) 241-288   doi: http://dx.doi.org/10.21741/9781644900130-7

treatment cost and the available experimental therapy is highly sophisticated, costly and shows unsatisfactory experimental outputs which further disqualifying them for preclinical trials. With the advancement of science remarkable progresses have been made in development of a therapy against cancer. Inhibition essential metabolic pathways linked with cell growth and division are the major strategies through which the present designed therapeutics causes cytotoxicity [133]. Some of these targeted therapeutics licensed for daily clinical use includes rituximab, trastuzumab, gefitinib, lapatinib, imatinib, bevacizumab and cetuximab [134]. But majority of the available chemotherapeutic agents are time-tested which shows good-disease free survival only for a limited time period.

Nanotechnological advancements has led to an revolution where it is possible to engineer a novel tunable nanoparticulated system loaded with natural or synthetic chemotherapeutic agents with anti-cancer activity [135]. These nanosystems can nullify the major obstacles of toxicity, drug resistance, unspecific delivery, low biocompatibility etc. [136]. Some of the nanotechnology base devices used in diagnosis against cancer include carbon nanotubes, quantum dots, paramagnetic NPs, liposomes, gold NPs, MRI contrast agents and nanotechnology based method for high-specificity detection of DNA and proteins [135]. There are many other hybrid nanoparticulated systems are currently being developed and are evaluated for anti-cancer activity (Table-1).

Neurodegenerative diseases

Neuronal stress induced regular loss of neurons in the central nervous system is the primary reason behind various chronic and progressive disorders in neurodegenerative diseases [137]. Alzheimer's disease (AD), Parkinson's disease (PD), Amyotrophic lateral sclerosis (ALS), Multiple sclerosis, frontotemporal dementia, Huntington's disease, spinocerebellar ataxias and multiple system atrophy are some of the progressive neurodegenerative diseases reported till now [138]. Out of the above mentioned diseases Alzheimer's disease and Parkinson's disease are more prevalent [139]. Although the degeneration of the neurons cannot be reversed completely by any drug, yet the early treatment of the disease can reduce the symptoms up to certain extent. The conventional therapy available shows a series of limitation and the major one is side-effects such as psychosis, confusion, hallucinations and the other is the incompetence to breach BBB [140].

Such limitations could be easily curtailed by the exploitation of nanotechnology, which could easily breach the BBB because of its small size and ameliorated side-effect because of its biocompatibility and biodegradability [141]. The BBB prevents the entry of molecules above 600 daltons into the central nervous system but nanoparticles can easily

Biosensors: Materials and Applications   Materials Research Forum LLC
Materials Research Foundations **47** (2019) 241-288   doi: http://dx.doi.org/10.21741/9781644900130-7

invade into the CNS by carrier mediated endocytosis or phagocytosis [142]. In addition to nano-size, functionalization of the nanoparticulated system with appropriate transporter ligands which mediates the endocytosis through BBB endothelial cells. Several nanotechnological tools are developed for the detection of AD by focusing on AD biomarkers. There is an increased expression of amyloid-β-derived diffusible ligands (ADDLs) in the brain of AD patients which is also found in CSF as well as blood. A localized surface plasmon resonance nanosensor conjugated to an anti-amyloid antibodies tom measeur ADDLs in CSF [143]. Another detection method, proposed by the Mirkin and associates is the bio-bar code assay that can measure the ADDLs level in attomolar range. In the bar-code, magnetic nanoparticles in conjugation with ADDL monoclonal antibody are used for the recognition and attachment of ADDL.

Nanotechnology has also provided drugs to reduce the effects of symptoms and behavioural changes appeared in case of AD. The drugs are carried to the target site by specific nanoparticles which protect the molecules from enzymatic degradation and do not cause any toxicity. These specific nanoparticles are liposomes, polymeric NPs, solid lipid NPs, inorganic NPs [144]. The drugs are entrapped inside a nanocore composed of biodegradable polymers like poly glycolic acid, chitosan, poly butyl cyanoacrylate (PBCA), poly caprolactone (PCL), etc. are best drug delivery systems for AD patients due to their high permeability to the BBB [145]. In the list of neurodegenerative diseases PD comes second to most sever where about 0.5-1% of people between the ages 65-69 years gets affected [146]. Though there is no permanent cure for it, many nanotechnology based treatments and detection methods are available for the systematic relief. PD is characterised by the depletion of dopamine due to the degeneration of the dopaminergic neurons [147], mitochondrial dysfunction and the aggregation of α-synuclein neuronal protein [148]. For the detection of the biomarkers of PD such as dopamine in micromolar, scientists have developed a nanoporous electrochemical biosensor, constructed by palladium nanoparticles over the nanoporous gold wire [149]. The level of α-synuclein is detected by the carbon-based nanomaterials such as grapheme quantum dots and grapheme oxide quantum dots [150]. Intranasal routes are basically preferred for efficient delivery of the therapeutics at the targeted site in PD patients. For example glial cell line-derived neurotrophic factor (GDNF) having neuroprotective effect was encapsulated in cationic liposomes formed of dioleoylphosphatidylcholine, cholesterol and stearylamine, and was effectively delivered through intranasal [151]. Recently for the treatment of PD, poly (D, L-lactide-co-glycolide) nanoparticles loaded with ropinirole is prepared by nanoprecipitation method [152]. And further research is still going on for the discovery of a novel nano-based therapeutic against these degenerative diseases some of these are listed in table no 1.

Microbial infections

Application of nanotechnology in amelioration of infectious diseases is the second most area of research after oncology [153]. The major huddle in the treatment or diagnosis of pathogenic infection is their location within cell/ intracellular form in an active or latent state, where the delivery of the drug is critical or out of the drug reach. And further if at all the drug is delivered, the inadequacy in the dose / efficiency or due to the exposed time period, the infectious pathogen develops a tendency to resist the particular drug [153,154]. Some of these pathogenic bacteria include *Mycobacterium tuberculosis*, *Staphylococcus aureus*, *Streptococcus pneumoniae*, Nontyphoidal *Salmonella*, *Entrococcus* sp., *Shigella* sp., *Klebsiella pneumonia*, *Pseudomonas aeruginosa*, *Actinetobacter baumannii* and *Neisseria gonorrhoeae* [153]. Invasive fungal infections caused by *Aspergillus* sp., Zygomycetes, *Fusarium* sp., *Scedosporium* sp., and non-*albicans Candida* sp. also represents a major risk as an opportunistic pathogen [155]. Viral infections on the other hand also possess noteworthy global impact which affects both health and socioeconomic developments. Specifically, human immunodeficiency virus (HIV), human papillomavirus (HPV), hepatitis B virus (HBV), hepatitis C virus (HCV), ebola virus disease (EVD), herpes simplex virus (HSV), zika virus (ZIKV) and influenza represent a major threat to human civilization, development of therapy against which is the primary task of present scientific community [156].

Though conventional antimicrobials such as bacterial cell wall inhibitors (fosfomycin, vancomycin etc.), protein biosynthesis inhibitors (tertacyclibnes, macrolides etc.), DNA synthesis inhibitors (4-quinolones), RNA synthesis inhibitors (rifampicin) and others such as ethambutol, isoniazid, streptomycin, kanamycin, capreomycin, amikacin, fluoroquinolones, para-amino salicylic acid, pyrazinamide, thioacetazone, rifabutin, ethionamide, linezolid and clofazimine have been proven to be beneficiary but still the adverse effect cannot be neglected [157]. Same is applicable for the anti-viral agents such as nucleoside reverse transcriptase inhibitors (zidovudine, didanosine etc.), nucleotide reverse transcriptase inhibitors (tenofovir disoproxil fumarate), non-nucleoside reverse transcriptase inhibitors (nevirapine, delavirdine etc.), protease inhibitors and viral entry inhibitors which also have certain demerits such as inefficiency to cross the blood-bone-barrier, targeting and low retention time [156,158].

These limitations could easily be compensated by the application of nanotechnology. Currently there are number of nano-based drug delivery vehicles have evolved mostly of polymeric or liposome origin which proves to be potential against these infectious diseases. For example, Gajendiran et al. [159] developed an *in vitro* tuberculosis drug release efficiency of silver NPs tartarate-linked poly (lactic-co-glycolic acid) (PLGA)-polyethylene glycol (PEG)-based multiblock copolymer loaded with rifampicin-, isoniazid- and pyrazinamid. Further, Costa-Gouveia et al. [160] developed a pulmonary nano-delivery system based on biodegradable β-cyclodextrin polymer, encapsulating a major second line anti-tuberculosis drug, enhionamide and its booster BDM41906. The efficiency of the nano-formulation as a chemotherapeutic against tuberculosis was evaluated *in vivo*. Toxicity and the inefficiency to breach the BBB are the major limitation of the presently available therapeutics. amphotericin B deoxycholate (AmB) is one of such therapeutic agent which is an antifungal agent and could be an efficient therapy against cryptococcal meningitis, but is not due to its inefficiency to cross the BBB. Xu et al. [161] formulated a nano formulation of AmB using α-butyl-cyanoacrylate against *Cryptococcus neoformans* 30629B induced experimental meningitis. The research team further showed that the nanoparticulated system encapsulating AmB could efficiently breach the BBB and were delectated in brain tissue after a span of 30 minutes of the injection, whereas the conventional AmB failed to do the same. Similarly, Roy et al. [162] developed surface unmodified and modified (-COOH and NH_2) nanodimonds with an ability to load an anti-HIV-1 drug efavirenz and evaluated its cytoxicity *in vitro*. In this comparative study they found that the drug loading capacity was higher in unmodified nano diamond than modified nano diamond along with minimum toxicity. Thus, the overall prospect of nanotechnology is highly dispersed as a therapeutic agent and further research should be carried out to cordially understand the durability of nano-based therapies.

Table 1: Type of nanoparticles and their evaluation procedure in biomedical analysis

Nanoparticulated systems	Model organism	Evaluation performed	Result	References
Obesity				
Superparamagnetic iron oxide nanoparticles (SPIONs)	human primary adipocytes	Effect of SPIONs on 22 and 29 risk genes (Based on gene wide association studies) for obesity and T2D in human adipocytes	mRNA of *GULP1, SLC30A8, NEGRI, SEC16B, MTCH2, MAF, MC4R*, and *TMEM195* were severely induced and *INSIG2, NAMPT, MTMR9, PFKP, KCTD15, LPL* and *GNPDA2* were down-regulated	[53]
Cerium oxide nanoparticles	3T3-L1 pre-adipocytes	Interfere of NPs with the adipogenic pathway	The NPs reduces the mRNA transcription of genes involved in adipogenesis, and by hindering the triglycerides accumulation	[54]

	Wistar rats	Toxicity and biochemical or metabolic changes	Negligible toxicity, efficient in reduction of weight gain and in lowering insulin, leptin, glucose and triglycerides level in plasma.	
Cubic phase nanoparticle/her bal extract mixture suspensions	3T3-L1 cells	Fluorescence activated cell sorting analysis and confocal laser scanning microscopy	NPs promotes incorporation of water souble dye calcein into adipocyte cell	[55]
	Wistar rats	Biochemical or physiological changes	NPs decreased the blood contents of aspartate aminotransferase, total cholesterol, triglyceride, urea nitrogen, and LDL and promoted the capability of the herbal extracts in suppressing the body weight gain and the liver weight in the animal model.	
Poly(lactide-co-glycolide) encapsulated Notch inhibitor (dibenzazepine)	Primary preadipocyt es from stromal vascular fraction (SVF) cells ofmale HFD-induced obese mice with the C57BL/6 background	Intracellular delivery of the NPs using fluorescence microscopy, endocytosis process by TEM and Notch targets and the browning markers by qPCR assay	Efficient intra cellular delivery (endolysosomal escape of NPs), expression of classical Notch target genes, *Hey1, HeyL,* and *Hes1,* was significantly inhibited and the mRNA levels of the browning markers such as *Ucp1, Cidea, Ppargc1a,* and *Dio2,* were elevated significantly as well as mRNA level of the cytochrome c oxidative subunit 5b (*Cox5b*)	[56]
Diabetes				
L-valine functionalized chitosan conjugated tripolyphosphat e NPs loaded with insulin	*HT-29 cells*	*In vitro cytotoxicity, pH and glucose triggered release and cellular uptake studies*	Negligible cytotoxicity, excellent stability against protein solution, efficient release and cellular ingestion was evaluated confocal laser scanning microscope	[57]

Insulin-loaded alginate/dextran sulphate NPs dual-coated with chitosan and albumin	Caco-2/HT29-MTX/Raji B cell monolayers	pH-sensitivity and mucoadhesivity, *in vitro* insulin *release and cellular uptake studies*	Sustained insulin release, efficient intestinal interactions, higher permeability and transport by clathrin-mediated endocytosis	[58]
Biomineralized insulin NPs	Insulin-resistant HepG2 cell	Effect on glucose metabolism, *cytotoxicity and cellular uptake studies,* glucose oxidase assay and, Glut-4, Glut-2, IRS-2 and IRS-1genes expression		[59]
	Diabetic KKAy mice		Durable upgrade in their glucose metabolisms	
Eprosartan mesylate loaded nano-bilosomes	Streptozotocin induced diabetes	Biochemical and molecular analysis	The NPs decreased the serum creatinine, urea, lactate dehydrogenase, total albumin, and malondialdehyde and expressions of Angiotensin II typ-1 receptor, inducible nitric oxide synthase, transforming growth factor-$\beta1$	[60]
Cardiovascular diseases				
Phe[D]-Pro-Arg-Chloromethylketone covalently linked to perfluorocarbon-core NP	-	Inhibition of thrombin were assessed via thrombin activity	Intrinsic activity against thrombin was observed	[61]
	Male C57BL/6 mice	Antithrombotic activity was assessed through intravenous	Efficient in inhibiting thrombosis	
Pitavastatin con jugated poly (DL-lactide-co-glycolide)	Human coronary artery smooth muscle cells (VSMC)	Cell proliferation and endothelial regeneration	Most potent effects on VSMC proliferation and endothelial regeneration was observed	[62]
	Domestic male pigs	Effect on endothelial healing	Effective endothelial healing effects were observed	

Collagen IV targeted poly[lactic-co-glycolic acid-b-poly(ethylene glycol)] NPs encapsulating amino acids 2–26 (Ac2-26)	Male Ldlr$^{-/-}$ mice	Atherosclerotic lesion analysis, LCM, RNA amplification, and RT-qPCR and Bone marrow transplantation	Prevention of oxidative stress, rise in the protective collagen layer overlying lesions, and a decrease in plaque necrosis	[63]
Cerium oxide NPs	Wistar rats	Cortisol and Aldosterone hormones were assessed in serum and oxidant-antioxidant parameters and histopathological examination in heart tissues were determined	Decrease in serum cardiac markers such as CK-MB, LDH, AST, ALT was observed by noticeable percent and increased tissues level of antioxidant enzymes such as catalase and superoxide dismutase	[64]
Liver fibrosis				
poly-ε-caprolactone encapsulated Silybin NPs	Male Wistar rats	Pharmacokinetics, pharmacodynamics and hepatoprotective activity	The NPs showedsuperior pharmacokinetic properties and hepatoprotective activity	[65]
Cerium oxide NPs	CCl4-treated rats	To estimate standard hepatic and renal function tests, evaluate steatosis, α-SMA expression, macrophage infiltration, apoptosis and mRNA expression of oxidative stress, inflammatory or vasoactive related genes, mean arterial pressure (MAP) and portal pressure (PP) were evaluated and serum samples obtained	Hepatic steatosis, ameliorated systemic inflammatory biomarkers were reduced and PP were improved by cerium NPs, located in liver, without affecting MAP, resulting the reduction in mRNA expression of inflammatory cytokines (TNF-α, IL1β, COX-2, iNOS), ET-1 and messengers related to oxidative (Epx, Ncf1, Ncf2) or Endoplasmic Reticulum (Atf3, Hspa5) stress signaling pathways was observed	[66]

Relaxin (RLN) loaded PEGylated iron oxide NPs	human Hepatic stellate cells (HSCs)	RLN conjugation NPs on human HSCs	RLN conjugation was verified, definite binding and uptake to TGFβ-activated human HSCs was shown by NPs	[67]
	CCl4-induced liver fibrosis mouse models and human liver cirrhosis tissues	RLN conjugation NPs on CCl4-induced advanced liver fibrosis mouse model	By slowing down HSC stimulation, ECM deposition and angiogenesis, both RLN conjugated NPs and free RLN actively disabled fibrosis but the release of Nitric oxide (NO) was only increased by RLN conjugated NPs through significant up-regulation of iNOS indicating supression of portal hypertension.	
Phosphatidylserine-modified nanostructured lipid carriers (mNLCs) containing curcumin	Male Sprague–Dawley rats	Severity of the disease was examined by both biochemical and histological methods	mNLCs were effective at reducing the liver damage and fibrosis, indicated by increase in liver enzymes and pro-inflammatory cytokines in the circulation, along with the least increase in collagen fibers and alpha smooth muscle actin and the most increased hepatocyte growth factors (HGF) and matrix metalloprotease (MMP) two in the livers	[68]
Cancer				
Doxorubicin loaded quantum dots-embedded mesoporous silica nanoparticles (Q-MS) as a core and poly(N-isopropylacrylamide (NIPAM))-graft-chitosan (CS) nanogels	Hep-G2	Tumor cell imaging was explored by confocal laser fluorescence microscope and thermo/pH-sensitive in vitro drug release and cytotoxicity was also determined	Nanospheresas were effective PL imaging in tumor cells, Loading content and embed efficiency of Dox into nanosphere carriers were markedly influenced by pH/thermo	[69]

Materials Research Forum LLC
doi: http://dx.doi.org/10.21741/9781644900130-7

Doxorubicin loaded cyclic Arg-Gly-Asp (cRGD) peptide conjugated to bovine serum albumin linked with cell-penetrating peptide (CCP) KALA	U87-MG and NIH 3T3 cells	Tumor targeting, cell-penetrating activity	Diverse actions of nanoformulation such as effective tumor targeting, cell-penetrating activity and endolysosomal and pH-responsive acivity was determined by using confocal laser scanning microscopy	[70]
By carboxylation, acylation, amination, PEGylation and conjugation with gemcitabine pristine single wall nanotube were employed	Human lung carcinoma cell line (A549), human pancreatic carcinoma cell line (MIA PaCa-2)	Cytotoxic activity by MTT assay	Effective cytotoxicity was observed	[71]
	B6 nude mice	Tumor growth inhibition activity	*Efficient inhibit tumor growth in nude mice was observed*	
Bcl-xL-specific shRNA and a very low doxorubicin content encapsulated by polyethylenimine, grafted through polyethylene glycol linker to carboxylated single-walled carbon nanotubes (SWCNT) and was covalently attached to AS1411 aptamer	L929, AGS cells	Evaluation of tumoricidal efficacy, shRNA-mediated gene-silencing strategy	The nanoformulation showed excellent tumoricidal efficacy which was demonstrated by MTT assay	[72]

Biosensors: Materials and Applications
Materials Research Foundations **47** (2019) 241-288

Materials Research Forum LLC
doi: http://dx.doi.org/10.21741/9781644900130-7

Neurodegenerative diseases				
Exosomal-based delivery system laden with catalase	Mouse macrophage cell line, Neuronal PC12 rat adrenal pheochromocytoma cell line	Exosomal uptake and cytotoxicity by the cell lines	Neurons were effectively imputed with exosomes that produce a better neuro-protective activity in response to oxidative Stress	[13]
	Female C57BL/6 mice	The exosomes's biodistribution in inflammed mouse brain	In mice with acute brain inflammation, catalase-filled exosomes defend *SNpc* neurons against oxidative stress	
L-Glutathione-decoration of poly(ethylene imine) loaded with plasmid DNA	hCMEC/D3 endothelial cell layer, L292, HEK	Bio- and hemocompatibility, cytotoxicity, as well as their ability to cross a hCMEC/D3 endothelial cell layer and cellular uptake	GSH-coupling represents a feasible and promising approach for the functionalization of nanocarriers intended to cross the BBB for drug-delivery while avoiding cellular toxicity	[17]
Angiopep-2 conjugated poly (L-lysine)-grafted polyethylenimine loaded with herpes simplex virus type I thymidine kinase NPs	MCF-7, U87MG, U87MG-LUC	*In vitro* anti-tumour activity cytotoxicity	Efficient in tumor killing effect and are biological safe	[73]
	BALB/c nude mice	*In vivo* anti-tumor effect and toxicity	The nanoformulation GCV system inhibited the orthotopic glioma growth *in vivo* by elevating the apoptosis and reducing the proliferation of the glioma cells simultaniously	
Zein encapsulated quercetin NPs	Male SAMP8 and SAMR1 mice	Effect of quercetin formulations in the spontaneous motor activity test and coordination	The nanoformulation improved the cognition and memory impairments characteristics of SAMP8 mice and promoted the oral absorption of quercetin	[14]

Microbial infection				
Aminosilane-coated magnetic nanoparticles functionalized with chlorhexidine	Human osteoblast cells hFOB 1.19	Anti-microbial activity against Methicillin-resistant *Staphylococcus aureus*, methicillin-sensitive *Staphylococcus aureus*, *Enterococcus faecalis*, *Escherichia coli*, *Pseudomonas aeruginosa* Xen 5, 4 different *Candida albicans, C. glabrata, C. tropicalis*,IL-8 concentration and cytotoxicity	Increase biocompatibility and efficient anti-microbial activity was observed	[74]
Vancomycin-loaded *N*-trimethyl chitosan NPs encapsulated with poly(trimethylene carbonate)		*In vitro* release activity, anti microbial activity against *S. aureus*, Osteoblast proliferation	The nanoformulation have elevated drug-loading ability and a stable, perishable, cytocompatible and bactericidal activity	[75]
	Male New Zealand White rabbits	Anti-microbial activity, bone repair activity	The nanoformulation were proved to be efficient in treatment of chronic osteomyelitis, and had excellent probability of promoting bone healing	
Vacuolar ATPase blocker, diphyllin encapsulated with poly(ethylene glycol)-block-poly(lactide-coglycolide)	fcwf-4 cells	Antiviral activity against the type II feline infectious peritonitis virus	Prominent antiviral effect against the feline coronavirus	[76]
	BALB/c mice	Blood chemistry	Total protein, blood urea nitrogen, creatinine were not affected, Aspartate Transaminase, alanine transaminase, were observed to be elevated and alkaline phosphatise level was decreased. NPs were well tolerated in mice following high-dose intravenous administration.	

Efavirenz-chitosan-g-hydroxypropyl β-cyclodextrin NPs	L929 cells	Drug release, permeability, CNS bioavailability, targeting efficiency, histocompatible, cytotoxicity, mucosal adhesive capacity	NPs showed active drug delivery, greater permeability, non-toxic	[77]
	Male Albino Wistar rats	Pharmacokinetic analysis, Gamma scintigraphy imaging and stability	Enhanced CNS bioavailability/ uptake, high drug-targeting percentage and drug-targeting index, elevated intranasal mucoadhesive	

5. Mechanism of action of nanotechnology based therapeutic agents

At the bottom line a therapeutic agent is preferable, if it harbours properties such as active solubility, targeted delivery, control release, preferable retention time, biocompatibility, low or negligible toxicity and biodegradability. Accompanying these properties size and structural makeup of nanotechnology based therapeutic agents contributes to their unique properties to breach the differential physiological barriers and reach the targeted site [163]. Again, the exploitation of functionalized nanoparticulated systems in *in vivo* manipulation of preferred receptors via various physiochemical tactics to control cellular signal transduction are being refined to perfection in the current era [164]. In the present nanotechnology based focused research, how a mechano sensitive receptor would respond to a targeted mechanical stimuli raised by a tuned nano particulated system at the molecular level to activated downstream signalling cascade is likely to be an anomaly [165]. The sophisticated and versatile magneto plasmonic nanoparticles-based platform which renders *in vivo* imaging, localizing and high spatiotemporal resolution of the targeted proteins activated by customized loading capabilities further contributes to enrich the molecular understanding interplayed by nanotechnology [165,166].

The diversely branched molecular cascades involved in either activation or inactivation of inflammatory molecules which may lead to apoptosis or initiate from apoptosis could be intensely regulated by a functionalized nano particulated system. Nanotechnology based innovations including these phenomenon are at the brim of present biomedical research, success in which may ultimately lead to the formulation of an effective nano-based therapy. Cancer therapy is one of the booming sector of biomedical science, and the research conducted are at such an extent where the experimental investigation are intertwined making it a multidisciplinary research.

Currently, it could be implicated that, in broad sense there are more than 277 diseases that can be regarded as cancer [167]. And for instance these 277 types of cancer my involve millions of specific signature molecules expressed by the local cell and billions of diversity of these signature molecules or receptors/ differentially expressed proteins (DEPs) (DEPs contributes to the onset and progression of cancer) specific to the cancer type, which may be regarded as molecular marker for that particular type of cancer. Currently, dbDEPC database contains 4,029 DEPs, which is from only 20 types of human cancer determined by 331 mass spectrometric experimentations [168]. Similarly there are numerous proteins or metabolic intermediates involved in onset or progress of inflammation and apoptosis in various infectious and non-infectious diseases and disorders. In the current decade these DEPs could be the molecular targets of the engineered nanoparticulated systems for the targeted delivery of the desired therapeutic agents.

Conserved sequences in the genome or suicide genes mRNAs on the other hand could also effectively be used for devising a nanoparticulated system. One of such system was developed by Davis for delivering siRNA. Here an approach was made to deliver siRNA encapsulated in cyclodextrin-PEG NPs to solid tumours. Likewise there are a number of nanoparticulated drug delivery systems being developed using epidermal growth factor receptors, RGD peptide (arginine, glycine, aspartic acid) and antibodies, which are under various phases of clinical trials [169]. In a nut shell it may be summed by stating that molecular interaction or targeting is at the base line is exploited nearly by every nanoparticulated system that displays good clinical results.

6.   Demerits of nanotechnology based therapeutics

Biomolecules are coated on nanoparticles for ease drug delivery, but its interaction leads to toxicological effects and injury to the living system and environment [170]. Cell cycle arrest or apoptosis are results of nanoparticle interaction with biomolecules due to increase in oxidative stress level [171]. The increase in oxidative stress leads to change in cell function which involves oxidative modification of protein, lipid peroxidation, mutatation, and modulation of inflammatory responses through signal transduction, these all factors leads to genotoxic effects resulting in cell death [172].

Nano-CuO is more toxic to cell by production of 8-hydroxy-20-deoxysuanosine (8-OHdG) whereas nano-TiO_2 induces a very lesser amount of (8-OHdG), thus it is least toxic to the cell [172]. While its interaction with protein may induce conformational changes and competitive binding [173]. Exposure of nanoparticles with immune cells during neonatal period may induce allergic inflammation at later stages of life [170]. Sometimes nanoparticles behaves as hapten which can change the protein structure and

Biosensors: Materials and Applications   Materials Research Forum LLC
Materials Research Foundations **47** (2019) 241-288   doi: http://dx.doi.org/10.21741/9781644900130-7

function causing autoimmune disorders as these NPs gets into the cell by endocytosis because of its nano structure [174]. The pores and channels produced by membrane proteins have structural closeness with nanoparticles due to which blocking of ion channels may occur. Carbon based nanoparticles with diameter of 0.9-1.3 nm can effectively block the transport of K^+ ions across the membrane that may lead to toxicity [175].

Interaction study of nanoparticles by Bressan et al. [176] proposed that gathering of nanosilver particles on the exterior of mitochondria obstructs the respiratory pathway and give rise to reactive oxygen species and oxygen stress followed by DNA damage. Nanosilver action on fibroblasts cell induces release of cytochrome c into cytosol and transfer of Bax to mitochondrio results in apoptosis through mitochondrial pathway. Nanosilver interaction with DNA induces G1 arrest and completely blocks S phase causing apoptosis [177]. Production of TGF-β1 and platelet derived growth factor are increased due to multi walled carbon nanotubes which is major reason for development of fibrosis [178].

Nanoparticles often imitate the characteristics of structural protein and are related with microtubules and centromere. This interaction may lead to mitotic spindle disruption, chromosome breakage/fragmentation, no/multinucleate cells and mutagenecity [175]. Reactive oxygen species formation due to interaction of nanoparticles with biomolecules like DNA, Protein possesses the highest one electron reduction potential. Understanding the mechanism of nanomaterials inducing toxic effect in the living system at molecular level will help to modify properties of nanomaterials for commercial use [172].

7.   Current nanotechnology based therapeutics for clinical trails

Improved solubility and pharmacokinetics, increased efficiency, ameliorated toxicity and increased biocompatibility and biodistribution are some of the basic physical and biological advantages of nanoformualted drugs when compared to conventional drugs [176-179]. Current FDA-approved bio-materials are mostly of polymeric, liposomal and nanocrystal, but an effort has been given towards the development of critically complex materials encompassing nano-materials such as micelles, proteins based NPs as well as variety of inorganic or metallic NPs for clinical trials [178,180]. The progress made in the design and usability of nanotechnology based therapeutic are best represented with the ones that are on the way to clinical trials or under clinical trials or finally got approved by the FDA after successful clinical results [181].

In the present scenario, there are a number of nanotechnology based therapeutics that are under different phases or on the way to clinical trials. For example, Xue et al. [81]

developed an anti-obesity nanoformulation which at the basic level is a peptide-functionalized NPs nanoparticulated system to deliver either PPARγ activator Rosi or 16,16-dimethyl PGE2 to adipose tissue vasculature to promote both transformation of WAT into BAT [81]. The nanoparticulated system has not been approved by the FDA, but in the near future its fate is awaited. Further a patient-friendly formulation "Afrezza" formulated by Mannkind Corporation and is approved by the FDA. Improvement in glycaemic control by the nano-formulated ultra-fast-acting inhalable insulin has been developed, which was efficient in lowering the level of glycated haemoglobin (HbA1c) in patients with type I and II diabetes [182]. Khaja et al. [183] developed a sterically stabilized phospholipid NPs encapsulating a siRNA suitable for liver and kidney fibrotic diseases. The transition of the nanoconstruct to clinics in a timely manner is under consideration. Li et al. [184] developed a nanotechnology based plasmonic gold chip which could simultaneously detect 10 cardiovascular autoantibodies targeting proteins related to myocardial structure, neurohormonal regulation, vascular proteins and apoptosis and coagulation associated proteins [184]. Observing the efficiency of this chip, it could be said that soon the chip will be on its way for clinical trials.

Nanotechnology based drugs are more extensively evaluated for cancer and some of the drugs have already been approved by the FDA. For example there are various liposomes, polymeric conjugates, polymeric NPs, polymeric micelles such as Myocet™, Doxil™, CRLX101, BIND-014, Genexol-PM™, etc. which are either under different phases of clinical trials or approved by the FDA [185]. For central nervous system the developed nano-based therapeutics includes Diprivan® with active ingredient propofol (marketed by Fresenius Kabi), Invega Sustenna® with active ingredient paliperidone palmitate (marketed by Janssen (Elan) [178]. Similarly, Donnellan et al. [186] developed an anti-tuberculosis solid drug NPs of two drugs namely rifampicine and isoniazid which is extensively investigated worldwide [186]. With further modification and experimentation the nanoformulation could be directed to clinical trials. Antifungal liposomal AmB formulation against Mucormycosis and Cryptococcal meningitis are under clinical trials and as well as certain nanoformulated antiviral agents Pegasys against hepatitis B and hepatitis C, Peglntron against hepatitis C have been approved by the FDA for clinical use [180]. Apart from these, there is other nano-based therapeutics which are under different phases of clinical trials.

8.   Future prospects of nanotechnology

In the current century nanotechnology based therapeutics and innovations have stunned the researches around the world by its unique potential application in the field of biomedical sciences. The first generation nano-therapeutics have proven their efficiency

Biosensors: Materials and Applications   Materials Research Forum LLC
Materials Research Foundations **47** (2019) 241-288   doi: http://dx.doi.org/10.21741/9781644900130-7

and provided a legitimate starting point for nano-therapeutics utility, but certain features could be further eliminated or modified in the next generation nano-therapeutics. The major focus of next generation nano-therapeutics would be to improve pharmacokinetic properties such as improved drug solubility, modifiable drug circulation time and controlled drug release at the targeted site. Altered biodistribution on the other hand would be profoundly benefiting to segregate targeted tissue from non-targeted tissue by availing varying degree of drug release. [187].

Some visible advancement includes the evaluation of hybrid nanocarriers, dual drug loaded nanoparticulated systems, high sensitive nano-based devices for detection and diagnosis, etc. For example different types of lipid-polymer core-shell systems have been developed such as polymer core-lipid shell, hollow core lipid-polymer-lipid systems, biomimetic LPH systems and polymer-caged liposomes [188]. A core shell lipid-polymer hybrid NPs with a core of poly (latic-co-glycolic acid) and an exterior of liped layer containing docetaxel and sphingosine kinase 1 FTY720 (fingolimod) inhibitors was developed by Wang and his team [189]. Re-sensitization of castrate resistant prostate cancer cells to the docetaxel was mediated by both free and conjugated FTY720 with nanoparticulated system was observed. The possibility of controlling or modulating these nano-based systems remotely is a result of brief understanding of molecular engineering of the system. This spectacular phenomenon or ability to be remotely triggered by exogenous stimuli have significantly intrigued the attention of current nano biotechnologists to peruse their research and contribute to this innovative science [190].

Khalid et al. [191] developed a unique nano-therapeutics in the form of nanosphere combining silk fibroin and nanodiamond loaded with anthracyline Doxorubicin for fluorescence tracking and control release. Further, Wang et al. [189] formulated a protein-based nanocomposite through nucleation and assembly of silica using self-assembled silk protein nanostructures as biotemplates to carry anti-cancerous drugs. For concurrent and extended intracellular imaging and ratiometric detection of hydrogen peroxide, Purdey et al. [192] devised a new hybrid nanometerial named peroxynanosensor which contained an organic fluorescent probe bound to a nanodiamond and was photostable in nature (earlier fluorescent probes lacked photostability). Cao et al. [193] further engineered a chelerythine loaded Fe_2O_3 gated multi-walled carbon nanotubenanoparticulation as a therapy against cancer [193]. Similarly, oxidized multi-walled carbon nanotube grafted with polyethylene glycol nano-system carrying etoposide (VP-16) and Bcl-2 phosphorothioate antisense deoxyoligonucleotides was developed by Heger et al. [194] for *in vitro* cytostastic efficiency [194].

Despite the momentous encroachments made in the field of biomedical sciences beyond its best predictable context, but still than there is a gap in the actual fruitfulness of the serviceability of nano-science. For example there have been numerous nano-based advancements made in engineering, diagnostics kits, therapeutic and contrasting agents, but only for limited number of diseases such as cancer and related diseases. Disorders like diabetes, neurodegenerative disorders, and other inflammatory diseases are not as much explored as cancer. There are numerous novel target sites and molecules which are slowly coming to light with a hope of future utilization. Further, various natural bioactive compounds with varying degree anti-diabetic, anti-cancer, liver protective, cardio protective efficiency are gradually accumulating after efficient *in vitro* and *in vivo* experimentations. These bioactive natural compounds could be further explored in conjugation with nanoparticulated system for their biomedical efficiency both *in vitro* and *in vivo* experimentations.In summation, it could be said that nanotechnology holds a world of opportunities for the researchers, who genuinely seek to make a difference.

Conclusion

Nanotechnology based theranostics are among the most innovative and strategic intervention that the scientific community had an opportunity to acknowledge. The extent to which the application could be exploited is beyond the comprehension of mankind as a whole. This expendable technology has much more to give than to take and should be deeply and critically verified in order to unravel the hidden mystery within. It is now possible to hope that the application of nanotechnology in diagnostics and therapeutics will greatly evolve to meet the present need. Utilization of nanoparticulated systems in engineering a novel experimental theranostics against obesity, diabetes, CVD, liver fibrosis, cancer, neurodegenerative diseases and microbial infections are the ongoing research. These nano-based next generation therapeutics are not far from eliminating the limitations of the current generation therapeutics. Intensive *in vivo* and *in vitro* experimentation and extensive clinical trials of this nano-based therapeutics are being backed up for their commercialization.

Moreover, the research made in this field could be accelerated as the innumerous bioactive compounds of diverse origin such as plant, microbial or animal in conjugation to this novel biodegradable, biocompatible, tunable and targeted nanoparticulated systems could breach almost every type of biological barriers with negligible toxicity be evaluated in near future. Thus, there is an extensive need of future research to properly understand the molecular notation of these nanoparticulated systems, how they interact inside the host tissue or cells and how specifically as well as precisely these are controlled or targeted. Then only an effective and novel nano-theranostic could be engineered with efficiency to serve all of mankind

Biosensors: Materials and Applications   Materials Research Forum LLC
Materials Research Foundations **47** (2019) 241-288   doi: http://dx.doi.org/10.21741/9781644900130-7

References

[1] M. Faraday, Experimental relations of gold (and other metals) to light, Phil. Trans. Roy. Soc. Lond. 147 (1857) 145-181. https://doi.org/10.1098/rstl.1857.0011

[2] B. Bodaiah, M.U. Kiranmayi, P. Sudhakar, A.R. Varma, K. Bhushanam, Insecticidal activity of green synthesized silver nanoparticles, Int. J. Recent Sci. Res. 7 (2016) 10652-10656.

[3] T. Cerna, T. Eckschlager, M. Stiborova, Targeted nanoparticles-a promising opportunity in cancer therapy-Review, J. Metallomics. Nanotechnol. 4 (2016) 6-11.

[4] T.C. Prathna, L. Mathew, N. Chandrasekaran, M.R. Ashok, A. Mukherjee, Biomimetic synthesis of nanoparticles: Science, technology & applicability, in (Ed.) M. Amitava, Biomimetics Learning from Nature (2010) 1-21.

[5] M. Bala, V. Arya, Biological synthesis of silver anoparticles from aqueous extract of endophytic fungus Aspergillus fumigatus and its antibacterial action, Int. J. Nanomater. Biostruct. 3 (2013) 37-41.

[6] R. Singh, S.K. Sahu, M. Thangaraj, Biosynthesis of silver nanoparticles by marine invertebrate (polychaete) and assessment of its efficacy against human pathogens, J. Nanoparticles 2014 (2014) 1-7. https://doi.org/10.1155/2014/718240

[7] W.H. Suh, K.S. Suslick, G.D. Stucky, Y.H. Suh, Nanotechnology, nanotoxicology, and neuroscience, Prog. Neurobiol. 87 (2009) 133-170. https://doi.org/10.1016/j.pneurobio.2008.09.009

[8] F. Fontana, D. Liu, J. Hirvonen, H.A. Santos, Delivery of therapeutics with nanoparticles: What's new in cancer immunotherapy? WIREs Nanomed. Nanobiotechnol. 9 (2017) 1-26.

[9] C. Saraiva, C. Praça, R. Ferreira, T. Santos, L. Ferreira, L. Bernardino, Nanoparticle mediated brain drug delivery: Overcoming blood-brain barrier to treat neurodegenerative diseases, J. Control Release 235 (2016) 34-47. https://doi.org/10.1016/j.jconrel.2016.05.044

[10] K. Kadota, A. Senda, H. Tagishi, J.O. Ayorinde, Y. Tozuka, Evaluation of highly branched cyclic dextrin in inhalable particles of combined antibiotics for the pulmonary delivery of anti-tuberculosis drugs, Int. J. Pharm. 517 (2017) 8-18. https://doi.org/10.1016/j.ijpharm.2016.11.060

[11] H.R. Lakkireddy, M. Urmann, M. Besenius, U. Werner, T. Haack, P. Brun, J. Alié, B. Illel, L. Hortala, R. Vogel, D. Bazile, Oral delivery of diabetes peptides-

comparing standard formulations incorporating functional excipients and nanotechnologies in the translational context, Adv. Drug Deliv. Rev. 106 (2016) 196-222. https://doi.org/10.1016/j.addr.2016.02.011

[12] D.B. Vieira, L.F. Gamarra, Advances in the use of nano carriers for cancer diagnosis and treatment, Einstein (Sao Paulo) 14 (2016) 99-103. https://doi.org/10.1590/S1679-45082016RB3475

[13] M.J. Haney, N.L. Klyachko, Y. Zhao, R. Gupta, E.G. Plotnikova, Z. He, T. Patel, A. Piroyan, M. Sokolsky, A.V. Kabanov, E.V. Batrakova, Exosomes as drug delivery vehicles for Parkinson's disease therapy, J. Control. Release. 207 (2015) 18-30. https://doi.org/10.1016/j.jconrel.2015.03.033

[14] L.C.G.E.I. Moreno, E. Puerta, J.E. Suárez-Santiago, N.S. Santos-Magalhães, M.J. Ramirez, J.M. Irache, Effect of the oral administration of nanoencapsulated quercetin on a mouse model of Alzheimer's disease, Int. J. Pharm. 517 (2017) 50-57. https://doi.org/10.1016/j.ijpharm.2016.11.061

[15] M.M. Wen, N. El-Salamouni, W.M. El-Refaie, H.A. Hazzah, M.M. Ali, G. Tosi, R.M. Farid, M.J. Blanco-Prieto, N. Billa, A.S. Hanafy, Nanotechnology-based drug delivery systems for Alzheimer's disease management: technical, industrial and clinical challenges, J. Control. Release. 245 (2017) 95-107. https://doi.org/10.1016/j.jconrel.2016.11.025

[16] Y.E. Choonara, P. Kumar, G. Modi, V. Pillay, Improving drug delivery technology for treating neurodegenerative diseases, Expert. Opin. Drug Deliv. 13 (2016) 1029-43. https://doi.org/10.1517/17425247.2016.1162152

[17] C. Englert, A.K. Trutzschler, M. Raasch, T. Bus, P. Borchers, A.S. Mosig, A. Traeger, U.S. Schubert, Crossing the blood-brain barrier: Glutathione-conjugated poly (ethyleneimine) for gene delivery, J. Control Release. 241 (2016) 1-14. https://doi.org/10.1016/j.jconrel.2016.08.039

[18] J.Y. Yhee, J. Im, R.S. Nho. Advanced therapeutic strategies for chronic lung disease using nanoparticle-based drug delivery, J. Clin. Med. 5 (2016) E82. doi: 10.3390/jcm5090082. https://doi.org/10.3390/jcm5090082

[19] S. Ranjita, A. Loaye, M. Khalil, Present status of nanoparticle research for treatment of tuberculosis, J. Pharm. Pharm. Sci.14 (2011) 100-116. https://doi.org/10.18433/J3M59P

Biosensors: Materials and Applications   Materials Research Forum LLC
Materials Research Foundations **47** (2019) 241-288   doi: http://dx.doi.org/10.21741/9781644900130-7

[20] N. Singh, R. Singh, An introduction to the approaches of novel drug delivery systems for acquired immune deficiency syndrome (AIDS), J. AIDS HIV Infections 1 (2016) 1-14.

[21] P.I. Siafaka, N.U. Okur, E. Karavas, D.N. Bikiaris, Surface modified multifunctional and stimuli responsive nanoparticles for drug targeting: Current status and uses, Int.J. Mol. Sci.17 (2016) E1440. doi: 10.3390/ijms17091440. https://doi.org/10.3390/ijms17091440

[22] S. Gupta, R. Bansal, S. Gupta, N. Jindal, A. Jindal, Nanocarriers and nanoparticles for skin care and dermatological treatments, Indian Dermatol Online J. 4 (2013) 267-272. https://doi.org/10.4103/2229-5178.120635

[23] A. Nasir, A. Kausar, A. Younus, A review on preparation, properties and applications of polymeric nanoparticle-based materials, Polymer-Plastics Tech. 54 (2015) 325-341. https://doi.org/10.1080/03602559.2014.958780

[24] K.S. Kavitha, S. Baker, D. Rakshith, H.U. Kavitha, H.C.Y. Rao, B.P. Harini, S. Satish, Plants as green source towards synthesis of nanoparticles, Int. Res. J. Bio. Sci. 2 (2013) 66-76.

[25] A. Watermann, J. Brieger, Mesoporous silica nanoparticles as drug delivery vehicles in cancer, Nanomaterials 7 (2017) E189. doi: 10.3390/nano7070189. https://doi.org/10.3390/nano7070189

[26] W. Wei, C. Xu, H. Wu, Magnetic iron oxide nanoparticles mediated gene therapy for breast cancer - an in vitro study, J. Huazhong Univ. Sci. Technol. Med. Sci. 26 (2006) 728-30. https://doi.org/10.1007/s11596-006-0628-y

[27] M.V. Yigit, D. Mazumdar, Y. Lu, MRI detection of thrombin with aptamer functionalized superparamagnetic ironoxide nanoparticles, Bioconjug.Chem.19 (2008) 412-7. https://doi.org/10.1021/bc7003928

[28] X.H. Pengm, X. Qian, H. Mao, A.Y. Wang, Z.G. Chen, S. Nie, Targeted magnetic iron oxide nanoparticles for tumorimaging and therapy, Int. J. Nanomed. 3 (2008) 311-21.

[29] B. Chertok, B.A. Moffat, A.E. David, F. Yu, C. Bergemann, B.D. Ross, Iron oxide nanoparticles as a drug delivery vehicle for MRI monitored magnetic targeting of brain tumors, Biomaterials 29 (2008) 487-96. https://doi.org/10.1016/j.biomaterials.2007.08.050

[30]   G. Weimuller, M. Zeisberger, K.M. Krishnan, Size-dependant heating rates of iron oxide nanoparticles for magnetic fluid hyperthermia, J. Magn. Magn. Mater. 321 (2009) 1947-50. https://doi.org/10.1016/j.jmmm.2008.12.017

[31]   G.K. Rout, H.S. Shin, S. Gouda, S. Sahoo, G. DaS, L.F. Fraceto, J.K. Patra, Current advances in nanocarriers for biomedical research and their applications, Artificial Cells, Nanomedicine, and Biotechnology 2018, doi.org/10.1080/21691401.2018.1478843. https://doi.org/10.1080/21691401.2018.1478843

[32]   S. Ahmed, M. Ahmad, B.L. Swami, S. Ikram, A review on plants extract mediated synthesis of silver nanoparticles for antimicrobial applications: A green expertise, J. Adv. Res.7 (2016) 17-28. https://doi.org/10.1016/j.jare.2015.02.007

[33]   D. Cabuzu, A. Cirja, R. Puiu, A.M. Grumezescu, Biomedical applications of gold nanoparticles, Curr. Top Med. Chem.15 (2015) 1605-1613. https://doi.org/10.2174/1568026615666150414144750

[34]   N.O. Mahmoodi, A. Ghavidast, N. Amirmahani, A comparative study on the nanoparticles for improved drug delivery systems, J. Photochem. Photobio.162 (2016) 681-693. https://doi.org/10.1016/j.jphotobiol.2016.07.037

[35]   S. Bhattacharyya, R.A. Kudgus, R. Bhattacharya, P. Mukherjee, Inorganic nanoparticles in cancer therapy, Pharma. Res.28 (2011) 237-259. https://doi.org/10.1007/s11095-010-0318-0

[36]   D.Y. Reddy, D. Dhachinamoorthi, K.B. Chandrasekhar, A brief review on polymeric nanoparticles for drug delivery and targeting, J. Med. Pharma. Innov.2 (2015) 19-32.

[37]   C. Carbone, S. Cupri, A. Leonardi, G. Puglisi, R. Pignatello, Lipid-based nanocarriers for drug delivery and targeting: A patent survey of methods of production and characterization, Pharma. Patent Analyst 2 (2013) 665-677. https://doi.org/10.4155/ppa.13.43

[38]   E. Abbasi, S.F. Aval, A. Akbarzadeh, M. Milani, H.T. Nasrabadi, S.W. Joo, Y. Hanifehpour, K. Nejati-Koshki, R. Pashaei-Asl, Dendrimers: synthesis, applications, and properties, Nanoscale Res. Lett. 9 (2014) 247. doi: 10.1186/1556-276X-9-247. https://doi.org/10.1186/1556-276X-9-247

[39]   I.F. Uchegbu, S.O. Vyas, Non-ionic surfactant based vesicles (Niosomes) in drug delivery, Int. J. Pharmaceu. 172 (1998) 33-70. https://doi.org/10.1016/S0378-5173(98)00169-0

[40] I.F. Uchegbu, A.T. Florence, Non-ionic surfactant vesicles (Niosomes): physical and pharmaceutical chemistry, Adv. Colloid. Interface. Sci. 58 (1995) 1-55. https://doi.org/10.1016/0001-8686(95)00242-I

[41] A.S. Magdum, Y.R. Hundekar, R.M. Chimkode, Niosomes: A promising vesicular drug delivery system for tuberculosis, Indo-Am. J. Pharma. Sci. 4 (2017) 2710- 23.

[42] U. Butt, A. ElShaer, L.A.S. Snyder, A.A. Al-Kinani, A. Le Gresley, R.G. Alany, Fatty acid based microemulsions to combat ophthalmia neonatorum caused by Neisseria gonorrhoeae and Staphylococcus aureus, Nanomaterials 8 (2018) E51. https://doi.org/10.3390/nano8010051

[43] X. Yu, I. Trase, M. Ren, K. Duval, X. Guo, Z. Chen, Design of nanoparticle-based carriers for targeted drug delivery, J. Nanomater 2016 doi.org/10.1155/2016/1087250. https://doi.org/10.1155/2016/1087250

[44] N. Mishra, P. Pant, A. Porwal, J. Jaiswal, Md. S Samad, S. Tiwari, Targeted drug delivery: A review, Am. J. Pharm. Tech. Res. 6 (2016). DOI: 10.21276/ajptr https://doi.org/10.21276/ajptr

[45] S. Gholizadeh, E.M. Dolman, R. Wieriks, R.W. Sparidans, W.E. Hennink, R.J. Kok, Anti-GD2 immunoliposomes for targeted delivery of the survivin inhibitor sepantronium bromide (YM155) to neuroblastoma tumor cells, Pharm. Res. 35 (2018) 85. doi: 10.1007/s11095-018-2373-x. https://doi.org/10.1007/s11095-018-2373-x

[46] L.Y. Chou, K. Ming, W.C. Chan, Strategies for the intracellular delivery of nanoparticles, Chem. Soc. Rev.40 (2011) 233-245. https://doi.org/10.1039/C0CS00003E

[47] C. Azevedo, M.H. Macedo, B. Sarmento, Strategies for the enhanced intracellular delivery of nanomaterials, Drug Discov. Today 23 (2018) 944-959. https://doi.org/10.1016/j.drudis.2017.08.011

[48] E. Ruoslahti, S.N. Bhatia, M.J. Sailor, Targeting of drugs and nanoparticles to tumors, J. Cell. Bio.188 (2010) 759-768. https://doi.org/10.1083/jcb.200910104

[49] N. Bertrand, J. Wu, X. Xu, N. Kamaly, O.C. Farokhzad, Cancer Nanotechnology: The impact of passive and active targeting in the era of modern cancer biology, Adv. Drug Deliv. Rev. 66 (2014) 2-25. https://doi.org/10.1016/j.addr.2013.11.009

[50] C. Foster, A. Watson, J. Kaplinsky, N. Kamaly, Improved targeting of cancers with nanotherapeutics, Methods Mol. Biol. 1530 (2017) 13-37. https://doi.org/10.1007/978-1-4939-6646-2_2

[51]   X. Huang, P.K. Jain, I.H. El-Sayed, M.A. El-Sayed, Plasmonic photothermal therapy (PPTT) using gold nanoparticles, Lasers Med. Sci.23 (2008) 217-28. https://doi.org/10.1007/s10103-007-0470-x

[52]   N.J. Song, S.H. Chang, D.Y. Li, C.J. Villanueva, K.W. Park, Induction of thermogenic adipocytes: molecular targets and thermogenic small molecules, Exp. Mol. Med. 49 (2017) 353. doi: 10.1038/emm.2017.70. https://doi.org/10.1038/emm.2017.70

[53]   S. Sharifi, S. Daghighi, M.M. Motazacker, B. Badlou, B. Sanjabi, A. Akbarkhanzadeh, A.T. Rowshani, S. Laurent, M.P. Peppelenbosch, F. Rezaee, Superparamagnetic iron oxide nanoparticles alter expression of obesity and T2D-associated risk genes in human adipocytes, Sci. Rep. 3 (2013) doi: 10.1038/srep02173. https://doi.org/10.1038/srep02173

[54]   A. Rocca, S. Moscato, F. Ronca, S. Nitti, V. Mattoli, M. Giorgi, G.Ciofani, Pilot in vivo investigation of cerium oxide nanoparticles as a novel anti-obesity pharmaceutical formulation, Nanomedicine11 (2015) 1725-34.

[55]   J.H. Lee, J.C. Kim, Effect of cubic phase nanoparticle on obesity-suppressing efficacy of herbal extracts, Biotechnol. Bioprocess. Eng. 20 (2015) 1005-1015. https://doi.org/10.1007/s12257-015-0417-1

[56]   C. Jiang, M.A. Cano-Vega, F. Yue, L. Kuang, N. Narayanan, G. Uzunalli, M.P. Merkel, S. Kuang, M. Deng, Dibenzazepine-loaded nanoparticles induce local browning of white adipose tissue to counteract obesity, Mol. Ther. 25 (2017) 1718-1729. https://doi.org/10.1016/j.ymthe.2017.05.020

[57]   L. Li, G. Jiang, W. Yu, D. Liu, H. Chen, Y. Liu, Z. Tong, X. Kong, J. Yao, Preparation of chitosan-based multifunctional nanocarriers overcoming multiple barriers for oral delivery of insulin, Mater. Sci. Eng. C Mater. Biol. Appl.70 (2017) 278-286. https://doi.org/10.1016/j.msec.2016.08.083

[58]   M. Lopes, N. Shrestha, A. Correia, M.A. Shahbazi, B. Sarmento, J. Hirvonen, F. Veiga, R. Seiça, A. Ribeiro, H.A. Santos. Dual chitosan/albumin-coated alginate/dextran sulfate nanoparticles for enhanced oral delivery of insulin, J. Control Release 232 (2016) 29-41. https://doi.org/10.1016/j.jconrel.2016.04.012

[59]   Y. Xiao, X. Wang, B. Wang, X. Liu, X. Xu, R. Tang, Long-term effect of biomineralized insulin nanoparticles on type 2 diabetes treatment, Theranostics, 7 (2017) 4301-4312. https://doi.org/10.7150/thno.21450

Biosensors: Materials and Applications
Materials Research Foundations **47** (2019) 241-288

Materials Research Forum LLC
doi: http://dx.doi.org/10.21741/9781644900130-7

[60]   A. Ahad, M. Raish, A. Ahmad, F.I. Al-Jenoobi, A.M. Al-Mohizea, Eprosartan
       mesylate loaded bilosomes as potential nano-carriers against diabetic nephropathy
       in streptozotocin-induced diabetic rats, Eur. J. Pharm. Sci.111 (2018) 409-417.
       https://doi.org/10.1016/j.ejps.2017.10.012

[61]   J. Myerson, L. He, G. Lanza, D. Tollefsen, S. Wickline, Thrombin-inhibiting
       perfluorocarbon nanoparticles provide a novel strategy for treatment and magnetic
       resonance imaging of acute thrombosis, J. Thromb. Haemost. 9 (2011) 1292-1300.
       https://doi.org/10.1111/j.1538-7836.2011.04339.x

[62]   N. Tsukie, K. Nakano, T. Matoba, S. Masuda, E. Iwata, M. Miyagawa, G. Zhao,
       W. Meng, J. Kishimoto, K. Sunagawa, K. Egashira, Pitavastatin-incorporated
       nanoparticle eluting stents attenuate in-stent stenosis without delayed endothelial
       healing effects in a porcine coronary artery model, J. Atheroscler. Thromb. 20
       (2013) 32-45. https://doi.org/10.5551/jat.13862

[63]   G. Fredman,N. Kamaly, S. Spolitu, J. Milton, D. Ghorpade, R. Chiasson, G.
       Kuriakose, M. Perretti, O. Farokzhad, I. Tabas, Targeted nanoparticles containing
       the proresolving peptide Ac2-26 protect against advanced atherosclerosis in
       hypercholesterolemic mice, Sci. Transl. Med. 18 (2015) doi:
       10.1126/scitranslmed.aaa1065. https://doi.org/10.1126/scitranslmed.aaa1065

[64]   S.S. El Shaer, T.A. Salaheldin, N.M. Saied, S.M. Abdelazim. In vivo ameliorative
       effect of cerium oxide nanoparticles in isoproterenol-induced cardiac toxicity,
       Exp. Toxicol.Pathol. 69 (2017) 435-441. https://doi.org/10.1016/j.etp.2017.03.001

[65]   C.R. Bonepally, S.J. Gandey, K. Bommineni, K.M. Gottumukkala, J. Aukunuru,
       Preparation, characterisation and in vivo evaluation of silybin nanoparticles for the
       treatment of liver fibrosis, Trop. J. Pharm. Res. 12 (2013) 1-6.
       https://doi.org/10.4314/tjpr.v12i1.1

[66]   D. Oró, T. Yudina, G. Fernández-Varo, E. Casals, V. Reichenbach, G. Casals, B.
       González de la Presa, S. Sandalinas, S. Carvajal, V. Puntes, W. Jiménez, Cerium
       oxide nanoparticles reduce steatosis, portal hypertension and display anti-
       inflammatory properties in rats with liver fibrosis, J. Hepatol. 64 (2016) 691-8.
       https://doi.org/10.1016/j.jhep.2015.10.020

[67]   R. Bansal, B. Nagórniewicz, G. Storm, J. Prakash, Relaxin-coated
       superparamagnetic iron oxide nanoparticles as a novel theranostic approach for the
       diagnosis and treatment of liver fibrosis, J. Hepatol. 66 (2017) S43.
       doi.org/10.1016/S0168-8278(17)30348-3. https://doi.org/10.1016/S0168-
       8278(17)30348-3

Biosensors: Materials and Applications   Materials Research Forum LLC
Materials Research Foundations **47** (2019) 241-288   doi: http://dx.doi.org/10.21741/9781644900130-7

[68] J. Wang, W. Pan, Y. Wang, W. Lei, B. Feng, C. Du, X.J. Wang, Enhanced efficacy of curcumin with phosphatidylserine-decorated nanoparticles in the treatment of hepatic fibrosis, Drug Deliv. 25 (2018) 1-11. https://doi.org/10.1080/10717544.2017.1399301

[69] R. Gui, Y. Wang, J. Sun, Embedding fluorescent mesoporous silica nanoparticles into biocompatible nanogels for tumor cell imaging andthermo/pH-sensitive in vitro drug release, Colloids Surf. B Biointerfaces116 (2014) 518-525. https://doi.org/10.1016/j.colsurfb.2014.01.044

[70] B. Chen, X.Y. He, X.Q. Yi, R.X. Zhuo, S.X. Cheng, Dual-peptide-functionalized albumin-based nanoparticles with pH dependent self-assembly behavior for drug delivery, ACS Appl. Mater. Interfaces 7 (2015) 15148-15153. https://doi.org/10.1021/acsami.5b03866

[71] A. Razzazan, F. Atyabi, B. Kazemi, R. Dinarvand, In vivo drug delivery of gemcitabine with PEGylated single-walled carbon nanotubes, Mater. Sci. Eng. C Mater. Biol. Appl. 62 (2016) 614-25. https://doi.org/10.1016/j.msec.2016.01.076

[72] T. Ganbold, G. Gerile, H. Xiao, H. Baigude, Efficient in vivo siRNA delivery by stabilized d-peptide-based lipid nanoparticles, RSC Advances 7 (2017) 8823-8831. https://doi.org/10.1039/C6RA25862J

[73] S. Gao, H. Tian, Z. Xing, D. Zhang, Y. Guo, Z. Guo, X. Zhu, X. Chen, A non-viral suicide gene delivery system traversing the blood brain barrier for non-invasive glioma targeting treatment, J. Control Release 243 (2016) 357-369. https://doi.org/10.1016/j.jconrel.2016.10.027

[74] G. Tokajuk, K. Niemirowicz, P. Deptuła, E. Piktel, M. Cieśluk, A.Z. Wilczewska, J.R. Dąbrowski, R. Bucki, Use of magnetic nanoparticles as a drug delivery system to improve chlorhexidine antimicrobial activity, Int. J. Nanomedicine, 12 (2017) 7833-7846. https://doi.org/10.2147/IJN.S140661

[75] Y. Zhang, R.J. Liang, J.J. Xu, L.F. Shen, J.Q. Gao, X.P. Wang, N.N. Wang, D. Shou, Y. Hu, Efficient induction of antimicrobial activity with vancomycin nanoparticle-loaded poly(trimethylene carbonate) localized drug delivery system, Int. J. Nanomed.12 (2017) 1201-1214. https://doi.org/10.2147/IJN.S127715

[76] C.M.J. Hu, W.S. Chang, Z.S. Fang, Y.T, Chen, W.L. Wang, H.H. Tsai, L.L. Chueh, T. Takano, T. Hohdatsu, H.W. Chen, Nanoparticulate vacuolar ATPase blocker exhibits potent host targeted antiviral activity against feline coronavirus, Sci. Rep.7 (2017) 13043. https://doi.org/10.1038/s41598-017-13316-0

[77]   A. Belgamwar, S. Khan, P. Yeole, Intranasal chitosan-g-HPβCD nanoparticles of
       efavirenz for the CNS targeting, Artif, Cells Nanomed, Biotechnol,46 (2018) 374-
       386. https://doi.org/10.1080/21691401.2017.1313266

[78]   C.F. Semenkovich, Insulin resistance and atherosclerosis, J. Clin. Invest.116
       (2006) 1813-1822. https://doi.org/10.1172/JCI29024

[79]   F. Louwen, A. Ritter, N.N. Kreis, J. Yuan, Insight into the development of obesity:
       functional alterations of adipose-derived mesenchymal stem cells, Obesity
       Reviews, 19 (2018) 888-904. https://doi.org/10.1111/obr.12679

[80]   EldawAbdellati, Nanotechnology in Elevation of the worldwide impact of obesity
       and obesity-related diseases: potential roles in human health and disease, J.
       Diabetes Sci. Technol. 5 (2011) 1005-1008.
       https://doi.org/10.1177/193229681100500424

[81]   Y. Xue, X. Xu, X.Q. Zhang, O.C. Farokhzad, R. Langer R. Preventing diet-
       induced obesity in mice by adipose tissue transformation and angiogenesis using
       targeted nanoparticles, Proc. Natl. Acad. Sci. USA. 113 (2016) 5552-5557.
       https://doi.org/10.1073/pnas.1603840113

[82]   J. Weiss, P. Takhistov, D.J. McClements, Functional materials in food
       nanotechnology. J. Food Sci. 71 (2006), doi.org/10.1111/j.1750-
       3841.2006.00195.x. https://doi.org/10.1111/j.1750-3841.2006.00195.x

[83]   H. Maeda, M. Hosokawa, T. Sashima, K. Funayama, K. Miyashita, Fucoxanthin
       from edible seaweed, Undaria pinnatifida, shows antiobesity effect through UCP1
       expression in white adipose tissues, Biochem. Bioph. Res. Comunn. 332 (2005)
       392-397. https://doi.org/10.1016/j.bbrc.2005.05.002

[84]   H. Maeda, M. Hosokawa,T. Sashima, K. Murakami-Funayama, K. Miyashita,
       Anti-obesity and anti-diabetic effects of fucoxanthin on diet-induced obesity
       conditions in a murine mode, Mol. Med. Rep., 2 (2009) 897-902.
       https://doi.org/10.3892/mmr_00000189

[85]   Z. Luo, L. Ma, Z. Zhao, H. He, D. Yang, X. Feng, S. Ma, X. Chen, T. Zhu, T.
       Cao,D. Liu, B. Nilius, Y. Huang, Z. Yan, Z. Zhu, TRPV1 activation improves
       exercise endurance and energy metabolism through PGC-1 alpha upregulation in
       mice, Cell Res, 22 (2012) 551-564. https://doi.org/10.1038/cr.2011.205

[86]   Z. Zhang, H. Zhang, B. Li, X. Meng, J. Wang, Y. Zhang, S. Yao, Q. Ma, L. Jin, J.
       Yang, W. Wang, G. Ning, Berberine activates thermogenesis in white and brown

adipose tissue, Nat. Commun. 5 (2014) 5493. doi: 10.1038/ncomms6493.
https://doi.org/10.1038/ncomms6493

[87]   Y. Song, Y. Li, Q. Xu, Z. Liu, Mesoporous silica nanoparticles for stimuli-
       responsive controlled drug delivery: advances, challenges and outlook, Int. J.
       Nanomed. 12 (2017) 87-110. https://doi.org/10.2147/IJN.S117495

[88]   K.E. Wellen, G.S. Hotamisligil, Inflammation, stress and diabetes, J. Clin. Invest,
       115 (2005) 1111-1119. https://doi.org/10.1172/JCI25102

[89]   M.A. Atkinson, G.S. Eisenbarth, A.W. Michels, Type 1 diabetes, Lancet, 383
       (2014) 69-82. https://doi.org/10.1016/S0140-6736(13)60591-7

[90]   M.A. Atkinson, The pathogenesis and natural history of type 1 diabetes,Cold
       Spring Harb Perspect Med, 2 (2012) doi: 10.1101/cshperspect.a007641.
       https://doi.org/10.1101/cshperspect.a007641

[91]   M. Van-Lummel, A. Zaldumbide, B.O. Roep, Changing faces, unmasking the
       beta-cell: post-translational modification of antigens in type 1 diabetes, Curr.
       Opin. Endocrinol Diabetes Obes., 20 (2013) 299-306.
       https://doi.org/10.1097/MED.0b013e3283631417

[92]   F.X. Mauvais, J. Diana, P. Van Endert, Beta cell antigens in type-1 diabetes:
       triggers in pathogenesis and therapeutic targets, F1000Res, Faculty Rev. 5 (2016)
       728. https://doi.org/10.12688/f1000research.7411.1

[93]   Y. Zhao, B. Krishnamurthy, Z.U. Mollah, T.W. Kay, H.E. Thomas, NF-kB in
       type-1 diabetes, Inflamm Allergy Drug Targets, 10(2011) 208-217.
       https://doi.org/10.2174/187152811795564046

[94]   C. Limbert, Type 1 diabetes – an auto-inflammatory disease: a new concept, new
       therapeutical strategies, J. Transl. Med. 10 (2012). doi:
       10.1101/cshperspect.a007641. https://doi.org/10.1101/cshperspect.a007641

[95]   H.E. Hohmeier, V.V. Tran, C.B. Newgard, Inflammatory mechanisms in diabetes:
       lessons from the β-cell, Int. J. Obesity 27 (2003) S12-S16.
       https://doi.org/10.1038/sj.ijo.0802493

[96]   A.R. Saltiel, J.E. Pessin, Insulin signaling pathways in time and space, Trends Cell
       Bio. 12 (2002) 65-71. https://doi.org/10.1016/S0962-8924(01)02207-3

[97]   A.J. King, The use of animal models in diabetes research, Br. J. Pharmacol.166
       (2012) 877-894. https://doi.org/10.1111/j.1476-5381.2012.01911.x

Biosensors: Materials and Applications
Materials Research Foundations 47 (2019) 241-288

Materials Research Forum LLC
doi: http://dx.doi.org/10.21741/9781644900130-7

[98] M.S. Alhadramy, Diabetes and oral therapies: A review of oral therapies for diabetes mellitus, Journal of Taibah University Medical Sciences 11 (2016) 317-329. https://doi.org/10.1016/j.jtumed.2016.02.001

[99] R. Gupta, Diabetes treatment by nanotechnology, J. Biotechnol.Biomater.7 (2017) 268. https://doi.org/10.4172/2155-952X.1000268

[100] J. Sheng, L. Han, J. Qin, G. Ru, R. Li, L. Wu, D. Cui, P. Yang, Y. He, J. Wang, N-trimethyl chitosan chloride-coated PLGA nanoparticles overcoming multiple barriers to oral insulin absorption, ACS. Appl. Mater. Interfaces. 7 (2015) 15430-41. https://doi.org/10.1021/acsami.5b03555

[101] Y. Shi, X. Sun, L. Zhang, K. Sun, K. Li, Y. Li, Q. Zhang, Fc-modified exenatide-loaded nanoparticles for oral delivery to improve hypoglycemic effects in mice, Sci. Rep. 8 (2018) 726. https://doi.org/10.1038/s41598-018-19170-y

[102] S.G. Chrysant, A new paradigm in the treatment of the cardiovascular disease continuum: focus on prevention, Hippokratia 15 (2011) 7-11.

[103] J. Stewart, G. Manmathan, P. Wilkinson, Primary prevention of cardiovascular disease: A review of contemporary guidance and literature, JRSM Cardiovasc. Dis. 6 (2017). doi: 10.1177/2048004016687211. https://doi.org/10.1177/2048004016687211

[104] W. Jiang, H. Liu, Nanocomposites for bone repair and osteointegration with soft tissues, (Ed) H. Liu, Nanocomposites musculoskelet Tissue Regeneration (2016) 241-257.

[105] J.W. Cassidy, Nanotechnology in the regeneration of complex tissues, Bone Tissue Regen. Insights 5 (2014) 25-35. https://doi.org/10.4137/BTRI.S12331

[106] N. Hao, L. Li, F. Tang, Roles of particle size, shape and surface chemistry of mesoporous silica nanomaterials on biological systems, Int. Mater. Rev.62 (2017) 57-77. https://doi.org/10.1080/09506608.2016.1190118

[107] R. Duivenvoorden, J. Tang, D.P. Cormode, A.J. Mieszawska, D. Izquierdo-Garcia, C. Ozcan, M.J. Otten, N. Zaidi, M.E. Lobatto, S.M. van Rijs, B. Priem, E.L. Kuan, C. Martel, B. Hewing, H. Sager, M. Nahrendorf, G.J. Randolph, E.S. Stroes, V. Fuster, E.A. Fisher, Z.A. Fayad, W.J. Mulder, A statin-loaded reconstituted high-density lipoprotein nanoparticle inhibits atherosclerotic plaque inflammation, Nat Commun. 5 (2014) 3065. doi: 10.1038/ncomms4065. https://doi.org/10.1038/ncomms4065

[108] S. Ahadian, S. Yamada, J. Ramón-Azcón, M. Estili, X. Liang, K. Nakajima, H. Shiku, A. Khademhosseini, T. Matsue, Hybrid hydrogel-aligned carbon nanotube scaffolds to enhance cardiac differentiation of embryoid bodies, Acta. Biomater.31 (2016) 134-143. https://doi.org/10.1016/j.actbio.2015.11.047

[109] S. Ahadian, L.D. Huyer, M. Estili, B. Yee, N. Smith, Z. Xu, Y. Sun, M. Radisic, Moldable elastomeric polyester-carbon nanotube scaffolds for cardiac tissue engineering, Acta. Biomater. 52 (2017) 81-91. https://doi.org/10.1016/j.actbio.2016.12.009

[110] H.A. Afdhal, D. Nunes, Evaluation of liver fibrosis: A concise review, Am. J. Gastroenterol. 99 (2004) 1160-74. https://doi.org/10.1111/j.1572-0241.2004.30110.x

[111] R.G. Wells, Cellular Sources of Extracellular Matrix in Hepatic Fibrosis, Clin. liver dis. 12 (2008) 759-768. https://doi.org/10.1016/j.cld.2008.07.008

[112] E. Arriazu, M. Ruiz de Galarreta, F.J. Cubero, M. Varela-Rey, M.P. Pérez de Obanos, T.M. Leung, A. Lopategi, A. Benedicto, I. Abraham-Enachescu, N. Nieto, Extracellular matrix and liver disease. Antioxid. Redox. Signal 21 (2014) 1078-1097. https://doi.org/10.1089/ars.2013.5697

[113] R. Anty, M. Lemoine, Liver fibrogenesis and metabolic factors, Clin. Res. Hepatol.Gastroenterol.35 (2011) S10-20. https://doi.org/10.1016/S2210-7401(11)70003-1

[114] K. Bettermann, T. Hohensee, J. Haybaeck, Steatosis and steatohepatitis: Complex disorders, Int. J. Mol. Sci.15 (2014) 9924-9944. https://doi.org/10.3390/ijms15069924

[115] P. Sorrentino, L. Terracciano, S. D'Angelo, U. Ferbo, A. Bracigliano, R. Vecchione, Predicting fibrosis worsening in obese patients with NASH through parenchymal fibronectin, HOMA-IR, and hypertension, Am. J. Gastroenterol. 105 (2010) 336-44. https://doi.org/10.1038/ajg.2009.587

[116] R. Bataller, D.A. Brenner, Hepatic stellate cells as a target for the treatment of liver fibrosis, Semin Liver Dis. 21 (2001) 437-51. https://doi.org/10.1055/s-2001-17558

[117] S.P. Surendran, R.G. Thomas, M.J. Moon, Y.Y. Jeong, Nanoparticles for the treatment of liver fibrosis, Int. J. Nanomed. 12 (2017) 6997-7006. https://doi.org/10.2147/IJN.S145951

Biosensors: Materials and Applications   Materials Research Forum LLC
Materials Research Foundations **47** (2019) 241-288   doi: http://dx.doi.org/10.21741/9781644900130-7

[118] V.G. Giby, T.A. Ajith, Role of adipokines and peroxisome proliferator-activated receptors in nonalcoholic fatty liver disease, World J. Hepatol. 6 (2014) 570-579. https://doi.org/10.4254/wjh.v6.i8.570

[119] D.R. Nelson, Z. Tu, C. Soldevila-Pico, M. Abdelmalek, H. Zhu, Y.L. Xu, R. Cabrera, C. Liu, G.L. Davis, Long-term interleukin 10 therapy in chronic hepatitis C patients has a proviral and anti-inflammatory effect, Hepatology 38 (2003) 859-68. https://doi.org/10.1002/hep.1840380412

[120] D. Tripathi, G. Therapondos, H.F. Lui, N. Johnston, D.J. Webb, P.C. Hayes, Chronic administration of losartan, an angiotensin II receptor antagonist, is not effective in reducing portal pressure in patients with preascitic cirrhosis, Am. J. Gastroenterol. 99 (2004) 390-4. https://doi.org/10.1111/j.1572-0241.2004.04051.x

[121] P.J. Pockros, L. Jeffers, N. Afdhal, Z.D. Goodman, D. Nelson, R.G. Gish, K.R. Reddy, R. Reindollar, M. Rodriguez-Torres, S. Sullivan, L.M. Blatt, S. Faris-Young, Final results of a double-blind, placebo-controlled trial of the antifibrotic efficacy of interferon-gamma1b in chronic hepatitis C patients with advanced fibrosis or cirrhosis, Hepatology 45 (2007) 569-78. https://doi.org/10.1002/hep.21561

[122] D.M. Mosser, J.P. Edwards, Exploring the full spectrum of macrophage activation, Nat. Rev. Immunol. 8 (2008) 958-69. https://doi.org/10.1038/nri2448

[123] M. Bartneck, K.T. Warzecha, F. Tacke, Therapeutic targeting of liver inflammation and fibrosis by nanomedicine, Hepatobiliary Surg. Nutr.3 (2014) 364-376.

[124] S.H. Zhang, K.M. Wen, W. Wu, W.Y. Li, J.N. Zhao, Efficacy of HGF carried by ultrasound microbubble-cationic nano-liposomes complex for treating hepatic fibrosis in a bile duct ligation rat model, and its relationship with the diffusion-weighted MRI parameters, Clin. Res. Hepatol. Gastroenterol. 37 (2013) 602-607. https://doi.org/10.1016/j.clinre.2013.05.011

[125] R.G Thomas, M.J. Moon, J.H. Kim, J.H. Lee, Y.Y. Jeong, Effectiveness of losartan-loaded hyaluronic acid (HA) micelles for the reduction of advanced hepatic fibrosis in C3H/HeN mice model, PLOS ONE 10 (2015) doi.org/10.1371/journal.pone.0145512. https://doi.org/10.1371/journal.pone.0145512

[126] American Cancer Society, cancer treatment & survivorship facts & figures 2016-2017, 2016: 1-40.

Biosensors: Materials and Applications   Materials Research Forum LLC
Materials Research Foundations **47** (2019) 241-288   doi: http://dx.doi.org/10.21741/9781644900130-7

[127] WHO cancer fact sheet 2017 france, world health organization.

[128] K.M. Debatin, Apoptosis pathways in cancer and cancer therapy, Cancer Immunol Immunother. 53 (2004) 153-9. https://doi.org/10.1007/s00262-003-0474-8

[129] A. Gross, J.M. McDonnell, S.J. Korsmeyer, BCL-2 family members and the mitochondria in apoptosis, Genes Dev.13 (1999) 1899-1911. https://doi.org/10.1101/gad.13.15.1899

[130] Y.Q. Xiao, K. Malcolm, G.S. Worthen, S. Gardai, W.P. Schiemann, V.A. Fadok, D.L. Bratton, P.M. Henson, Cross-talk between ERK and p38 MAPK mediates selective suppression of pro-inflammatory cytokines by transforming growth factor-β, J. Bio. Chem. 277 (2002) 14884-14893.

[131] K.N. Kropp, S. Maurer, K. Rothfelder, B.J. Schmied,K.L. Clar, M. Schmidt, B. Strunz, H.G. Kopp, A. Steinle, F. Grünebach, S.M. Rittig, H.R. Salih, D. Dörfel, The novel deubiquitinase inhibitor b-AP15 induces direct and NK cell-mediated antitumor effects in human mantle cell lymphoma. Cancer Immunol. Immunother. 67 (2018) 935-947. https://doi.org/10.1007/s00262-018-2151-y

[132] Q. Yue, G. Gao, G. Zou, H. Yu, X. Zheng, Natural products as adjunctive treatment for pancreatic cancer: Recent trends and advancements, Bio. Med. Res. Int. (2017) 2017. doi.org/10.1155/2017/8412508 https://doi.org/10.1155/2017/8412508

[133] A.S. Narang, D.S. Desai, Anticancer drug development unique aspects of pharmaceutical development. Lu Y, Mahato RI (eds.), Pharmaceutical perspectives of cancer therapeutics, Springer-Verlag New York, 694 (2009) 31.

[134] K. Sikora, The impact of future technology on cancer care, Clin. Med.2 (2002) 560-568. https://doi.org/10.7861/clinmedicine.2-6-560

[135] N.R. Jabir, S. Tabrez, G.M. Ashraf, S. Shakil, G.A. Damanhouri, M.A. Kamal, Nanotechnology-based approaches in anticancer research, Int. J. Nanomedicine 7 (2012) 4391-4408.

[136] R Ranganathan, S Madanmohan, A Kesavan, G Baskar, YR Krishnamoorthy, R Santosham, D Ponraju, SK Rayala, G Venkatraman, Nanomedicine: towards development of patient-friendly drug-delivery systems for oncological applications, Int. J. Nanomedicine 7 (2012) 1043-1060.

[137] S. Saxena, P. Caroni, Selective neuronal vulnerability in neurodegenerative diseases: from stressor thresholds to degeneration, Neuron 71 (2011) 38-48. https://doi.org/10.1016/j.neuron.2011.06.031

[138] A.D. Gitler, P. Dhillon, J. Shorter, Neurodegenerative disease: models, mechanisms, and a new hope, Dis. Model Mech. 10 (2017) 499-502. https://doi.org/10.1242/dmm.030205

[139] A. Xie, J. Gao, L. Xu, D. Meng, Shared mechanisms of neurodegeneration in Alzheimer's disease and Parkinson's disease, Biomed. Res. Int. 2014 (2014). doi: 10.1155/2014/648740. https://doi.org/10.1155/2014/648740

[140] G. Soursou, A. Alexiou, G.M. Ashraf, A.A. Siyal, G. Mushtaq, M.A. Kamal, Applications of nanotechnology in diagnostics and therapeutics of Alzheimer's and Parkinson's disease, Curr. Drug Metab. 16 (2015) 705-712. https://doi.org/10.2174/1389200216081511071250 49

[141] J. Wen, K. Yang, Y. Xu, H. Li, F. Liu, S. Sun, Construction of a triple-stimuli responsive system based on cerium oxide coated mesoporous silica nanoparticles Sci. Rep. 6 (2016) doi.org/10.1038/srep38931 https://doi.org/10.1038/srep38931

[142] C. Spuch, O. Saida, C. Navarro, Advances in the treatment of neurodegenerative disorders employing nanoparticles, Recent Pat. Drug Deliv. Formul. 6 (2012) 2-18. https://doi.org/10.2174/187221112799219125

[143] K.M. Jaruszewski, R.S. Omtri, K.K. Kandimalla, Role of nanotechnology in the diagnosis and treatment of Alzheimer's disease, Curr. Adv. Med. Appl. Nanotechnol. 18 (2012) 107-124. https://doi.org/10.2174/97816080513111120 1010107

[144] G. Karthivashan, P. Ganesan, S. Park, J. Kim, D. Choi, Therapeutics strategies and nano-drug delivery applications in management of ageing Alzheimer's disease, Drug Delivery 25 (2018) 307-320. https://doi.org/10.1080/10717544.2018.1428243

[145] D. Hadavi, A.A. Poot, Biomaterials for the treatment of Alzheimer's disease, Front Bioeng. Biotechnol. 4 (2016). doi: 10.3389/fbioe.2016.00049. https://doi.org/10.3389/fbioe.2016.00049

[146] R.L. Jayaraj, V. Chandramohan, E. Namasivayam, Nanomedicine for Parkinson disease: current status and future perspective, Int. J. Pharm. Bio. Sci.4 (2013) 692-704.

[147] D. Martinez-Fong, M.J. Bannon, L. Trudeau, J.A. Gonzalez-Barrios, M.L. Arango Rodriguez, N.G. Hernandez-Chan, D. Reyes-Corona, J. Armendariz-Borunda, I. Navarro-Quiroga, NTS-Polyplex: a potential nanocarrier for neurotrophic therapy

of Parkinson's disease, Nanomedicine: NBM 8 (2012) 1052-1069.
https://doi.org/10.1016/j.nano.2012.02.009

[148] A.R. Esteves, D.M. Arduino, D.F.F Silva, C.R. Oliveira, S.M. Cardoso,
Mitochondrial dysfunction: the road to alpha-synuclein oligomerization in PD,
Parkinsons Disease (2011) doi.org/10.4061/2011/693761.
https://doi.org/10.4061/2011/693761

[149] X. Yi, Y. Wu, G. Tan, P. Yu, L. Zhou, Z. Zhou, J. Chen, Z. Wang, J. Pang, C.
Ning, Palladium nanoparticles entrapped in a self-supporting nanoporous gold
wire as sensitive dopamine biosensor, Sci. Rep.7 (2017).
https://doi.org/10.1038/s41598-017-07909-y

[150] S. Mohammadi, M. Nikkhah, S. Hosseinkhani, Investigation of the effects of
carbon-based nanomaterials on A53T alpha-synucleinaggregation using a whole-
cell recombinant biosensor, Int. J. Nanomedicine 12 (2017) 8831-8840.
https://doi.org/10.2147/IJN.S144764

[151] M.M. Migliore, R. Ortiz, S. Dye, R.B. Campbell, M.M. Amiji, B.L. Waszczak,
Neurotrophic and neuroprotective efficacy of intranasal GDNF in a rat model of
Parkinson's disease, Neuroscience 274 (2014) 11-23.
https://doi.org/10.1016/j.neuroscience.2014.05.019

[152] E. Barcia, L. Boeva, L. Garcia-Garcia, K. Slowing, A. Fernandez-Carballido, Y.
Casanova, S. Negro, Nanotechnology-based drug delivery of ropinirole for
Parkinson's disease, Drug Delivery 24 (2017) 1112-1123.
https://doi.org/10.1080/10717544.2017.1359862

[153] H. Zazo, C.I. Colino, J.M. Lanao, Current applications of nanoparticles in
infectious diseases, J. Control Release 224 (2016) 86-102.
https://doi.org/10.1016/j.jconrel.2016.01.008

[154] P. Sendi, R.A. Proctor, Staphylococcus aureus as an intracellular pathogen: the
role of small colony variants, Trends Microbiol. 17 (2009) 54-58.
https://doi.org/10.1016/j.tim.2008.11.004

[155] G.M. Soliman, Nanoparticles as safe and effective delivery systems of antifungal
agents: Achievements and challenges, Int. J. Pharm. 523 (2017) 15-32.
https://doi.org/10.1016/j.ijpharm.2017.03.019

[156] L. Singh, H.G. Kruger, G.E.M. Maguire, T. Govender, R. Parboosing, The role of
nanotechnology in the treatment of viral infections, Ther. Adv. Infectious Dis.4
(2017) 105-131. https://doi.org/10.1177/2049936117713593

Materials Research Forum LLC
doi: http://dx.doi.org/10.21741/9781644900130-7

[157] T. Yokota, Kinds of antimicrobial agents and their mode of actions, Nihon Rinsho 55 (1997) 1155-60.

[158] E. De Clercq, Antiviral drugs in current clinical use, J. Clin.Virol.30 (2004)115-133. https://doi.org/10.1016/j.jcv.2004.02.009

[159] M. Gajendiran, P. Balashanmugam, P.T. Kalaichelvan, S. Balasubramanian, Multi-drug delivery of tuberculosis drugs by π-back bonded gold nanoparticles with multiblock copolyesters.Mater. Res. Express 3 (2016). DOI:10.1088/2053-1591/3/6/065401. https://doi.org/10.1088/2053-1591/3/6/065401

[160] J. Costa-Gouveia, E. Pancani, S. Jouny, A. Machelart, V. Delorme, G. Salzano, R. Iantomasi, C. Piveteau, C.J. Queval, O.R. Song, M. Flipo, B. Deprez, J.P. Saint-André, J. Hureaux, L. Majlessi, N. Willand, A. Baulard, P. Brodin, R. Gref, Combination therapy for tuberculosis treatment: pulmonary administration of ethionamide and booster co-loaded nanoparticles. Sci. Rep. 7 (2017) 5390. https://doi.org/10.1038/s41598-017-05453-3

[161] N. Xu, J. Gu, Y. Zhu, H. Wen, Q. Ren, J. Chen, Efficacy of intravenous amphotericin B-polybutylcyanoacrylate nanoparticles against cryptococcal meningitis in mice, Int. J. Nanomedicine 6 (2011) 905-913. https://doi.org/10.2147/IJN.S17503

[162] U. Roy, V. Drozd, A. Durygin, J. Rodriguez, P. Barber, V. Atluri, X. Liu, T.G. Voss, S. Saxena, M. Nair, Characterization of nanodiamond-based anti-HIV drug delivery to the brain, Sci. Rep. 8 (2018). https://doi.org/10.1038/s41598-017-16703-9

[163] S. Ojha, B. Kumar. A review on nanotechnology based innovations in diagnosis and treatment of multiple sclerosis, J. Cellular Immunother. 4 (2018) 56-64. https://doi.org/10.1016/j.jocit.2017.12.001

[164] R.J. Mannix, S. Kumar, F. Cassiola, M. Montoya-Zavala, E. Feinstein , M. Prentiss, D.E. Ingber, Nanomagnetic actuation of receptor-mediated signal transduction, Nat. Nanotechnol.3 (2008) 36-40. https://doi.org/10.1038/nnano.2007.418

[165] D. Yang, W.P. Wong, Small but Mighty: Nanoparticles Probe Cellular Signaling PathwayS, Dev.Cell.37 (2016) 397-398. https://doi.org/10.1016/j.devcel.2016.05.021

[166] D. Seo, K.M. Southard, J.W. Kim, H.J. Lee, J. Farlow, J.U. Lee, D.B. Litt, T. Haas, A.P. Alivisatos, J. Cheon, Z.J. Gartner, Y.W. Jun, A mechanogenetic toolkit

Biosensors: Materials and Applications   Materials Research Forum LLC
Materials Research Foundations **47** (2019) 241-288   doi: http://dx.doi.org/10.21741/9781644900130-7

for interrogating cell signaling in space and time, Cell165 (2016) 1507-1518.
https://doi.org/10.1016/j.cell.2016.04.045

[167] S.H. Hassanpour, M. Dehghani, Review of cancer from perspective of molecular,
J. Cancer Res. Practice 4 (2017) 127-129.
https://doi.org/10.1016/j.jcrpr.2017.07.001

[168] A. Pavlopoulou, Da Spandidos, I. Michalopoulos, Human cancer databases, Oncol.
Rep.33 (2015) 3-18. https://doi.org/10.3892/or.2014.3579

[169] B. Li, Q. Li, J. Mo, H. Dai, Drug-loaded polymeric nanoparticles for cancer stem
cell targeting, Front Pharmacol. 8 (2017). doi: 10.3389/fphar.2017.00051.
https://doi.org/10.3389/fphar.2017.00051

[170] B. Viswanath, S. Kim, Influence of nanotoxicity on human health and
environment: The alternative strategies. de Voogt P. (Ed.) Reviews of
environmental contamination and toxicology, 242 (2016) 61-104.

[171] Y.W. Huang, M. Cambre, H.J. Lee, The Toxicity of nanoparticles depends on
multiple molecular and physicochemical mechanisms, Int. J. Mol. Sci.18 (2017)
2702. https://doi.org/10.3390/ijms18122702

[172] P.P. Fu, Q. Xia, H.M. Hwang, P.C. Ray, H. Yu, Mechanisms of nanotoxicity:
Generation of reactive oxygen species, J. Food Drug Anal.22 (2014) 64-75.
https://doi.org/10.1016/j.jfda.2014.01.005

[173] C.A. Jimenez-Cruz, S. Kang, R. Zhou, Large scale molecular simulations of
nanotoxicity. WIREs Syst. Biol. Med. 6 (2014) 329-343.
https://doi.org/10.1002/wsbm.1271

[174] A. Elsaesser, C.V. Howard, Toxicology of nanoparticles, Adv. Drug Deliv.Rev.64
(2012) 129-37. https://doi.org/10.1016/j.addr.2011.09.001

[175] N. Yanamala, V.E. Kagan, A.A. Shvedova, Molecular modeling in structural
nano-toxicology: Interactions of nano-particles with nano-machinery of cells, Adv.
Drug Deliv. Rev. 65 (2013) 2070-2077. https://doi.org/10.1016/j.addr.2013.05.005

[176] E. Bressan, V. Vindigni, L. Ferroni, W. Cairns, C. Gardin, C. Rigo, B. Zavan, M
Stocchero, Silver Nanoparticles and Mitochondrial Interaction, Int. J. Dentistry.
(2013) 2013 1-8. https://doi.org/10.1155/2013/312747

[177] D. McShan, P.C. Ray, H. Yu, Molecular toxicity mechanism of nanosilver, J. Food
Drug Anal. 22 (2014) 116-127. https://doi.org/10.1016/j.jfda.2014.01.010

[178] J.P. Ryman-Rasmussen, E.W. Tewksbury, O.R. Moss, M.F. Cesta, B.A. Wong , J.C. Bonner, Inhaled multiwalled carbon nanotubes potentiate airway fibrosis in murine allergic asthma, Am. J. Respir. Cell. Mol. Boil., 40 (2009) doi.org/10.1165/rcmb.2008-0276OC. https://doi.org/10.1165/rcmb.2008-0276OC

[179] J.M. Caster, A.N. Patel, T. Zhang, A. Wang, Investigational nanomedicines in 2016: a review of nanotherapeutics currently undergoing clinical trials, Wiley Interdiscip. Rev. Nanomed.Nanobiotechnol. 9 (2017). doi: 10.1002/wnan.

[180] C.L. Ventola, Progress in Nanomedicine: Approved and investigational nanodrugs, Pharm. Ther. 42 (2017) 742-755.

[181] A.C. Anselmo, S. Mitragotri, Nanoparticles in the clinic, Bioengg. Translational Med.1 (2016) 10-29. https://doi.org/10.1002/btm2.10003

[182] O. Veiseh, B.C. Tang, K.A. Whitehead, D.G. Anderson, R. Langer, Managing diabetes with nanomedicine: challenges and opportunities, Nat. Rev. Drug Discov. 14 (2015) 45-57. https://doi.org/10.1038/nrd4477

[183] F. Khaja, D. Jayawardena, A. Kuzmis, H. Önyüksel, Targeted sterically stabilized phospholipid siRNA nanomedicine for hepatic and renal fibrosis, Nanomaterials 6 (2016). doi: 10.3390/nano6010008. https://doi.org/10.3390/nano6010008

[184] X. Li, T. Kuznetsova, N. Cauwenberghs, M. Wheeler, H. Maecker, J.C. Wu, F. Haddad, H. Dai, Autoantibody profiling on a plasmonic nano-gold chip for the early detection of hypertensive heart disease, Proc. Natl. Acad. Sci. USA 114 (2017) 7089-7094. https://doi.org/10.1073/pnas.1621457114

[185] J.I. Hare, T. Lammers, M.B. Ashford, S. Puri, G. Storm, S.T. Barry, Challenges and strategies in anti-cancer nanomedicine development: An industry perspective, Adv. Drug Deliv. Rev. 108 (2017) 25-38. https://doi.org/10.1016/j.addr.2016.04.025

[186] S. Donnellan, V. Stone, H. Johnston, M. Giardiello, A. Owen, S. Rannard, G. Aljayyoussi, B. Swift, L. Tran, C. Watkins, K. Stevenson, Intracellular delivery of nano-formulated antituberculosis drugs enhances bactericidal activity, J. Interdiscip.Nanomed.2 (2017) 146-156. https://doi.org/10.1002/jin2.27

[187] www.cancer.gov/sites/nano/research/plan.

[188] T. Date, V. Nimbalkar, J. Kamat, A. Mittal, R.I. Mahato, D. Chitkara, Lipid-polymer hybrid nanocarriers for delivering cancer therapeutics, J. Control Release 271 (2018) 60-73. https://doi.org/10.1016/j.jconrel.2017.12.016

[189] J. Wang, S. Yang, C. Li, Y. Miao, L. Zhu, C. Mao, M. Yang, Nucleation and assembly of silica into protein-based nanocomposites as effective anticancer drug carriers using self-assembled silk protein nanostructures as biotemplates, ACS Appl. Mater. Interfaces 9(2017) 22259-22267. https://doi.org/10.1021/acsami.7b05664

[190] Z. Li, E. Ye, David, R Lakshminarayanan, X.J. Loh, Recent advances of using hybrid nanocarriers in remotely controlled therapeutic delivery, Small 12 (2016) 4782-4806. https://doi.org/10.1002/smll.201601129

[191] A. Khalid, A.N. Mitropoulos, B. Marelli, S. Tomljenovic-Hanic, F.G. Omenetto, Doxorubicin loaded nanodiamond-silk spheres for fluorescence tracking and controlled drug release, Biomed. Opt. Express.7 (2016) 132-147. https://doi.org/10.1364/BOE.7.000132

[192] M.S. Purdey, P.K. Capon, B.J. Pullen, P. Reineck, N. Schwarz, P.J. Psaltis, S.J. Nicholls, B.C. Gibson, A.D. Abell, An organic fluorophore-nanodiamond hybrid sensor for photostable imaging and orthogonal, on-demand biosensing, Sci. Rep.7 (2017) doi.org/10.1038/s41598-017-15772-0 https://doi.org/10.1038/s41598-017-15772-0

[193] L. Cao, Y. Liang, F. Zhao, X. Zhao, Z. Chen, Chelerythrine and Fe3O4 loaded multi-walled carbon nanotubes for targeted cancer therapy, J. Biomed.Nanotech.12 (2016) 1312-1322. https://doi.org/10.1166/jbn.2016.2280

[194] Z. Heger, H. Polanska, S. Krizkova, J. Balvan, M. Raudenska, S. Dostalova, A. Moulick, M. Masarik, V. Adam, Co-delivery of VP-16 and Bcl-2-targeted antisense on PEG-grafted oMWCNTs for synergistic in vitro anti-cancer effects in non-small and small cell lung cancer, Colloids Surf. B Biointerfaces. 150 (2017) 131-140. https://doi.org/10.1016/j.colsurfb.2016.11.023

Biosensors: Materials and Applications
Materials Research Foundations 47 (2019) 289-316

Materials Research Forum LLC
doi: http://dx.doi.org/10.21741/9781644900130-8

Chapter 8

Enzymatic Biosensor for *in vivo* Applications

Pedro Salazar*, Miriam Martín, José Luis González–Mora

Laboratory of Sensors, Biosensors and Advanced Materials (Neurochemistry and Neuroimaging Group), Faculty of Medical Sciences, University of La Laguna, Campus de Ofra s/n, Tenerife, 38071 La Laguna, Spain

Abstract

A large number of analytical devices have recently been developed for detecting biomarkers without the need for laboratory handling procedures. At present, there is a growing demand for developing minimal–or non–invasive implantable devices able to measure in real–time with high spatial resolution. Among the different approaches, implantable biosensors are the most common approach thanks to their high sensitivity and selectivity, the simplicity of starting materials and their cost-effectiveness because they can be easily miniaturized for the manufacture of implantable devices. This chapter reviews the main issues (sensitivity, selectivity, tolerance to fluctuations in oxygen levels, biocompatibility, long-term functional stability, etc.) involved in the design phase and the application of biosensors for *in vivo* monitoring of key molecules in health applications.

Keywords

Biosensor, *in vivo*, Enzyme, Amperometric Detection, Biocompatibility, Point–of–Care, Glucose, Lactate, Glutamate, Diabetes

Contents

1.   Introduction

There is a growing demand for new analytical tools in health, clinical monitoring, pharmaceutical, agro-alimentary and environmental applications. Among these applications, medical diagnostics has driven scientists to develop new sophisticated methods and technologies able to measure directly in biological samples (blood, serum, saliva, cerebrospinal fluid, etc.) and tissues (brain, skin, oral cavity, retina, subcutaneous space, etc.). In this regard, analytical devices that can simultaneously measure multiple biomarkers in physiological fluids or tissues, without the need of laboratory handling procedures, would be an important improvement for clinical diagnosis and health care applications in the near future. In addition, there is an increasing demand for developing minimal– or non–invasive analytical methods with high sensitivity and selectivity, able to measure in real–time with high spatial resolution. Therefore, scientists are working on developing new technologies, materials and methods to develop Lab–on–a–Chips (LOC) devices that allow the simultaneous and easy determination of multiple biomarkers [1–4]. Nevertheless, current devices adapted for multiplex detection in *in vivo* applications are either bulky or not biocompatible, making them unsuitable for implantation [4], which will need a lot of intensive work from the scientific community in the near future. Fortunately, the ultimate goal is providing point–of–care (POC) testing devices for common metabolites (such as glucose, lactate, and cholesterol) to patients today. In addition, advances in nanotechnology, miniaturization and microfluidics, along with developments in cloud-connected POC diagnostic technologies, are pushing the frontiers of POC towards wearable [5], user–friendly and low–cost devices. Therefore, the future combination of various biosensing platforms within smartphone–integrated electronic readers will provide accurate on–site and on–time diagnostics [6–8]. The main advantage of the present approach is that patients can continuously measure and control their own metabolite levels at home, using an easy and painless method. In this context, biosensors are ideally small and portable devices, allowing the selective quantification of chemical and biochemical analytes in a complex matrix without the need for separation techniques before analyzing. Today, they are replacing other more sophisticated techniques such as spectroscopic and chromatographic methods or coupling with others such as micro-

Biosensors: Materials and Applications   Materials Research Forum LLC
Materials Research Foundations **47** (2019) 289-316   doi: http://dx.doi.org/10.21741/9781644900130-8

dialysis to develop minimally invasive devices for real-time determination of different metabolites. In this regard, biosensors are a good example of an interdisciplinary practice where many areas of the science (analytical chemistry, surface and material sciences, biology, biochemistry, electrochemistry, device fabrications, etc.) converge. Unlike "labels" or "imaging", *in vivo* biosensors are designed for the continuous monitoring of the target analyte in real biological systems with sufficient sensitivity, selectivity, biocompatibility and long-term stability.

Without any doubt, glucose biosensors have received more attention than other devices due to the fact that *"Diabetes"* is a serious global problem resulting from westernized lifestyles (with high–fat diets and decreased exercise). The diagnosis and management of *Diabetes* requires a strict monitoring of blood glucose levels. Thus, over the last few decades, the main objective of many types of research has been to develop a method that provides a reliable and strict glycemic control. Recent studies revealed the increasing prevalence of *Diabetes* worldwide, where approximately 360 million people had *Diabetes* in 2011, of whom more than 95% had *type–2–Diabetes* [9]. Furthermore, this number is expected to increase to 552 million by 2030 and it is thought that about half of these patients will be unaware of their condition. Additionally, it is estimated that another 300 million people had symptoms indicating future risk of developing *type–2–Diabetes*. Thanks to this huge effort, a large number of commercial devices, industrial developments and patents have been developed with this purpose [10–16]. A good example of the great interest in developing analytical devices for the management of *Diabetes* is the commercial glucometer named *Glucoday* (marketed in ten European countries) and developed jointly by the Tor Vergata University and A. Menarini Diagnostics. Under the present approach, they coupled a microdialysis probe with a flow cell containing an electrochemical glucose biosensor, allowing continuous glucose monitoring. In addition, this collaboration has continued with the development of a new generation microdialysis–based device, the GlucoMen®Day, a smaller and more compact instrument, where a new screen–printed glucose biosensor continuously measures the subcutaneous glucose concentration for a week [17–20]. Even, *"artificial pancreas"* have been reported in the literature and commercial devices are available for patients [21,22], combining an intelligent insulin pump that is able to release insulin in response to changing glucose levels in a similar way to a human pancreas.

2.   Biosensors: definition and classification

The first biosensor was first described by Clark and Lyons in a Conference at a Symposium in the New York Academy of Sciences in 1962 when the term enzyme–electrode was adopted [23]. Such authors coupled an enzyme (glucose oxidase, Gox) to an amperometric oxygen electrode and, found that the increase of the glucose concentration (in the solution test) lowered the O_2 concentration in the proximity of the electrode surface. Subsequently, this design was implemented by Updike and Hicks (1967) to decrease the matrix effect due to O_2 fluctuation in the working solution [24]. Later on, Guilbault and Lubrano (1973) developed an enzymatic electrode to determine blood glucose by amperometric detection of the H_2O_2 generated during the enzymatic reaction [25]. Finally, the term biosensor was adopted in 1975 when Divies developed the first device to determine ethanol using microorganisms as the recognition element [26]. The term *"bio"* was adopted to emphasise the biological nature of the recognition element. At present, and an according to the International Union of Pure and Applied Chemistry (IUPAC), a biosensor is defined as a self–contained integrated device, which is capable of providing specific quantitative or semi-quantitative analytical information using a biological recognition element, which is retained in direct spatial contact with a transduction element [27]. Based on the previous definition some important criteria can be used to classify different kinds of biosensors as follows: (1) the nature of the biological recognition element (*biosensor, immunosensor, aptasensor, genosensor*) or (2) the method used to transform this biological reaction into a detectable signal on the transducer surface (*electrochemical, optical, piezoelectric, acoustic, thermal*). In general, a biosensor is composed of five elements: the bioreceptor (1) that binds or catalyzes a specific reaction with the target analyte, an interface (2) where the bioreceptor is retained and the specific biological event takes place and gives rise to a signal picked up by (3) the transducer element that converts the specific biological reaction into an electrical signal that can be amplified and finally, the signal processor (4) and an appropriate display (5) allows the filtering and/or processing of the electrical signal and visualization of the result (e.g., analyte concentration) respectively.

One of the most critical steps during biosensor assemble is the immobilization of the bioreceptor on the transducer surface. Among the most common different immobilization methods are adsorption, physical entrapment, affinity interactions, cross-linking and covalent immobilization [28,29]. The choice of the most appropriate immobilization technique depends on various factors such as the nature of the biomolecule and transducer, the detection method, etc. Moreover, the analytical performance of the biosensors (sensitivity, linear range, selectivity and biocompatibility) depends on the immobilization technique used. Therefore, much effort has been made to develop successful immobilization strategies in order to assure greater sensitivity and stability of biosensors in biomedical applications, including the use of nanomaterials and biocompatible materials to ensure the bioreceptor structure is not altered.

Biosensors: Materials and Applications
Materials Research Foundations **47** (2019) 289-316

Materials Research Forum LLC
doi: http://dx.doi.org/10.21741/9781644900130-8

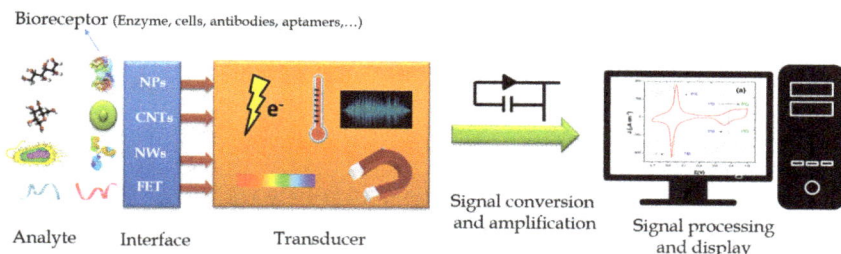

Figure 1. Schematic representation of different biosensor components.

Among the different approaches, the most common devices used in *in vivo* applications are electrochemical enzymatic biosensors (named biosensors). The main advantages in the use of such an approach are their high sensitivity and selectivity, the simplicity of starting materials and their cost-effectiveness because they can be easily miniaturized for the manufacture of implantable biosensors. In addition, the possibility to develop a user–friendly and ready–to–use devices with real-time output results is of great interest in POC applications, where patients can easily measure the target analyte in real–time.

3.   Biosensors: Michaelis–Menten model in amperometric biosensors

Of the different biosensor configurations, enzymatic amperometric biosensors are preferred for *in vivo* applications due to their easy implementation, the simplicity of material starting, the commercial ability of a great number of enzymes and their autocatalytic and reusable properties. The design of electrochemical biosensors is predicated firstly on the target analyte and the choice of appropriate biological element. The vast majority of enzyme-based amperometric biosensors exploit the biocatalytic oxidation of analyte, using oxidase enzymes containing the prosthetic group, flavin adenine dinucleotide (FAD); see **Eqs. 1**and **2**. Despite this common foundation, a wide range of oxidase–based devices have been designed which make use of different signal transduction mechanisms, materials and electrochemical methods.

For an oxidase enzyme (E) with FAD as a prosthetic group, the catalytic conversion of substrate (S) and co-substrate (O_2) to products (P) and H_2O_2 can be written as **Eqs. 1** and **2**:

$$S + E/FAD \rightarrow P + E/FADH_2 \qquad\qquad (1)$$

$$E/FADH_2 + O_2 \rightarrow E/FAD + H_2O_2 \qquad\qquad (2)$$

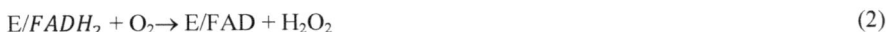

Referring to **Eqs. 1–2**, where S and O_2 are considered the substrate and co-substrate, respectively, two–substrate forms of the Michaelis–Menten equation for the overall rate of reaction contains two enzyme affinity terms $K_M(S)$ and $K_M(O_2)$. However, if the concentration of the co-substrate is large and constant, this reduces to the well–known one–substrate Michaelis–Menten equation. The biosensor responses usually need complex mathematical simulations due to the fact that enzymes are incorporatedinto a thick membrane layer, basically to include the diffusion phenomena of the analyte in thick films where the enzyme is retained. However, in thin layers of ca. 10–30 nm or when the enzyme is directly immobilized on the electrode surface via covalent bonding, the basic Michaelis–Menten model provides more readily accessible insights into factors affecting the responsiveness of biosensors. Such a simplification is justified because the substrate diffusion is not limiting for this thin film layer [30]. According to the Michaelis–Menten model, the enzymatic reaction can be expressed as **Eqs. 5–6**:

$$E + S \underset{k_{-1}}{\overset{k_1}{\longleftrightarrow}} ES \overset{k2}{\rightarrow} E + P \qquad\qquad (5)$$

$$E + S \overset{K_M}{\longleftrightarrow} E + P \qquad\qquad (6)$$

$$K_M = \frac{k_{-1}k_2}{k_1} \qquad\qquad (7)$$

where the true Michaelis constant for one–substrate, K_M, is defined (**Eq. 7**) in terms of the rateconstants for the generalized reaction describing the overall conversion of substrate to product, catalyzed by the enzyme. In addition, k_1 is the substrate binding to the enzyme, k_{-1} is its reversal and k_2 is the catalytic rate [31]. Therefore, the overall rate of the enzyme reaction according to the more commonly used one–substrate Michaelis–Menten model is described as follows in **Eq. 8**:

$$V_S = \frac{V_{max}}{1 + \frac{K_M(S)}{[S]}} \qquad\qquad (8)$$

Where V_{max} is the V_S value at enzyme saturation. Finally, converting the rate (**Eq. 8**) to current (**Eq.9**) we obtain the electrochemical version of the Michaelis–Menten equation [32]:

$$J_S = \frac{J_{max}}{1 + \frac{K_M(S)}{[S]}} \tag{9}$$

where J_S is the current density for S and represents the overall rate of the enzyme reaction, and J_{max} is the J_S value at enzyme saturation. K_M can be phenomenologically defined as the concentration of substrate that gives half the J_{max} response (see **Figure 2**). Thus changes in K_M are sensitive to the binding constant (k_1) and often reflect changes in the access/binding of substrate to the enzyme, furthermore, variations in K_M have been interpreted in terms of diffusion barriers to substrate/enzyme binding [32]. The Michaelis constant for S is also useful for defining the slope in the linear response region (LRS, linear region slope). The slope of the tangent of a hyperbola, as the independent variable approaches zero, is given by J_{max}/K_M; see **Figure 2** and **Eq. 10** [31]:

$$\underset{[S]\to 0}{Lt} J_s = \frac{J_{max}}{1 + K_M/[S]} = \frac{J_{max}[S]}{[S] + K_M} \approx \frac{J_{max}}{K_M}[S] \approx LRS \cdot [S] \tag{10}$$

Of course, the response of a biosensor following Michaelis–Menten kinetics is never truly linear. Nevertheless, for low concentrations of substrate, the analytical response is 'practically' linear (see **Figure 2, *inset***), and all that remains is to define the substrate concentration range for which the linearity of the response is analytically acceptable. In practice, the range of the linear response to S is usually considered to extend up to $\sim\frac{1}{2}K_M$ (see **Figure 2, *inset***).

4.   Biosensors in *in vivo* applications: important issues

The successful application of implantable biosensors in complex media, such as biological tissues and fluids, demands that strict criteria be addressed during the biosensor assembly [33]. The main factors that affect the performance of implantable amperometric biosensors including target analyte sensitivity, tolerance to fluctuations in oxygen concentration, selectivity against common endogenous reducing agents, biocompatibility, minimization of tissue disruption and foreign body response and long-term stability will be discussed below.

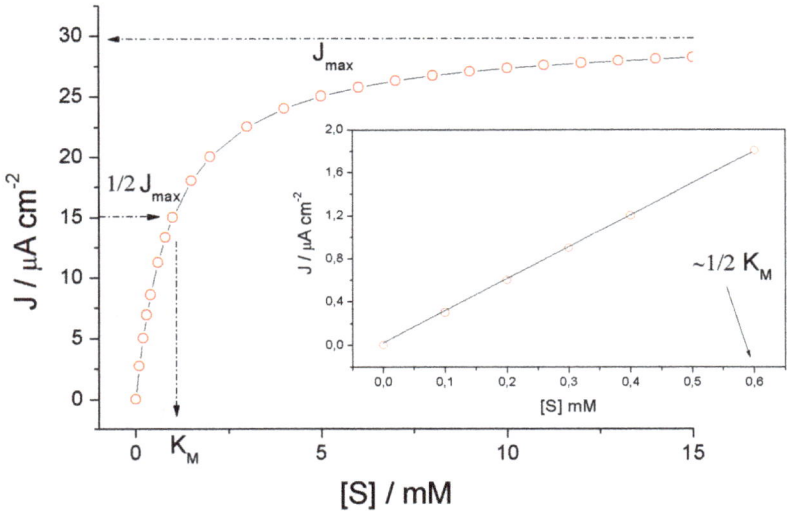

Figure 2. Generalized example of a biosensor calibration analysis for enzyme substrate, S. The Michaelis-Menten curve is generated by fitting the calibration data to Eqn. 4. The apparent Michaelis constant, K_M, is the concentration of S, [S], that yields a biosensor response equal to half the Jmax value. Inset: Calibration data for [S] values up to approx. half the K_M value fit a linear regression analysis well to provide the main analytical parameter used to compare the sensitivity of different biosensor designs to substrate, viz. the linear region slope or LRS.

5. Sensitivity, limit of detection, limit of quantification and linear range

One of the most important issues for biosensing is to develop analytical devices with enough sensitivity and selectivity to detect the target analyte with an acceptable uncertainty grade, especially when the analysis at a very low concentration in presence of interference. The term sensitivity refers to the magnitude of its response, while selectivity addresses its ability to respond only to changes of the target analyte in the presence of other compounds. Both sensitivity and selectivity are of vital importance for implantable biosensors. In addition, both are interrelated based on the fact that improving sensitivity of a biosensor usually reduces the selectivity and vice versa. Therefore, the use of an inner or outer membrane, for eliminating interference, decreases the sensitivity. Moreover, the addition of perm-selective membranes such as Nafion®, cellulose acetate, polyester sulfonic, etc. on the top of the working electrode comes at the expense of the response time because of diffusional effects. Such an approach may be used, as commented below, to mitigate the oxygen deficit as the flux of analyte across this membrane is reduced by a factor of nearly 100, whereas the flux of oxygen in not highly affected [34].

The main factors affecting biosensor sensitivity are size and geometry; the nature of the working electrode and its surface activity, the specific activity of the enzyme; the enzymatic loading, the success in enzyme immobilization preserving as much activity as possible; the presence of an inner polymeric layer to immobilize the enzyme or to eliminate interference; and the presence of outer layer to (1) alleviate the oxygen dependence, (2) interference artefacts, (3) increase the operative linear range or prevent (4) biofouling and (5) biocompatibility of the biosensor surface [35].

The *first* generation amperometric biosensors based on the detection of enzymatically generated H_2O_2, are preferred for implantable devices due to their easy and practical implementation [32]. *First* amperometric biosensors are based on the measurement of the current resulting from the electrochemical oxidation or reduction of electroactive species generated or consumed during the enzymatic reaction (see **Figure 3**). Amperometry is usually performed by maintaining a constant potential at a Pt, Au or C based working electrode with respect to a reference electrode, which may also serve as the auxiliary electrode if currents are low (<1 pA), in other cases an additional auxiliary electrode is needed to avoid the polarization of the reference electrode. Therefore, the resulting steady-state current is directly correlated to the bulk concentration of the electroactive species in the proximity of the biosensor surface and hence with the bulk analyte concentration when reaction rates are first-order–dependent [34].

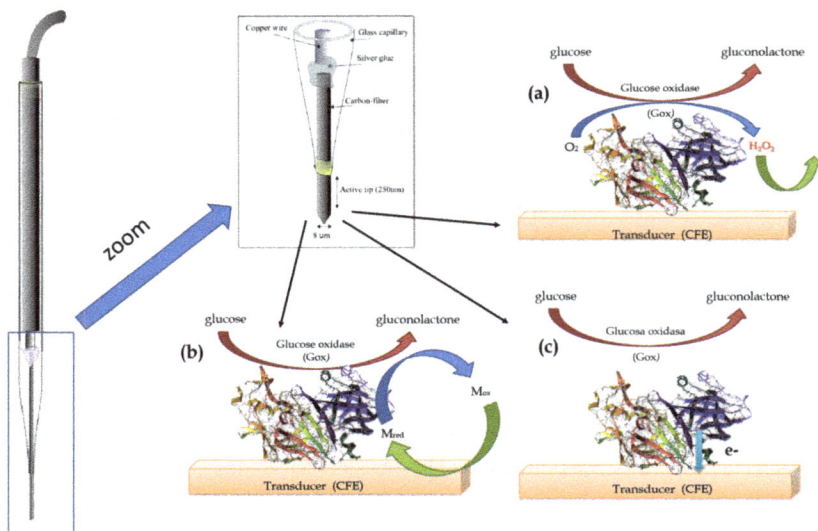

*Figure 3. Detection schemes for different glucose microbiosensor configurations in which glucose is converted to gluconolactone, catalyzed by the enzyme glucose oxidase. Secondary to this reaction, one of the following may occur: the H_2O_2 generated by reaction of the reduced enzyme with ambient oxygen, may be detected directly using first-generation approaches (**a**); an artificial redox mediator may be used in the second-generation biosensors for monitoring the enzymatic reaction (**b**); or electrons may be directly transferred between the glucose oxidase to the electrode surface in third-generation biosensors (**c**).*

Platinum (Pt), thanks to its outstanding electrocatalytic properties and commercial availability in different configurations, dimensions and alloy compositions, is the most common material for electrochemical detection of H_2O_2 [32,36–40]. H_2O_2 oxidation is catalyzed by the presence of platinum oxides (Pt–$(OH)_2$) on the surface of the electrode according to **Eqs. 3** and **4** [41] at ca. 700 mV:

$$H_2O_2 + Pt(OH)_2 \rightarrow Pt + 2\,H_2O + O_2 \tag{3}$$

$$Pt + 2\,H_2O \rightarrow Pt(OH)_2 + 2\,H^+ + 2\,e^- \tag{4}$$

Biosensors: Materials and Applications   Materials Research Forum LLC
Materials Research Foundations **47** (2019) 289-316   doi: http://dx.doi.org/10.21741/9781644900130-8

Microsensors based on carbon fiber electrodes (CFEs) with very low dimensions ($\phi \sim 10$ μm) have been used in neuroscience applications to detect several electroactive neurotransmitters and their metabolites such as dopamine (DA), norepinephrine (NA), adrenaline (A) and serotonin (5–HT). However, CFEs and other carbonaceous materials are not optimal for H_2O_2 detection in biosensor applications. In order to combine their very low dimensions, with the electrocatalytic properties of platinum for H_2O_2 detection, CFEs have been properly modified with Pt nanostructures to increase their sensitivity for H_2O_2 in biosensor devices [42]. Thus, platinum nanoparticles with different growth densities, black platinum, *etc.* have been electrodeposited onto the CFE surface from acid solutions of H_2PtCl_6 using different strategies [43,44]. Thanks to this hybridization, Pt-modified electrodes show higher nanoscale roughness, enhancing their surface area and their electrocatalytic activity.

More recently,other noble metals (Ag, Au, Pd), metals (Ni, Cu, Mn) and metal oxides (NiO, CuO, ZnO), nanoparticles (NPs) and nanowires (NWs) as well as carbon nanotubes (CNTs), graphene (Gr), composites and hybrid materials are being used to construct more efficient electrochemical and optical biosensors with improved analytical performance. The use of such nanostructured materials allows the detection H_2O_2 at lower over potentials where interference is alleviated.

Among different electrocatalytic materials, Prussian Blue (PB) and its analogues are oneof the most promising materials. PB can be considered as the oldest known hexacyanometallatecompound and has attracted the attention of many scientists seeking an understanding of its formation, composition, structure and physical properties and more importantly thanks to its electrocatalytic properties for H_2O_2 detection (sensitivity ca. 1–1.5 A M^{-1} cm^{-2}) [39]. The latter property is important when we consider that more than 90% of commercially available enzyme-based biosensors contain oxidase enzymes as terminal enzymes responsible for the generation of the analytical signal. Due to its high activity and selectivity at mild applied potentials (ca. 0mV) [39,45], PB has been denoted as an "*artificial peroxidase*" (see **Figure 4**). Moreover, thanks to the low potential needed for H_2O_2 detection PB–modified biosensors present lower artifacts because of biological reducing agents and biofouling phenomena in biological matrixes [46]. Thanks to its excellent analytical characteristics, PB has been exploited in the development of commercial biosensor devices for diabetes monitoring [17–20], and more recently, PB–modified CFEs ($\phi \sim 10$ μm) have been reported for developing glucose, glutamate and lactate biosensors for neuroscience applications with excellent *in vitro* and *in vivo* results [47–50].

Biosensors: Materials and Applications   Materials Research Forum LLC

Materials Research Foundations **47** (2019) 289-316   doi: http://dx.doi.org/10.21741/9781644900130-8

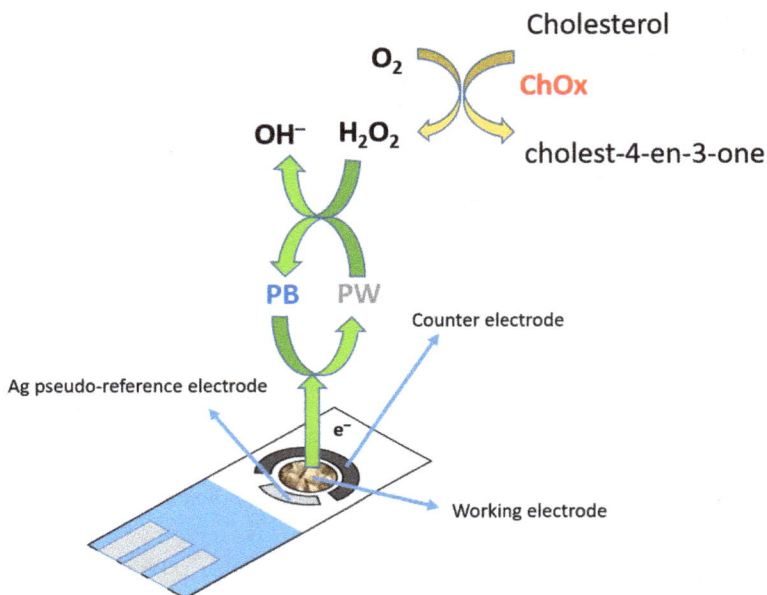

*Figure 4. Detection scheme for a cholesterol biosensor based on a PB-modified
electrode. Cholesterol is converted to cholest-4-en-3-one catalyzed by cholesterol
oxidase (ChOx) immobilized on the electrode surface. Secondary to this reaction, the
production of H_2O_2 can be amperometrically detected at low applied overpotentials,
electrocatalyzed by the PB layer.*

Biosensors: Materials and Applications   Materials Research Forum LLC
Materials Research Foundations **47** (2019) 289-316   doi: http://dx.doi.org/10.21741/9781644900130-8

Limit of detection (LOD) of a biosensor refers to the lowest change in analyte concentration that can be detected. LOD is a critical parameter for monitoring analytes present at ultra-low concentrations in biological fluid or tissues. LOD can be lowered using nanostructured biosensors or by using nano/microelectrodes thanks to their hemispherical diffusion profiles [51]. To solve the low current generated by such individual nano/micro electrodes, sensors can be assembled in array configurations in order to accumulate the generated current and amplify the resulting signal. Glucose and lactate typically measured in brain, subcutaneous tissue and interstitial fluids are present at high enough concentrations (mM) and do not present detection limit problems. Nevertheless, previous studies have shown that glutamate biosensors with a sensitivity of ca. 10 A mM^{-1} cm^{-2} are not adequate for neuro chemical applications [35,45] due to their low concentrations ($\sim\mu M$) in brain extracellular fluids (ECF). In order to improve this low sensitivity nanomaterials–modified biosensors have been developed to increase the surface–to–volume ratio and the electrocatalytic properties of the working electrode. Another promising strategy is to increase the enzymatic loading on the biosensing layer by incorporating a polycation (polyethyleneimine, PEI) to counterbalance the electrostatic repulsion between adjacent enzyme molecules and the anionic substrate, glutamate (Glu). Therefore, under the present approach glutamate biosensors with sensitivities of ca. 100 A mM^{-1} cm^{-2} have been developed for neuroscience applications [35,45].

Linear range is defined as the range of analyte concentrations for which the biosensor response changes linearly with the concentration. Therefore, the biosensor response can be mathematically represented as $y=mc+y_0$, where c is the concentration of the analyte, y is the output signal, m is the sensitivity of the biosensor and y_0 is a constant that describes the background. Consequently, the LOD and the limit of quantification (LOQ) can be defined by the expressions $LOD=3xSD/m$ and $LOQ=10xSD/m$, with SD being the standard deviation of a blank (solution without analyte). Based on previous definitions, the linear range is defined as that between the LOQ and the point where a plot of concentration versus response goes non–linear (limit of linearity, LOL). For a biosensor that follows the Michaelis–Menten mechanism, the range of the linear response is usually considered to extend up to $\sim\frac{1}{2}K_M$. From a practical point of view, the LOL for implantable devices can be extended using an outer membrane to restrict the diffusion of the analyte that reaches the enzymatic layer (decreasing its sensitivity and increasing its response time). Such membranes can be simultaneously used for improving the oxygen dependence and biocompatibility of the biosensors surface too. An example of this approach is found in [52] where the authors used an outer membrane based on Poly(Vinyl

Alcohol) (PVA) hydrogels for improving the sensor linearity (ca. 60%) and oxygen dependence (twofold factor).

6. Selectivity: interference of endogenous reducing agents

One of the most important issues for implantable enzymatic biosensors is their selectivity. Thanks to the high specificity of enzymes, biosensors have a high degree of selectivity for the target analyte. The chemical specificity of biosensors can be very high thanks to the use of high specificity enzymes, even to the level of stereoselectivity – for example, the use of L–glutamate oxidase (L–GluOx).Nevertheless, other enzymes such as alcohol, sugar groups or amino–acid oxidases, tyrosinase, laccase and peroxidases are non-selective enzymes and can be used to determinate different analytes [27]. On the other hand, electroactive species presented in the target medium such as ascorbic acid (AA), uric acid (UA), dopamine (DA), 3,4–dihydroxyphenylacetic acid (DOPAC), etc can react with the electrochemical transducer surface, compromising the selectivity of the device. Therefore, the appropriate choice of the biological receptor and transducer will be determinant on the selectivity properties of the enzyme-based amperometric biosensors. This problem is particularly evident for biosensors implanted in biological tissues because separation techniques cannot be used to eliminate the interference.

There are two recommended methods for quantifying biosensor selectivity that depends on the aim of measurement [27]. The first one consists of measuring the biosensor response, at the same concentration, for individual interference and the target analyte. Therefore, the sensitivity for interference (S_R) is compared against the sensitivity for the analyte (S_A) and selectivity is expressed as the S_A/S_R ratio. In the second approach, interfering substances are added, at their expected concentration, into the measuring cell, already containing the usual analyte concentration, at the mid-range of their expected value. The selectivity is then expressed as the percentage of variation of the biosensor response. Although the latter is easier than the calibration curve comparison, it has a more restricted significance because selectivity depends on the analyte and interference concentration range used. Nevertheless, when interfering substances are well identified they turn out to be more convenient.

As commented above, Pt transducers are still preferred because of their outstanding catalytic activity. Despite the excellent sensitivity for H_2O_2, the main problem of Pt transducers for oxidase-based biosensors is the high overpotential needed to detect H_2O_2 (often ca. + 700 mV) [39]. At this high potential, many substances, including ascorbic and uric acids, in the biosensor target medium (blood, fat, neural tissues, *etc.*) also are oxidized, thus interfering with the biosensor signal [53,54]. Over the last twenty years, many approaches have been proposed to solve alleviate interference, with the most common proceduresbeing: the use of inner and outer permselective polymer layers [55–57], the introduction of *second* and *third* generation biosensors [58–63] (*see below*), the incorporation of a "blank" or "sentinel" sensors (containing all the design elements of the biosensor, but omitting the enzyme) to removethe interference response, the introduction of ascorbate oxidase in the sensing layer [58,59], and the use of electrocatalytic (PB) [64–66] and nanostructured compounds (CNTs, Gr, NWs NPs) [67–70] to improve the analytical signal and reduce the potential detection window to name a few.

7.   Oxygen deficit

During the design of the biosensor it is vital to consider: (1) the stoichiometric problem generated due to the natural oxygen fluctuation in the sampling site, (2) the very low oxygen concentration in some tissues or (3) the high analyte concentration in the area of interest that increases the oxygen demand during the catalytic reaction (see **Eq. 2**) [31]. This is an important issue in neurophysiological studies because the mean oxygen concentration in ECF has been reported to be close to 50 μM. Moreover, the relatively high concentration of some important analytes such as glucose and lactate (mM) with respect to the natural enzymatic cofactor can accentuate the oxygen demand. Finally, oxygen fluctuations under induced physiological conditions,electrical stimulation, anaesthesia, pharmacological intervention and even, the foreign body response afterimplantationneed to be taken into account.

As mentioned above, biosensors designed to monitor glucose in the blood, interstitial fluids or subcutaneous tissues need a significantly greater oxygen tolerance compared to other biosensor devices (e.g., glutamate) because glucose levels are an order of magnitude higher in these peripheral tissues (5–10 mM), thereby increasing the oxygen demand. Such an imbalance can lead to a severe stoichiometric limitation leading to a loss of sensitivity and a decrease in the operative linear range. To solve this problem, the use of *second* and *third* generation biosensors has been introduced in implantable devices. However, the more widely used strategy for solving the oxygen deficit is the use of flux limiting outer membranes or inner membranes that act as oxygen reservoirs.

Second generation biosensors [71] use an artificial redox mediator instead of oxygen (see **Figure 3)**. Therefore, such a redox mediator is able to reoxidize the prosthetic group of the enzyme and it finally is detected on the surface of the transducer. Some soluble low molecular weight metal complexes are used for this purpose, e.g., ruthenium and osmium–complexes, ferrocene derivatives, etc. However, the main disadvantage of such an approach is the leakage of diffusible mediators which induce sample contamination, instability of the biosensors response and toxicity. It is usually assumed that redox mediator eliminates the dependence of the biosensor response on oxygen. However, oxygen is always present in interstitial fluids and tissues and is free diffusing, in contrast to the redox mediator which is usually immobilized to reduce its toxicity. Alternatively, *third* generation biosensors [72] are able to establish the direct electrochemical communication between the prosthetic group of the enzyme with the electrode surface (see **Figure 3).** This is the simplest and most ideal method to solve the oxygen demand and the use of an artificial inter medium via self–exchange reactions. However, such an approach is only available for veryfew enzymes such as peroxidases, whose active site is close to their surface.

As commented above, the most commonly used strategy to overcome the oxygen limitation is the use of outer polymeric membranes torestrict the access of the substrate to the enzymatic layer (increasing the linear range) without affecting or minimizing the impact of oxygen flow through this outer membrane. The main inconvenience of such an approach is the increase in the response time of the implantable devices and the possibility that outer membranes can be prone to degradation via oxidative degradation, calcification, delamination, etc. and cause a foreign body response. To overcome such limitations, biocompatible membranes and nanomaterials, surfactant agents, lipids, anti–inflammatory drugs and other tissue response–modifying agents can be used to modify the biosensor surface [72].

A more functional approach to solve the oxygen deficit without affecting the biosensor response (response time and sensitivity) is the incorporation of an oxygen reservoir layer in the biosensor design [73]. Fluorocarbon–derived materials such as poly(chlorotrifluorethylene) (Kel–F) oil and H200, H700 and H1000 oils and several silicone oils, such as poly(dimethylsiloxane) (PDMS) have been used because they have a remarkable solubility for oxygen. Finally, the addition of such an inner oxygen reservoir is able to prolong the enzymatic reaction for extended periods, even in media severely depleted of oxygen [74].

8.   Biocompatibility and long-term stability

One of the most critical issues for implantable devices, including biosensors, artificial organs, scaffolds for tissue engineering, drug-eluting stents, etc [75–77] is their biocompatibility and foreign body response. Such a response will affect the long-term stability and functionality of the biosensor, but more importantly, the foreign body reaction will pose a significant potential risk for patients. The inflammatory response occurs after device implantation due to tissue injury as well as the continual presence of the device in the body [78]. Following the first stage (acute inflammatory response) responsible for provisional matrix formation and cleaning of the wound site, chronic inflammation and the formation of granulomatous tissue involving the implantable device

Biosensors: Materials and Applications   Materials Research Forum LLC
Materials Research Foundations **47** (2019) 289-316   doi: http://dx.doi.org/10.21741/9781644900130-8

occur. Therefore, a polymer–blood interface is created and nonspecific absorption of tissue fluid proteins and blood onto the surface of the device is induced. Following nonspecific protein absorption, immune and inflammatory cells such as leukocytes, monocytes and platelets intervene to protect the body from the foreign object. Finally, the implantable device is walled off by a collagenous fibrous capsule to prevent it from interacting with surrounding tissues [78,79]. As mentioned above, the functional definition of biocompatibility means that the interaction between the biosensor and the surrounding tissue must be minimal. Therefore, biosensor degradation and changes in the proximity of the implanted devices, that affect the diffusion of the analyte toward the biosensor surface, must be avoided. Nevertheless, a more general definition takes into consideration a second criterion related to the interaction between both elements and how the biosensor can modify or perturb the anatomical and/or physiological functions (homeostasis) of the surrounding tissue. In this regard, studies have shown how brain extracellular uric acid concentration increases when using chronically implanted carbon paste electrodes and microdialysis probe of different sizes [80,81], demonstrating that special caution must be taken into account during biosensor implantation. After twenty years of studying the main causes affecting the degree of the foreign body reaction, some factors have been identified such as shape, roughness, porosity, size, implantation time, the toxicity of materials used, surface chemistry and degradation of the implanted device [82]. In this respect, carbon paste electrodes (ϕ ~100–500 μm), used in the early 80's in the field of neuroscience, have been substituted over the last ten years by carbon fiber electrodes (ϕ ~5–10 μm) with lower dimensions, lower tissue injury and betterspatial resolution [47,50].

The surface chemistry composition of implantable biosensors is of vital importance for improving the biocompatibility of such devices. General approaches used to overcome the *in vivo* instability of implantable devices include (a) biocompatible material coatings, (b) steroidal and nonsteroidal anti-inflammatory drugs, and (c) angiogenic drugs [82]. The most commonly used method to improve the biocompatibility properties of implantable biosensors is the use of an outer membrane that decreases the foreign body response. For this purpose, natural, synthetic and semisynthetic materials such as collagen, alginate, hyaluronan, chitosan, dextran and other polymers includingpoly(lactic acid), poly(lactic–co–glycolic acid), poly(ethylene glycol), Nafion®, 2–hydroxy ethyl methacrylate, polyethylenimine and poly(vinyl alcohol) are used.

The most common functionalization methodologies are photo– and electro–polymerization, chemical and physical vapour deposition, layer–by–layer deposition, plasma polymerizationand self–assembled monolayer. Such modifications allow the control of surface properties and the addition of new functionalities such as antimicrobial

Biosensors: Materials and Applications   Materials Research Forum LLC
Materials Research Foundations **47** (2019) 289-316   doi: http://dx.doi.org/10.21741/9781644900130-8

activity, biocompatibility, chemical stability, etc. Nevertheless, these methods can be complex, time consuming, laborious, sometimes poorly reproducible and limited to certain surfaces. Therefore, new green functionalization methods that allow easy and reproducible surface modifications have been recently proposed. In this regard, one promising material is poly(dopamine) (PDA), a mussel-inspired coating, which has recently attracted considerable attention of researchers [83–86]. PDA–modified surfaces have been used for enzyme [87,88], antibody [89] and DNA immobilization for biosensing applications [83,85] efficient and reliable manipulation of human neural stem cells (NSCs), growth factor immobilization, hematopoietic cell adhesion, antibacterial surface preparations and hydroxyapatite crystallization [83]. In addition, recent studies have shown that PDA–modified surfaces can serve as an effective strategy to form ultra-stable coatings on nanoparticles (NPs) in *in vivo* applications, which can improve the intracellular delivery capacity and biocompatibility of NPs for biomedical applications [90].

Conclusions

Over the last ten years, biosensors have become an important analytical tool for health care applications, especially for implantable devices. Such implantable biosensors (cutaneous, subcutaneous) or hybrid approaches (e.g., microdialysis coupled with an amperometric biosensor) are replacing classical glucometers with enzyme–modified strips, allowing, for example, the continuous monitoring of glucose levels. The success of amperometric biosensors for *in vivo* monitoring depends on several factors that have been explained in the present chapter in terms of biosensor design features that provide adequate sensitivity, linear range, limit of detection and selectivity. Moreover, different strategies to overcome certain limitations including oxygen deficit, biofouling, biocompatibility, biosensor stability and lifetime are discussed. Therefore, the introduction of perm–a selective inner or outer layer, advanced materials– and/or nanomaterials–modified electrode surfaces with better physicochemical properties, more efficient immobilization of bioreceptors and fast occurrence of the electrochemical event have beenintroduced and discussed. Finally, it is possible to conclude, that in the near future, the key challenge to achieving better implantable devices will be the tailor–made the preparation of advanced functional materials–modified biosensor with novel and well–controlled defined properties, which is adapted for biological matrixes and has low discomfort.

Biosensors: Materials and Applications
Materials Research Foundations 47 (2019) 289-316

Materials Research Forum LLC
doi: http://dx.doi.org/10.21741/9781644900130-8

References

[1]   G. Luka, A. Ahmadi, H. Najjaran, E. Alocilja, M. DeRosa, K. Wolthers, A. Malki, H. Aziz, A. Althani, M. Hoorfar, Microfluidics integrated biosensors: a leading technology towards lab-on-a-chip and sensing applications, Sensors. 15 (2015) 30011–30031. https://doi.org/10.3390/s151229783

[2]   J.P. Lafleur, A. Jönsson, S. Senkbeil, J.P. Kutter, Recent advances in lab-on-a-chip for biosensing applications, Biosens. Bioelectron. 76 (2016) 213–233. https://doi.org/10.1016/j.bios.2015.08.003

[3]   J.-Y. Yoon, Lab-on-a-Chip Biosensors, in: Introd. to Biosens., Springer International Publishing, Cham, 2016: pp. 257–297. https://doi.org/10.1007/978-3-319-27413-3_14

[4]   A. Malima, S. Siavoshi, T. Musacchio, J. Upponi, C. Yilmaz, S. Somu, W. Hartner, V. Torchilin, A. Busnaina, Highly sensitive microscale in vivo sensor enabled by electrophoretic assembly of nanoparticles for multiple biomarker detection, Lab Chip. 12 (2012) 4748. https://doi.org/10.1039/c2lc40580f

[5]   X. Zhu, W. Liu, S. Shuang, M. Nair, C.-Z. Li, Intelligent tattoos, patches, and other wearable biosensors, in: Med. Biosens. Point Care Appl., Elsevier, 2017: pp. 133–150. https://doi.org/10.1016/B978-0-08-100072-4.00006-X

[6]   D. Xu, X. Huang, J. Guo, X. Ma, Automatic smartphone-based microfluidic biosensor system at the point of care, Biosens. Bioelectron. 110 (2018) 78–88. https://doi.org/10.1016/j.bios.2018.03.018

[7]   M. Zarei, Advances in point-of-care technologies for molecular diagnostics, Biosens. Bioelectron. 98 (2017) 494–506. https://doi.org/10.1016/j.bios.2017.07.024

[8]   S. Kanchi, M.I. Sabela, P.S. Mdluli, Inamuddin, K. Bisetty, Smartphone based bioanalytical and diagnosis applications: A review., Biosens. Bioelectron. 102 (2018) 136–149. https://doi.org/10.1016/j.bios.2017.11.021

[9]   L. Ryden, P.J. Grant, S.D. Anker, C. Berne, F. Cosentino, N. Danchin, C. Deaton, J. Escaned, H.-P. Hammes, H. Huikuri, M. Marre, N. Marx, L. Mellbin, J. Ostergren, C. Patrono, P. Seferovic, M.S. Uva, M.-R. Taskinen, M. Tendera, J. Tuomilehto, P. Valensi, J.L. Zamorano, ESC Guidelines on diabetes, pre-diabetes, and cardiovascular diseases developed in collaboration with the EASD, Eur. Heart J. 35 (2014) 1824–1824. https://doi.org/10.1093/eurheartj/ehu076

[10]   D. Bruen, C. Delaney, L. Florea, D. Diamond, Glucose sensing for diabetes
       monitoring: recent developments, Sensors. 17 (2017) 1866.
       https://doi.org/10.3390/s17081866

[11]   P. Salazar, V. Rico, A.R. González-Elipe, Nickel/Copper Bilayer-modified Screen
       Printed Electrode for Glucose Determination in Flow Injection Analysis,
       Electroanalysis. 30 (2018) 187–193. https://doi.org/10.1002/elan.201700592

[12]   V.R. and A.R.G. Pedro Salazar, Non–enzymatic glucose sensors based on nickel
       nanoporous thin films prepared by physical vapor deposition at oblique angles for
       beverage industry applications, J. Electrochem. Soc. 163 (2016) B704–B709.
       https://doi.org/10.1149/2.1241614jes

[13]   G.S. Wilson, R. Gifford, Biosensors for real-time in vivo measurements, Biosens.
       Bioelectron. 20 (2005) 2388–2403. https://doi.org/10.1016/j.bios.2004.12.003

[14]   E.-H. Yoo, S.-Y. Lee, Glucose biosensors: an overview of use in clinical practice,
       Sensors. 10 (2010) 4558–4576. https://doi.org/10.3390/s100504558

[15]   V. Scognamiglio, Nanotechnology in glucose monitoring: Advances and
       challenges in the last 10 years, Biosens. Bioelectron. 47 (2013) 12–25.
       https://doi.org/10.1016/j.bios.2013.02.043

[16]   J. Zhang, W. Hodge, C. Hutnick, X. Wang, Noninvasive diagnostic devices for
       diabetes through measuring tear glucose, J. Diabetes Sci. Technol. 5 (2011) 166–
       172. https://doi.org/10.1177/193229681100500123

[17]   F. Ricci, F. Caprio, A. Poscia, F. Valgimigli, D. Messeri, E. Lepori, G. Dall'Oglio,
       G. Palleschi, D. Moscone, Toward continuous glucose monitoring with planar
       modified biosensors and microdialysis, Biosens. Bioelectron. 22 (2007) 2032–
       2039. https://doi.org/10.1016/j.bios.2006.08.041

[18]   F. Ricci, D. Moscone, C.S. Tuta, G. Palleschi, A. Amine, A. Poscia, F. Valgimigli,
       D. Messeri, Novel planar glucose biosensors for continuous monitoring use,
       Biosens. Bioelectron. 20 (2005) 1993–2000.
       https://doi.org/10.1016/j.bios.2004.09.010

[19]   A. Poscia, M. Mascini, D. Moscone, M. Luzzana, G. Caramenti, P. Cremonesi, F.
       Valgimigli, C. Bongiovanni, M. Varalli, A microdialysis technique for continuous
       subcutaneous glucose monitoring in diabetic patients (part 1), Biosens.
       Bioelectron. 18 (2003) 891–898. https://doi.org/10.1016/S0956-5663(02)00216-6

[20]   F. Lucarelli, F. Ricci, F. Caprio, F. Valgimigli, C. Scuffi, D. Moscone, G.
       Palleschi, GlucoMen day continuous glucose monitoring system: a screening for

Biosensors: Materials and Applications   Materials Research Forum LLC
Materials Research Foundations **47** (2019) 289-316   doi: http://dx.doi.org/10.21741/9781644900130-8

enzymatic and electrochemical interferents, J. Diabetes Sci. Technol. 6 (2012) 1172–1181. https://doi.org/10.1177/193229681200600522

[21] H. Blauw, P. Keith-Hynes, R. Koops, J.H. DeVries, A review of safety and design requirements of the artificial pancreas, Ann. Biomed. Eng. 44 (2016) 3158–3172. https://doi.org/10.1007/s10439-016-1679-2

[22] S.C. Christiansen, A.L. Fougner, Ø. Stavdahl, K. Kölle, R. Ellingsen, S.M. Carlsen, A review of the current challenges associated with the development of an artificial pancreas by a double subcutaneous approach, diabetes Ther. 8 (2017) 489–506. https://doi.org/10.1007/s13300-017-0263-6

[23] L.C. Clark, C. Lyons, Electrode systems for continuous monitoring in cardiovascular surgery, Ann. N. Y. Acad. Sci. 102 (2006) 29–45. https://doi.org/10.1111/j.1749-6632.1962.tb13623.x

[24] S J Updike, G P Hicks, The enzyme electrode, Nature. 214 (1967) 986–988

[25] G.G. Guilbault, G.J. Lubrano, An enzyme electrode for the amperometric determination of glucose, Anal. Chim. Acta. 64 (1973) 439–455. https://doi.org/10.1016/S0003-2670(01)82476-4

[26] C.A. Corcoran, G.A. Rechnitz, Cell-based biosensors, Trends Biotechnol. 3 (1985) 92–96. https://doi.org/10.1016/0167-7799(85)90091-5

[27] D.R. Thévenot, K. Toth, R.A. Durst, G.S. Wilson, Electrochemical biosensors: recommended definitions and classification1International Union of Pure and Applied Chemistry: Physical Chemistry Division, Commission I.7 (Biophysical Chemistry); Analytical Chemistry Division, Commission V.5 (Electroanalytical, Biosens. Bioelectron. 16 (2001) 121–131. https://doi.org/10.1016/S0956-5663(01)00115-4

[28] S.A. Ansari, Q. Husain, Potential applications of enzymes immobilized on/in nano materials: A review, Biotechnol. Adv. 30 (2012) 512–523. https://doi.org/10.1016/j.biotechadv.2011.09.005

[29] A. Sassolas, L.J. Blum, B.D. Leca-Bouvier, Immobilization strategies to develop enzymatic biosensors, Biotechnol. Adv. 30 (2012) 489–511. https://doi.org/10.1016/j.biotechadv.2011.09.003

[30] C.P. McMahon, R.D. O'Neill, Polymer–enzyme composite biosensor with high glutamate sensitivity and low oxygen dependence, Anal. Chem. 77 (2005) 1196–1199. https://doi.org/10.1021/ac048686r

Biosensors: Materials and Applications   Materials Research Forum LLC
Materials Research Foundations **47** (2019) 289-316   doi: http://dx.doi.org/10.21741/9781644900130-8

[31]   C.P. McMahon, G. Rocchitta, S.M. Kirwan, S.J. Killoran, P.A. Serra, J.P. Lowry, R.D. O'Neill, Oxygen tolerance of an implantable polymer/enzyme composite glutamate biosensor displaying polycation-enhanced substrate sensitivity, Biosens. Bioelectron. 22 (2007) 1466–1473. https://doi.org/10.1016/j.bios.2006.06.027

[32]   R.D. O'Neill, G. Rocchitta, C.P. McMahon, P.A. Serra, J.P. Lowry, Designing sensitive and selective polymer/enzyme composite biosensors for brain monitoring in vivo, TrAC Trends Anal. Chem. 27 (2008) 78–88. https://doi.org/10.1016/j.trac.2007.11.008

[33]   G. Rocchitta, A. Spanu, S. Babudieri, G. Latte, G. Madeddu, G. Galleri, S. Nuvoli, P. Bagella, M. Demartis, V. Fiore, R. Manetti, P. Serra, Enzyme biosensors for biomedical applications: strategies for safeguarding analytical performances in biological fluids, Sensors. 16 (2016) 780. https://doi.org/10.3390/s16060780

[34]   G.S. Wilson, M. Ammam, In vivo biosensors, FEBS J. 274 (2007) 5452–5461. https://doi.org/10.1111/j.1742-4658.2007.06077.x

[35]   C.P. McMahon, G. Rocchitta, P.A. Serra, S.M. Kirwan, J.P. Lowry, R.D. O'Neill, The efficiency of immobilised glutamate oxidase decreases with surface enzyme loading: an electrostatic effect, and reversal by a polycation significantly enhances biosensor sensitivity, Analyst. 131 (2006) 68–72. https://doi.org/10.1039/B511643K

[36]   P. Salazar, M. Martín, R. Roche, R.D. O'Neill, J.L. González-Mora, Prussian Blue-modified microelectrodes for selective transduction in enzyme-based amperometric microbiosensors for in vivo neurochemical monitoring, Electrochim. Acta. 55 (2010) 6476–6484. https://doi.org/10.1016/j.electacta.2010.06.036

[37]   J. Wang, Electrochemical glucose biosensors, Chem. Rev. 108 (2008) 814–825. https://doi.org/10.1021/cr068123a

[38]   P. Salazar, M. Martín, R.D. O'Neill, R. Roche, J.L. González-Mora, Biosensors based on prussian blue modified carbon fibers electrodes for monitoring lactate in the extracellular space of brain tissue, Int. J. Electrochem. Sci. 7 (2012) 5910–5926

[39]   G.J. Salazar P, Martín M, O'Neill RD, Lorenzo-Luis P, Roche R, Prussian blue and analogues: biosensing applications in health care, in: T. Ashutosh, A.N. Nordin (Eds.), Adv. Biomater. Biodevices, 2014: pp. 423–450

Biosensors: Materials and Applications   Materials Research Forum LLC
Materials Research Foundations **47** (2019) 289-316   doi: http://dx.doi.org/10.21741/9781644900130-8

[40]   P. Salazar, M. Martín, R.D. O'Neill, R. Roche, J.L. González-Mora, Surfactant-promoted Prussian Blue-modified carbon electrodes: Enhancement of electro-deposition step, stabilization, electrochemical properties and application to lactate microbiosensors for the neurosciences, Colloids Surfaces B Biointerfaces. 92 (2012) 180–189. https://doi.org/10.1016/j.colsurfb.2011.11.047

[41]   S.B. Hall, E.A. Khudaish, A.L. Hart, Electrochemical oxidation of hydrogen peroxide at platinum electrodes. Part 1. An adsorption-controlled mechanism, Electrochim. Acta. 43 (1998) 579–588. https://doi.org/10.1016/S0013-4686(97)00125-4

[42]   P. O'Connell, C. O'Sullivan, G. Guilbault, Electrochemical metallisation of carbon electrodes, Anal. Chim. Acta. 373 (1998) 261–270. https://doi.org/10.1016/S0003-2670(98)00414-0

[43]   S. Domínguez-Domínguez, J. Arias-Pardilla, Á. Berenguer-Murcia, E. Morallón, D. Cazorla-Amorós, Electrochemical deposition of platinum nanoparticles on different carbon supports and conducting polymers, J. Appl. Electrochem. 38 (2008) 259–268. https://doi.org/10.1007/s10800-007-9435-9

[44]   Eiichi Tamiya, I. Karube, Development of micro-biosensors for brain research, in: T. Yoshida, R.D. Tanner (Eds.), Bioprod. Bioprocesses 2, Springer-Verlag, 1991: pp. 169–175

[45]   P. Salazar, M. Martín, R.D. O'Neill, J.L. González-Mora, Glutamate microbiosensors based on Prussian Blue modified carbon fiber electrodes for neuroscience applications: In-vitro characterization, Sensors Actuators B Chem. 235 (2016) 117–125. https://doi.org/10.1016/j.snb.2016.05.057

[46]   A.A. Karyakin, Prussian blue and its analogues: electrochemistry and analytical applications, Electroanalysis. 13 (2001) 813–819

[47]   P. Salazar, M. Martín, R. Roche, J.L. González–Mora, R.D. O'Neill, Microbiosensors for glucose based on Prussian Blue modified carbon fiber electrodes for in vivo monitoring in the central nervous system, Biosens. Bioelectron. 26 (2010) 748–753. https://doi.org/10.1016/j.bios.2010.06.045

[48]   R. Roche, P. Salazar, M. Martín, F. Marcano, J.L. González-Mora, Simultaneous measurements of glucose, oxyhemoglobin and deoxyhemoglobin in exposed rat cortex, J. Neurosci. Methods. 202 (2011) 192–198. https://doi.org/10.1016/j.jneumeth.2011.07.003

Biosensors: Materials and Applications   Materials Research Forum LLC
Materials Research Foundations **47** (2019) 289-316   doi: http://dx.doi.org/10.21741/9781644900130-8

[49]   P. Salazar, M. Martín, R.D. O'Neill, R. Roche, J.L. González-Mora, Improvement and characterization of surfactant-modified Prussian blue screen-printed carbon electrodes for selective H_2O_2 detection at low applied potentials, J. Electroanal. Chem. 674 (2012) 48–56. https://doi.org/10.1016/j.jelechem.2012.04.005

[50]   P. Salazar, R.D. O'Neill, M. Martín, R. Roche, J.L. González-Mora, Amperometric glucose microbiosensor based on a Prussian Blue modified carbon fiber electrode for physiological applications, Sensors Actuators B Chem. 152 (2011) 137–143. https://doi.org/10.1016/j.snb.2010.11.056

[51]   S. Vaddiraju, I. Tomazos, D.J. Burgess, F.C. Jain, F. Papadimitrakopoulos, Emerging synergy between nanotechnology and implantable biosensors: A review, Biosens. Bioelectron. 25 (2010) 1553–1565. https://doi.org/10.1016/j.bios.2009.12.001

[52]   S.S. Vaddiraju, H. Singh, D.J. Burgess, F.C. Jain, F. Papadimitrakopoulos, Enhanced glucose sensor linearity using poly(vinyl alcohol) hydrogels, J. Diabetes Sci. Technol. 3 (2009) 863–874. https://doi.org/10.1177/193229680900300434

[53]   P. D'Orazio, Biosensors in clinical chemistry, Clin. Chim. Acta. 334 (2003) 41–69. https://doi.org/10.1016/S0009-8981(03)00241-9

[54]   P. D'Orazio, Biosensors in clinical chemistry -2011 update, Clin. Chim. Acta. 412 (2011) 1749–1761. https://doi.org/10.1016/j.cca.2011.06.025

[55]   G. Calia, P. Monti, S. Marceddu, M.A. Dettori, D. Fabbri, S. Jaoua, R.D. O'Neill, P.A. Serra, G. Delogu, Q. Migheli, Electropolymerized phenol derivatives as permselective polymers for biosensor applications, Analyst. 140 (2015) 3607–3615. https://doi.org/10.1039/C5AN00363F

[56]   S.A. Rothwell, R.D. O'Neill, Effects of applied potential on the mass of non-conducting poly(ortho-phenylenediamine) electro-deposited on EQCM electrodes: comparison with biosensor selectivity parameters, Phys. Chem. Chem. Phys. 13 (2011) 5413. https://doi.org/10.1039/c0cp02341h

[57]   S.A. Rothwell, C.P. McMahon, R.D. O'Neill, Effects of polymerization potential on the permselectivity of poly(o-phenylenediamine) coatings deposited on Pt–Ir electrodes for biosensor applications, Electrochim. Acta. 55 (2010) 1051–1060. https://doi.org/10.1016/j.electacta.2009.09.069

[58]   W.H. Oldenziel, G. Dijkstra, T.I.F.H. Cremers, B.H.C. Westerink, In vivo monitoring of extracellular glutamate in the brain with a microsensor, Brain Res. 1118 (2006) 34–42. https://doi.org/10.1016/j.brainres.2006.08.015

Biosensors: Materials and Applications   Materials Research Forum LLC
Materials Research Foundations **47** (2019) 289-316   doi: http://dx.doi.org/10.21741/9781644900130-8

[59]   W.H. Oldenziel, M. van der Zeyden, G. Dijkstra, W.E.J.M. Ghijsen, H. Karst,
       T.I.F.H. Cremers, B.H.C. Westerink, Monitoring extracellular glutamate in
       hippocampal slices with a microsensor, J. Neurosci. Methods. 160 (2007) 37–44.
       https://doi.org/10.1016/j.jneumeth.2006.08.003

[60]   J. Castillo, S. Gáspár, S. Leth, M. Niculescu, A. Mortari, I. Bontidean, V.
       Soukharev, S.A. Dorneanu, A.D. Ryabov, E. Csöregi, Biosensors for life quality -
       design, development and applications, Sensors Actuators, B Chem. 102 (2004)
       179–194. https://doi.org/10.1016/j.snb.2004.04.084

[61]   J.J. Mitala, A.C. Michael, Improving the performance of electrochemical
       microsensors based on enzymes entrapped in a redox hydrogel, Anal. Chim. Acta.
       556 (2006) 326–332. https://doi.org/10.1016/j.aca.2005.09.053

[62]   M. Cano, J. Luis Ávila, M. Mayén, M.L. Mena, J. Pingarrón, R. Rodríguez-
       Amaro, A new, third generation, PVC/TTF–TCNQ composite amperometric
       biosensor for glucose determination, J. Electroanal. Chem. 615 (2008) 69–74.
       https://doi.org/10.1016/j.jelechem.2007.11.032

[63]   G. Sánchez-Obrero, M. Cano, J.L. Ávila, M. Mayén, M.L. Mena, J.M. Pingarrón,
       R. Rodríguez-Amaro, A gold nanoparticle-modified PVC/TTF-TCNQ composite
       amperometric biosensor for glucose determination, J. Electroanal. Chem. 634
       (2009) 59–63. https://doi.org/10.1016/j.jelechem.2009.07.017

[64]   F. Ricci, G. Palleschi, Sensor and biosensor preparation, optimisation and
       applications of Prussian Blue modified electrodes, Biosens. Bioelectron. 21 (2005)
       389–407. https://doi.org/10.1016/j.bios.2004.12.001

[65]   A. Karyakin, E. Karyakina, L. Gorton, Prussian-Blue-based amperometric
       biosensors in flow-injection analysis, Talanta. 43 (1996) 1597–1606.
       https://doi.org/10.1016/0039-9140(96)01909-1

[66]   D. Moscone, D. D'Ottavi, D. Compagnone, G. Palleschi, A. Amine, Construction
       and analytical characterization of prussian blue-based carbon paste electrodes and
       their assembly as oxidase enzyme sensors, Anal. Chem. 73 (2001) 2529–2535.
       https://doi.org/10.1021/ac001245x

[67]   B.D. Malhotra, M.A. Ali, Nanostructured biomaterials for in vivo biosensors, in:
       Nanomater. Biosens., 2018: pp. 183–219. https://doi.org/10.1016/B978-0-323-
       44923-6.00007-8

[68]   S. Kumar, W. Ahlawat, R. Kumar, N. Dilbaghi, Graphene, carbon nanotubes, zinc
       oxide and gold as elite nanomaterials for fabrication of biosensors for healthcare,

Biosens. Bioelectron. 70 (2015) 498–503.
https://doi.org/10.1016/j.bios.2015.03.062

[69]   S.C. Ray, N.R. Jana, Application of carbon-based nanomaterials as biosensor, in:
       Carbon Nanomater. Biol. Med. Appl., Elsevier, 2017: pp. 87–127.
       https://doi.org/10.1016/B978-0-323-47906-6.00003-5

[70]   G. Maduraiveeran, M. Sasidharan, V. Ganesan, Electrochemical sensor and
       biosensor platforms based on advanced nanomaterials for biological and
       biomedical applications, Biosens. Bioelectron. 103 (2018) 113–129.
       https://doi.org/10.1016/j.bios.2017.12.031

[71]   F. Wang, S. Liu, M. Lin, X. Chen, S. Lin, X. Du, H. Li, H. Ye, B. Qiu, Z. Lin, L.
       Guo, G. Chen, Colorimetric detection of microcystin-LR based on disassembly of
       orient-aggregated gold nanoparticle dimers, Biosens. Bioelectron. 68 (2015) 475–
       480. https://doi.org/10.1016/j.bios.2015.01.037

[72]   N. Wisniewski, M. Reichert, Methods for reducing biosensor membrane
       biofouling, Colloids Surfaces B Biointerfaces. 18 (2000) 197–219.
       https://doi.org/10.1016/S0927-7765(99)00148-4

[73]   M. Martin, R.D. O'Neill, J.L. Gonzalez-Mora, P. Salazar, The use of
       fluorocarbons to mitigate the oxygen dependence of glucose microbiosensors for
       neuroscience applications, J. Electrochem. Soc. 161 (2014) H689–H695.
       https://doi.org/10.1149/2.1071410jes

[74]   J. Wang, L. Fang, Oxygen rich oxidase enzyme electrodes for operation in
       oxygen-free solutions, J. Am. Chem. Soc. 120 (1998) 1048–1050 ST–Oxygen rich
       oxidase enzyme electro

[75]   Y. Fang, Y. Ni, G. Zhang, C. Mao, X. Huang, J. Shen, Biocompatibility of CS–
       PPy nanocomposites and their application to glucose biosensor,
       Bioelectrochemistry. 88 (2012) 1–7.
       https://doi.org/10.1016/j.bioelechem.2012.05.006

[76]   C. Sun, L. Gao, D. Wang, M. Zhang, Y. Liu, Z. Geng, W. Xu, F. Liu, H. Bian,
       Biocompatible polypyrrole-block copolymer-gold nanoparticles platform for
       determination of inosine monophosphate with bi-enzyme biosensor, Sensors
       Actuators B Chem. 230 (2016) 521–527. https://doi.org/10.1016/j.snb.2016.02.111

[77]   C. Menti, M. Beltrami, A.L. Possan, S.T. Martins, J.A.P. Henriques, A.D. Santos,
       F.P. Missell, M. Roesch-Ely, Biocompatibility and degradation of gold-covered

Biosensors: Materials and Applications   Materials Research Forum LLC
Materials Research Foundations **47** (2019) 289-316   doi: http://dx.doi.org/10.21741/9781644900130-8

magneto-elastic biosensors exposed to cell culture, Colloids Surfaces B Biointerfaces. 143 (2016) 111–117. https://doi.org/10.1016/j.colsurfb.2016.03.034

[78]   Y. Onuki, U. Bhardwaj, F. Papadimitrakopoulos, D.J. Burgess, A review of the biocompatibility of implantable devices: current challenges to overcome foreign body response, J. Diabetes Sci. Technol. 2 (2008) 1003–1015. https://doi.org/10.1177/193229680800200610

[79]   Y. Wang, S. Vaddiraju, B. Gu, F. Papadimitrakopoulos, D.J. Burgess, Foreign body reaction to implantable biosensors, J. Diabetes Sci. Technol. 9 (2015) 966–977. https://doi.org/10.1177/1932296815601869

[80]   R.D. O'Neill, J.P. Lowry, On the significance of brain extracellular uric acid detected with in-vivo monitoring techniques: a review, Behav. Brain Res. 71 (1995) 33–49. https://doi.org/10.1016/0166-4328(95)00035-6

[81]   A. Duff, R.D. O'Neill, Effect of probe size on the concentration of brain extracellular uric acid monitored with carbon paste electrodes, J. Neurochem. 62 (1994) 1496–1502. https://doi.org/10.1046/j.1471-4159.1994.62041496.x

[82]   J.M. Morais, F. Papadimitrakopoulos, D.J. Burgess, Biomaterials/tissue interactions: possible solutions to overcome foreign body response, AAPS J. 12 (2010) 188–196. https://doi.org/10.1208/s12248-010-9175-3

[83]   Y. Liu, K. Ai, L. Lu, Polydopamine and its derivative materials: synthesis and promising applications in energy, environmental, and biomedical fields, Chem. Rev. 114 (2014) 5057–5115. https://doi.org/10.1021/cr400407a

[84]   Y.H. Ding, M. Floren, W. Tan, Mussel-inspired polydopamine for bio-surface functionalization, Biosurface and Biotribology. 2 (2016) 121–136. https://doi.org/10.1016/j.bsbt.2016.11.001

[85]   M.E. Lynge, P. Schattling, B. Städler, Recent developments in poly(dopamine)-based coatings for biomedical applications, Nanomedicine. 10 (2015) 2725–2742. https://doi.org/10.2217/nnm.15.89

[86]   M. Liu, G. Zeng, K. Wang, Q. Wan, L. Tao, X. Zhang, Y. Wei, Recent developments in polydopamine: an emerging soft matter for surface modification and biomedical applications, Nanoscale. 8 (2016) 16819–16840. https://doi.org/10.1039/C5NR09078D

[87]   M. Martín, P. Salazar, R. Álvarez, A. Palmero, C. López-Santos, J.L. González-Mora, A.R. González-Elipe, Cholesterol biosensing with a polydopamine-modified nanostructured platinum electrode prepared by oblique angle physical

Biosensors: Materials and Applications   Materials Research Forum LLC
Materials Research Foundations **47** (2019) 289-316   doi: http://dx.doi.org/10.21741/9781644900130-8

vacuum deposition, Sensors Actuators B Chem. 240 (2017) 37–45.
https://doi.org/10.1016/j.snb.2016.08.092

[88]   M. Martín, P. Salazar, S. Campuzano, R. Villalonga, J.M. Pingarrón, J.L.
       González-Mora, Amperometric magnetobiosensors using poly(dopamine)-
       modified Fe 3 O 4 magnetic nanoparticles for the detection of phenolic
       compounds, Anal. Methods. 7 (2015) 8801–8808.
       https://doi.org/10.1039/C5AY01996F

[89]   M. Martín, P. Salazar, C. Jiménez, M. Lecuona, M.J. Ramos, J. Ode, J. Alcoba, R.
       Roche, R. Villalonga, S. Campuzano, J.M. Pingarrón, J.L. González-Mora, Rapid
       Legionella pneumophila determination based on a disposable core–shell Fe 3 O 4
       @poly(dopamine) magnetic nanoparticles immunoplatform, Anal. Chim. Acta.
       887 (2015) 51–58. https://doi.org/10.1016/j.aca.2015.05.048

[90]   X. Liu, J. Cao, H. Li, J. Li, Q. Jin, K. Ren, J. Ji, Mussel-inspired polydopamine: a
       biocompatible and ultrastable coating for nanoparticles in *vivo*, ACS Nano. 7
       (2013) 9384–9395. https://doi.org/10.1021/nn404117j

Keyword Index

About the Editors

Dr. Inamuddin is currently working as Assistant Professor in the Chemistry Department, Faculty of Science, King Abdulaziz University, Jeddah, Saudi Arabia. He is a permanent faculty member (Assistant Professor) at the Department of Applied Chemistry, Aligarh Muslim University, Aligarh, India. He obtained Master of Science degree in Organic Chemistry from Chaudhary Charan Singh (CCS) University, Meerut, India, in 2002. He received his Master of Philosophy and Doctor of Philosophy degrees in Applied Chemistry from Aligarh Muslim University (AMU), India, in 2004 and 2007, respectively. He has extensive research experience in multidisciplinary fields of Analytical Chemistry, Materials Chemistry, and Electrochemistry and, more specifically, Renewable Energy and Environment. He has worked on different research projects as project fellow and senior research fellow funded by University Grants Commission (UGC), Government of India, and Council of Scientific and Industrial Research (CSIR), Government of India. He has received Fast Track Young Scientist Award from the Department of Science and Technology, India, to work in the area of bending actuators and artificial muscles. He has completed four major research projects sanctioned by University Grant Commission, Department of Science and Technology, Council of Scientific and Industrial Research, and Council of Science and Technology, India. He has published 133 research articles in international journals of repute and eighteen book chapters in knowledge-based book editions published by renowned international publishers. He has published forty two edited books with Springer, United Kingdom, Elsevier, Nova Science Publishers, Inc. U.S.A., CRC Press Taylor & Francis Asia Pacific, Trans Tech Publications Ltd., Switzerland and Materials Research Forum LLC, U.S.A. He is the member of various editorial boards of the journals and serving as associate editor for journals such as Environmental Chemistry Letter, Applied Water Science, Euro-Mediterranean Journal for Environmental Integration, Springer-Nature, Frontiers Section Editor of Current Analytical Chemistry, published by Bentham Science Publishers, editorial board member for Scientific Reports-Nature and editor for Eurasian Journal of Analytical Chemistry. He has attended as well as chaired sessions in various international and national conferences. He has worked as a Postdoctoral Fellow, leading a research team at the Creative Research Initiative Center for Bio-Artificial Muscle, Hanyang University, South Korea, in the field of renewable energy, especially biofuel cells. He has also worked as a Postdoctoral Fellow at the Center of Research Excellence in Renewable Energy, King Fahd University of Petroleum and Minerals, Saudi Arabia, in the field of polymer electrolyte membrane fuel cells and computational fluid dynamics of polymer electrolyte membrane fuel cells. He is a life member of the Journal of the Indian

Chemical Society. His research interest includes ion exchange materials, a sensor for heavy metal ions, biofuel cells, supercapacitors and bending actuators.

Dr. Tauseef Ahmad Rangreez is working as postdoctoral fellow at National Institute of Technology, Srinagar, India. He completed his Ph.D in Applied Chemistry, from Aligarh Muslim University, Aligarh, India on the topic "Development of Nanostructure Organic-Inorganic Composite Materials based Sensors for Inorganic Pollutants". He worked as a Project Fellow under the UGC Funded Research Project entitled "Development of Nanostructured Conductive Organic Inorganic Composite Materials based sensors Functionalities for Organic and Inorganic Pollutants". He completed his Masters in Chemistry (Industrial Applications) from Jamia Hamdard, New Delhi. He has published several research articles of international repute. He has edited one book with Springer. His research interest includes ion exchange chromatography and biosensor.

Mohd Imran Ahamed is a Research Scholar at Department of Chemistry, Aligarh Muslim University, Aligarh, India. He is working towards his Ph.D. thesis entitled Synthesis and characterization of inorganic-organic composite heavy metals selective cation-exchangers and their analytical applications. He has published several research and review articles in the journals of international recognition. He has also edited three books published by Springer and Materials Science Forum, U.S.A. He has completed his Bachelor of Science (Chemistry) from Aligarh Muslim University, Aligarh, India, and Masters in Chemistry (Organic Chemistry) from Dr. Bhimrao Ambedkar University, Agra, India. His research work includes ion exchange chromatography, wastewater treatment, and analysis, bending actuator and electrospinning.

Prof. Abdullah M. Asiri is the Head of the Chemistry Department at King Abdulaziz University since October 2009 and he is the founder and the Director of the Center of Excellence for Advanced Materials Research (CEAMR) since 2010 till date. He is the Professor of Organic Photochemistry. He graduated from King Abdulaziz University (KAU) with B.Sc. in Chemistry in 1990 and a Ph.D. from University of Wales, College of Cardiff, U.K. in 1995. His research interest covers color chemistry, synthesis of novel photochromic and thermochromic systems, synthesis of novel coloring matters and dyeing of textiles, materials chemistry, nanochemistry and nanotechnology, polymers and plastics. Prof. Asiri is the principal supervisors of more than 20 M.Sc. and six Ph.D. theses. He is the main author of ten books of different chemistry disciplines. Prof. Asiri is the Editor-in-Chief of King Abdulaziz University Journal of Science. A major achievement of Prof. Asiri is the discovery of tribochromic compounds, a new class of compounds which change from slightly or colorless to deep colored when subjected to small pressure or when grind. This discovery was introduced to the scientific community as a new terminology published by IUPAC in 2000. This discovery was awarded a patent from European Patent office and from UK patent. Prof. Asiri involved in many committees at the KAU level and on the national level. He took a major role in the advanced materials committee working for KACST to identify the national plan for science and technology in 2007. Prof. Asiri played a major role in advancing the chemistry education and research in KAU. He has been awarded the best researchers from KAU for the past five years. He also awarded the Young Scientist Award from the Saudi Chemical Society in 2009 and also the first prize for the distinction in science from the Saudi Chemical Society in 2012. He also received a recognition certificate from the American Chemical Society (Gulf region Chapter) for the advancement of chemical science in the Kingdome. He received a Scopus certificate for the most publishing scientist in Saudi Arabia in chemistry in 2008. He is also a member of the editorial board of various journals of international repute. He is the Vice- President of Saudi Chemical Society (Western Province Branch). He holds four USA patents, more than one thousand publications in international journals, several book chapters and edited books.

www.ingramcontent.com/pod-product-compliance
Lightning Source LLC
Chambersburg PA
CBHW071325210326
41597CB00015B/1351

* 9 7 8 1 6 4 4 9 0 0 1 2 3 *